PROGRESS
MI(

PROGRESS IN ENVIRONMENTAL MICROBIOLOGY

MYUNG-BO KIM
EDITOR

Nova Biomedical Books
New York

For permission to use material from this book please contact us:
Telephone 631-231-7269; Fax 631-231-8175
Web Site: http://www.novapublishers.com

Library of Congress Cataloging-in-Publication Data

Progress in environmental microbiology / Myung-Bo Kim, editor.
p. ; cm.
Includes bibliographical references and index.
ISBN-13: 978-1-60021-940-5 (hardcover)
ISBN-10: 1-60021-940-3 (hardcover)
1. Microbial ecology. I. Kim, Myung-Bo.
[DNLM: 1. Environmental Microbiology. QW 55 P964 2008]
QR100.P73 2008
579'.17--dc22 2007032468

Published by Nova Science Publishers, Inc. ✦ New York

Contents

Preface

This book presents new and important research on environmental microbiology which is area of interaction that studies the interaction of microorganisms with the environment. It includes the structure, activities and communal behaviour of microbial communities, microbial interactions and interactions with plants, animals and non-living environmental factors, population biology and clonal structure microbes and surfaces, adhesion and biofouling responses to environmental signals and stress factors growth and survival, modelling and theory development, microbial community genetics and evolutionary processes, microbial physiological, metabolic and structural diversity, pollution microbiology, extremophiles and life in extreme and unusual little-explored habitats, primary and secondary production, element cycles and biogeochemical processes and microbially-influenced global changes.

Expert Commentary A - Since the early years of hygiene and microbiology it is known that epidemics can be related to drinking water supplies. A famous example are numerous outbreaks of cholera in major European cities in the nineteenth century, e. g. the Munich epidemic in the year 1854 (von Pettenkofer, 1855). In the same period, John Snow studied the epidemiology of cholera in several cities of England. He could trace back the famous London outbreak to a contaminated drinking water pump (on Broad Street). As a consequence of these investigations, modern public drinking water supply systems as well as sewage systems were built in the following decades. On the basis of scientific expeditions in the year 1883 to cholera outbreaks in Egypt and India, Robert Koch and colleges were able to obtain pure cultures of *Vibrio cholerae* (Sack et al., 2004; Thompson et al., 2004). About the same time Koch also demonstrated the presence of heterotrophic bacteria in tap-water (Exner et al., 2005).

Expert Commentary B - In nature, microorganisms exist by interacting with each other. Microbiology of pure culture is not enough to describe their behaviors. Several microbial interactions have been recognized to affect the growth or metabolism of others; e.g., syntrophic co-metabolism, competition for substrate or space, production of inhibitors or activators, and predation. A harmonious balance of negative and positive interactions is found in stable microbial communities. In addition, third-party organisms easily affect the two-species relationships. For example, a third-party will cancel an inhibitory interaction by removing an inhibitory chemical. Therefore, a network of these relationships as a total system should be depicted for a complete understanding. We should begin the challenging study by detecting all members of a microbial community. However, unlike flora and fauna, we can now detect only a part of a dominant species and can characterize limited species by isolation. Thus, a defined mixed culture system stably comprising four to five strains will serve as a model community. The relationships obtained by comprehensive experiments will be evaluated with the help of bioinformatics and systems biology.

Chapter I - The molecular and population mechanisms involved in the evolution of bacteria, beneficial in symbioses with plants are reviewed. These bacteria possess the complicated genomes characterized by enormous plasticity (due to saturation with mobile DNA elements) and usually differentiated into several large replicons. In root nodule bacteria (rhizobia), evolution of symbiotic (*sym*) gene networks occurs under the impacts of partners' feedbacks, which may be either negative (at the early nodulation stages controlled by *nod*/*nol*/*noe* genes encoding for signaling and host penetration) or positive (at the late stages controlled by *nif* genes encoding for N_2 fixation). At the population level, these feedbacks are organized into "Infection and Release" Cycle (IRC), which includes: competition among virulent strains for host infection; *in planta* propagation of the winners; release of endosymbiotic bacteria into soil and their interactions with resident strains. The resulted microevolutionary factors include Darwinian selection (at all stages of IRC), frequency dependent selection and genetic drift (during inter-strain competition for plant nodulation), group selection (due to preferential *in planta* propagation of the host-beneficial clones) and population waves (after the release of bacteria into environment). Co-ordinated operation of these factors is responsible for the enormous polymorphism and for the panmictic structures in bacterial populations reflecting the crucial roles of intra-genome rearrangements and of horizontal gene transfer in their evolution. Circulation of rhizobia in host-environment systems results in the "Gain-and-Loss" dynamics of *sym* genes leading to: (i) their rapid evolution due to gene recruiting from different metabolic/regulatory pathways; (ii) expansion of *sym* genes within plant-associated bacterial communities, which may be stimulated greatly by plant invasions into novel environments. Evolution of mutualistic (host-beneficial) traits in bacteria is discussed in the terms of inter-species (reciprocal) altruism. Inter-deme and kin selection pressures are addressed as the major forces, which ensure the evolution of *nif* genes responsible for mutualistic (late) interactions, while *nod*/*nol*/*noe* genes responsible for pathogenic-like (early) interactions may evolve under the impacts of individual selection.

Chapter II - In this review, the possibility of the use of mixtures or combinations of active microorganisms as a more consistent and effective method of disease control than the application of a single biocontrol agent (BCA) is discussed. The growing pollution of the environment, the general concern of harmful residues in food, and resistance of numerous pathogens to commercial pesticides have induced researchers to find an alternative and nature-safe method of crop protection. During recent decades, numerous bacteria and fungi were isolated and tested for their effectiveness as soil, seed, root and tuber inoculants in control of plant pathogens. However, the commercial use of the biocontrol preparations in practical agriculture is still limited. Single BCA typically has a relatively narrow spectrum of activity compared with synthetic pesticides, and it is strongly affected by various biotic and abiotic factors under natural field conditions. Thus, while effective in the laboratory or in controlled field experiments, BCAs rarely give consistent and satisfying results in practice.

Despite the problems and limitations, the researchers spare no pain to find new active microorganisms and to develop the most effective methods of their application. However, studying past and present efforts in BCA's evaluation, it seems that a new outlook on biocontrol is needed. The natural environment is a very complicated and changeable system, therefore, an application of a single, even very active strain of the antagonist will never give as satisfactory result as a more condition-independent pesticide. Integration of several, complementary methods, e.g. application of BCA supported by favourable-for-microorganisms agrotechnic practices or organic amendments, could provide more reliable effects in plant protection against pathogens. Recently, the possible enhancement of the efficacy of BCAs by their combination was studied in many scientific laboratories. There are examples that some bacteria and fungi may interact with each other stimulating some beneficial aspects of their physiology. Moreover, bioprotection observed in naturally suppressive soils is usually attributed to the general activity of diverse indigenous microorganisms existing in these soils. Therefore, it is more likely that a community of several compatible microorganisms with multiple mechanisms of disease suppression and different requirements for growth conditions may broaden the spectrum of their activity and enhance the efficacy of biocontrol. The use of microbial mixtures would more closely mimic the situation in suppressive soils, and under natural changeable conditions one mechanism may compensate for the lack of activity of the other resulting in an additive or synergistic effect.

The review presents hitherto existing studies documenting the enhanced protection of plants treated with combined microorganisms, even against multiple pathogens. However, the reports describing a lack or negative effect of the microbial mixtures on plant development and health are also shown. The strategies in selection of the microorganisms for use in the mixtures, their possible formulation and methods of the application are considered. Also discussed are the problems resulting from the production and registration process of such multiple preparations and some potential areas for future research.

Chapter III - Anthropic activities have been responsible for the introduction of increasing amounts of heavy metals in the environment. Metal production, leather and tanning processes, gas and electricity production, sewage and waste disposal and related activities, contribute to the presence of copper, cadmium, zinc, lead, chromium and nickel in soil and surface and ground waters if waste products are not properly treated before discharged. Exposition to heavy metals causes irreversible damage to living organisms; their presence above certain limits is a potential risk to the environment and human health. In order to evaluate this risk, total metal concentration is a poor indicator because reactivity, bioavailability and toxicity depend on the distribution of the different metal species in that particular environment. A physicochemical understanding of metal speciation is required.

Microorganisms from different habitats have developed several strategies in order to cope with metal toxicity. Thorough studies on microbes-metal interactions can help to understand detoxifying mechanisms that can be applied to wastewater treatment. An important advantage of these innovative metal removal technologies, particularly if they are to be employed in developing countries, is the cost-effectiveness of using autochthonous bacteria, since they may be isolated from local polluted environments.

Buenos Aires Metropolitan Area presents one of the most polluted watersheds in Argentina: the Reconquista River. It receives high amounts of both faecal and industrial wastes without previous treatment, leading to high loads of pathogen microorganisms and metals in sediments, surface and pore waters of Reconquista basin. Autochthonous microorganisms, able to grow in the presence of copper, zinc, cadmium and chromium, were isolated from water and sediment samples taken from this basin, and used in metal biosorption studies under different experimental conditions to improve metal retention. Cadmium has been chosen as model metal because its toxicity limits bacterial growth. Cadmium complexing capacity (CC) of culture media and electroplating effluents was evaluated in terms of total ligand concentration (L_t) and conditional stability constants (K_f'), assuming 1:1 Cd-ligand complexes are formed. In these systems total ligand concentration is in the μM range, far from the typical results obtained for seawater (nM), where most speciation studies were performed. As only moderate strength ligands were detected ($4<\log K_f'<8$), it is expected that the isolated bacteria would compete favorably with the ligands for metal retention.

The aim of this article is to describe how understanding of both, the physicochemistry of aqueous systems and microbial strategies to survive in polluted environments, contribute to a proper design of wastewater treatment process.

Chapter IV - Management practices have a significant impact on soil organism populations and their activity. Thus, microbiological properties have been considered as interesting attributes which are highly sensitive to changes produced by soil management. To be considered sustainable, management practices should promote fertility and productivity in the long term and preserve soil microbial diversity.

Soil microbial community dynamics is directly involved with ecosystem functions since it regulates fundamental processes such as organic matter decomposition, nutrient cycling or xenobiotic degradation. Changes in the composition or activity of microbial communities may have short or long-term effects in the ecosystem functioning. Mainly due to this fact, an increasing interest has arisen in the study of microbial communities in the recent years. Different approaches have been developed, based on physiological traits, biochemical analysis of cell constituents or genetic procedures.

By applying the commercial Biolog GN microplates originally designed for Gram negative isolate identification, Garland and Mills (1991) developed a rapid community-level technique based on sole carbon source oxidation for estimating microbial community diversity. Each GN microplate consists of 95 wells with a carbon source incorporated to a basal medium as sole C-source and one control without any C-source; the redox dye tetrazolium violet colorimetrically indicates the utilization of the carbon substrates.

Since then, this approach which was later called community level physiological profile (CLPP) has been increasingly applied and proved to be useful to detect relative changes in microbial communities.

However, as all methods available up to now for describing microbial communities as regards structure and/or function, CLPP has advantages and limitations. Selective enrichment resulting from the high substrate concentrations employed and relatively long incubation periods have to be taken into account in order to prevent over-interpretation of the information obtained.

In this chapter the authors summarize some of our findings working on CLPP from Biolog GN™ and from the later developed Biolog EcoPlate™ assay (which allows testing a more reduced number of substrates and a larger number of replicates), in field and greenhouse experiments. The suitability of CLPP to be used for monitoring soil quality is also analyzed.

Critical considerations quoted in recent literature as regards interpretation of CLPP are discussed, and also alternative approaches directly measuring oxygen consumption or carbon dioxide production due to the carbon substrate catabolism more recently developed in order to provide a more accurate and ecologically relevant overview on soil functional microbial diversity.

Chapter V - The selective extraction procedure was further developed and perfected newly in the present study in order to understand more mechanisms of heavy metals adsorption onto the surficial sediments (SSs) and a kind of special surficial sediments: natural surface coating samples (NSCSs). Then the extraction of endogenic Cu and Zn in company with the extraction of Fe, Mn oxides and organic materials (OMs) and adsorption of anthropogenic (added) Cu and Zn onto the solid particles before and after the extractions were investigated using the selective extraction- (metal adsorption)-statistical analysis methods. The results indicate that 0.1 mol/L $NH_2OH{\cdot}HCl$ + 0.1 mol/L HNO_3 has great potential for removing Mn oxides, $(NH_4)_2C_2O_4$ (0.2 mol/L) + $H_2C_2O_4$ (pH 3.0) for extracting both Fe and Mn hydrous oxides, and H_2O_2 (30%) for selectively extracting OMs. And the statistical analyses of experimental results suggest that both for endogenic and anthropogenic heavy metals, the adsorption capacities of NSCSs were higher than those of SSs, implying that the role of the NSCSs in the transformation and cycling of heavy metals in aquatic environments was more important than that of SSs. For the two sorts of metal pollutants, the relative adsorption roles of non-residual and residual fractions of solid particles were different from each other, i.e. non-residual fraction contributed more roles in anthropogenic Cu adsorption to the solid particles, and the contributions of non-residual and residual fractions to the anthropogenic Zn adsorption were comparative; but for endogenic pollutants, the contribution of non-residual fraction was significant more than that of residual fraction for Zn, and the role of non-residual fraction was similar to that of residual fraction for Cu. However, the relative roles of Fe, Mn oxides and OMs in the non-residual fraction of the particles were similar to each other for endogenic and anthropogenic metals, and the greatest contribution to metals adsorption on a molar basis was from Mn oxides. Metals adsorption capacities of Mn oxides exceeded those of Fe oxides by one order of magnitude, fewer roles were found attributing to adsorption by OMs. These results reveal that Mn oxides in the non-residual fractions were the most important component in controlling heavy metals in aquatic environments.

Chapter VI - Pneumonia with *Legionella spp.* presents a public health challenge especially because fatal outcomes still remain frequent. *Pseudomonas aeruginosa* is a significant source of hospital-acquired pneumonia and can also cause devastating chronic infections in compromised hosts, for example respiratory infections in cystic fibrosis patients. The aim of the authors studies was to describe the abundance and epidemiology of *Legionellaceae* in the man-made environment. In total, water systems of 70 different buildings in the German town of Göttingen (Lower Saxony) were examined for the presence of *Legionella* in two sampling cycles. Of these 22 (31%) had the bacterium in at least one water sample. *Legionella pneumophila* serogroups 1, 3, 4, 5 and 6 could be identified in the water samples. Most of the buildings were colonized solely by one distinct strain, as proven by PCR typing. Some buildings contained more than one PCR type or even more than one serogroup. Additionally the colonization of greenhouse misting systems with *Legionella spp.* and *Pseudomonas spp.* was studied in 20 different greenhouse misting systems located in Northern Germany. In total 80 water samples were collected. Each system was tested on two different occasions. Water was drawn at a central tap and at the outlet of spray nozzles. Sampled greenhouses were used to cultivate various plants and trees for commercial, recreational or scientific reasons, some of them in tropical conditions. *Legionella spp.* was detected in 10% of the systems (two systems), but only in low numbers. *Pseudomonas spp.* was recovered from 70% of the greenhouse watering systems (fourteen systems), occasionally at counts greater than 10,000 CFU/100 ml. Each colonized greenhouse had one or several individual strains of *Legionella* and *Pseudomonas* that could not be detected in any other system. This was demonstrated by a random amplified polymorphic DNA typing method. The possible health hazard caused by these water systems for both genera of bacteria is evaluated and discussed.

Chapter VII - Maturation ponds are commonly used as a treatment method for improving or polishing effluent from secondary biological processes, activated sludge, trickling filters or facultative ponds. Methods for removing fecal coliforms (FC) have been studied by many authors, and all indicate that sunlight, temperature and retention time are the principal factors that cause FC reduction. Pond geometry affects retention time, and this has been demonstrated on both pilot- and real-scales by several authors. The objective of this chapter is to demonstrate the effect of sunlight exposure time on FC removal in maturation ponds on a pilot- and full-scale basis. The chapter will also explain how to take advantage of this important resource, by improving the geometric configuration of the system, to increase retention time and, therefore, natural disinfection. A case study developed in tertiary treatment with maturation ponds, located after a conventional percolating filter plant, is used as an example. The use of baffles to change pond configuration is the best way to control, handle or manipulate pond hydraulic behavior, and thus provide a series of benefits directly related to sunlight intensity. The two parameters, retention time and sunlight exposure time, constitute the principal binomial for FC removal in maturation ponds as tertiary treatment after conventional wastewater treatment.

Chapter VIII - A microcosm studies was carried out to determine the effect of *Aspergillus niger*-treated mixture of two agro-industrial waste (AIW) materials, dry olive wastes and sugar beet press mud, on tomato (*Lycopersicon esculentum*) plants grown in a soil inoculated with *Fusarium oxysporum* f. sp. *lycopersici* (*Fol*). These waste materials were selected as they constitute a major environmental problem especially for Mediterranean countries. Agro-wastes were treated in conditions of solid-state fermentation in the presence of Morocco apatite (RP) and further applied at a rate of 50 g/kg soil. Soil-plant systems were additionally inoculated or not with the arbuscular mycorrhizal (AM) fungus *Glomus intraradices*. Plant growth and nutrition, symbiotic developments and soil enzymatic activities were stimulated in *Fol*-free soil supplemented with treated agro-wastes and significantly greater in treatments where AM fungus was introduced compared with the non-amended control. The introduction of *Fol* into the soil-plant system adversely influenced all studied parameters. AM fungus alone reduced the effect of the plant pathogen on tomato plants and significantly decreased (by $2.3x10^3$) the number of *Fol*-CFU compared with the non-mycorrhizal control grown in non-amended soil. Higher levels of pathogen control were achieved in *A. niger*/AIW/RP-amended mycorrhized plant-soil system. In separate, *in vitro* studies, *A. niger* demonstrated a strong suppressive effect on *Fol* with a 5-fold reduction of the pathogen colony diameter compared to the control. Production of siderophore-like metabolites was also detected in chrome azurol S plate assays with *A. niger*.

Chapter IX - Present methods for the detection of micro-organisms in the aquatic environment are slow, inefficient and often unreliable. Furthermore, direct microscopic enumeration has shown that numbers of bacteria capable of forming colonies on "nonselective" media are usually several orders of magnitude fewer than numbers actually present and metabolically active in freshwater and marine environments. The development of molecular techniques, in particular in situ hybridization offers the potential for rapid and specific assays. The fluorescent in situ hybridization (FISH) technique uses rRNA- targeted fluorescent nucleic acid probes, and over the last decade has become an important tool for microbial ecologists. In short, cells are fixed (i.e., they are not viable anymore and the status quo of their DNA and RNA is preserved), permeabilized to facilitate access of the probe to the target site and then hybridized with nucleic acid probes. The probes are labeled with a fluorochrome, and the samples can then be analysed by epifluorescence. The classic FISH technique rely on (usually 16S) rRNA as a probe target. In this mini review the authors will summarise methods and applications for FISH in aquatic environment and we will highlight in particular their advantages and limitations.

In: Progress in Environmental Microbiology
Editor: Myung-Bo Kim, pp. 1-6
ISBN: 978-60021-940-5

Expert Commentary A

Complex Ecology of Microbial Biofilm Communities in Drinking Water Supply Systems

B. P. Zietz*

Medical Institute of General Hygiene and Environmental Health, University of Göttingen, Germany

Since the early years of hygiene and microbiology it is known that epidemics can be related to drinking water supplies. A famous example are numerous outbreaks of cholera in major European cities in the nineteenth century, e. g. the Munich epidemic in the year 1854 (von Pettenkofer, 1855). In the same period, John Snow studied the epidemiology of cholera in several cities of England. He could trace back the famous London outbreak to a contaminated drinking water pump (on Broad Street). As a consequence of these investigations, modern public drinking water supply systems as well as sewage systems were built in the following decades. On the basis of scientific expeditions in the year 1883 to cholera outbreaks in Egypt and India, Robert Koch and colleges were able to obtain pure cultures of *Vibrio cholerae* (Sack et al., 2004; Thompson et al., 2004). About the same time Koch also demonstrated the presence of heterotrophic bacteria in tap-water (Exner et al., 2005).

In July 1976, an epidemic of acute pneumonia took place at an American Legion convention in Philadelphia (Fraser et al., 1977). *Legionella pneumophila* was recognized as the aetiological agent of this outbreak (McDade et al., 1977). Later research showed that *Legionella spp.* are common bacteria in the aquatic environment. They can colonize especially domestic warm water installations. Thermally altered aquatic environments can favour a rapid multiplication of *Legionella* (Fields et al., 2002; WHO, 2007).

* Correspondence concerning this article should be addressed to Dr. B. P. Zietz, MPH; Medical Institute of General Hygiene and Environmental Health, University of Göttingen, Lenglerner Str. 75, D-37079 Göttingen, Germany. Tel.: +49 551 5007886-1; fax: +49 551 5007886-3. E-mail: bjoern.zietz@nlga.niedersachsen.de

Pathogens like *Legionella spp.* are known to have free-living amoebal hosts such as *Acanthamoeba,* which are widely distributed in the environment including domestic water systems (Fields et al., 2002; Marciano-Cabral and Cabral, 2003). For example in an investigation of six hospital warm water systems, amoebae were detected in 29 of 56 (52%) water samples. They could be identified as *Hartmannella vermiformis*, *Echinamoebae spp.*, *Saccamoebae spp.*, and *Vahlkampfia spp.* (Rohr et al., 1998). Several free-living amoebae can itself cause diseases like amoebic encephalitis or keratitis (by *Acanthamoeba*, primarily associated with contact-lens usage) (Schuster and Visvesvara, 2004; Marciano-Cabral and Cabral, 2003). Additionally amoebae can be naturally infected with other amoeba-resisting bacteria that include potential pathogens, such as *Pseudomonas aeruginosa*, *Vibrio cholerae*, *Helicobacter pylori*, or *Mycobacterium avium* (Marciano-Cabral and Cabral, 2003; Alsam et al., 2006). These facts suggest that both amoebae and bacteria are involved in complex ecological interactions.

Another important human (opportunistic) pathogen that may be present in drinking water distribution systems is the genera *Pseudomonas*, which may cause, for example, pneumonia, dermatitis or otitis media. Infections with these bacteria predominantly affect predisposed persons. *Pseudomonads* and especially *Pseudomonas aeruginosa* are significant nosocomial pathogens, which for example cause wound infections and bacteraemia in burn victims or urinary tract infections in catheterized patients. (Garau and Gomez, 2003; Stover et al., 2000). Tap-water can be among the nosocomial infection sources (Trautmann et al., 2006; Exner et al., 2005).

Pathogenic environmental mycobacteria can also be part of biofilm communities in drinking water supply systems. The WHO has classified these bacteria as an emerging cause of waterborne disease. This is especially true for the *Mycobacterium avium* complex (WHO, 2004a).

Generally the bacterial communities in drinking water distribution systems found the basis of a food chain including the grazing activity of protozoa and especially amoebae (Sibille et al., 1998). Sessile and planktonic microorganisms in these water systems fill different and complex environmental habitats (Donlan and Costerton, 2002). Floating cell aggregates ("flocs") can share many of their characteristics with biofilms (Hall-Stoodley et al., 2004). It has been proposed that a basic community structure of living, fully hydrated biofilms is composed of cells (±15% by volume) and of matrix material (±85% by volume), and the cells are located in matrix-enclosed "towers", "mushrooms" or filamentous "streamers" (Donlan and Costerton, 2002; Hall-Stoodley et al., 2004). There are indications that biofilm formation in drinking water supply systems is based on long-term successional developments extending over many months (Martiny et al., 2003). Even in the last years new bacterial species from drinking water biofilms were isolated (Rickard et al., 2005; Kalmbach et al., 1999).

Studies indicate that the growth of biofilms in drinking water systems is influenced by factors like nature of the pipe material, the hydraulic conditions and the physical and chemical characteristics of the water (WHO, 2004b). An overview of important factors influencing microbial communities in drinking water systems can be found in Figure 1.

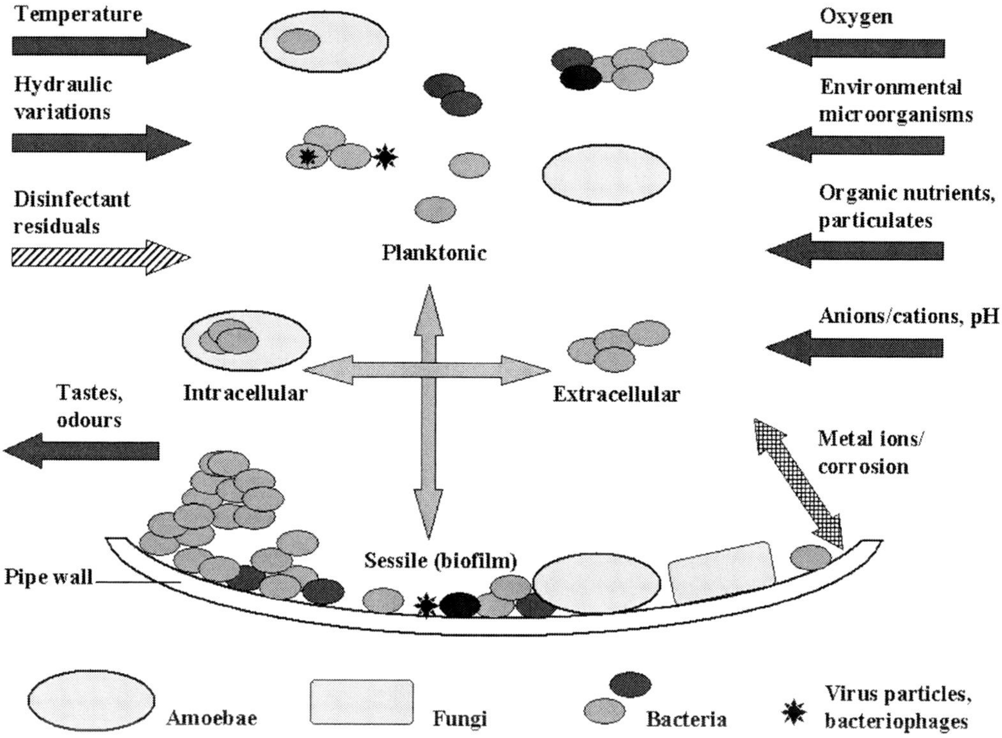

Figure 1. Schematic depiction of microbial communities in drinking water distribution systems with metal pipes and important influence factors and interactions therein (excluding algae, cyanobacteria and helminths).

One important factor favouring biofilm growth seems to be biodegradable organic matter (Volk and LeChevallier, 1999). Pipe materials can also have an influence on biofilm formation and microbial colonization. Certain pipe materials may stimulate growth by releasing the bioavailable nutrients iron and phosphorus or neutralize the disinfectant residual. On the other hand leached metals such as copper from copper pipes may slow biofilm growth (Berry et al., 2006).

Studies indicate that chemical disinfection trends to effect planktonic bacteria more than biofilm bacteria. Multispecies biofilms may also be more resistant to biocides than single-species biofilms. Additionally pathogenic bacteria hosted in amoebae seem to have an increased resistance to disinfection (Berry et al., 2006; Exner et al., 2005).

Due to the complexity of microbial communities and influencing factors in drinking water supply systems, potentially hosting many different human pathogens, an integrated control strategy should be implemented. The third edition of the WHO Guideline for drinking water includes a substantive revision of approaches to ensure microbial safety. It requests comprehensive system-specific water safety plans. These plans use hazard identification and risk assessment as a starting point for system management. All steps in water supply from water catchment to consumer's taps should be included. In these assessments, risks, for example, can be ranked in a matrix using the factor's severity of consequences and likelihood of event. Water safety plans can vary in complexity, as appropriate for the situation (WHO, 2006).

Microbial control measures in the user's installations should co-ordinate, for example, temperature control (heating, thermal insulation of pipes), proper dimensioning, choice of installation materials, control of water chemistry, regular cleaning and maintenance, and in some cases additional disinfection or UV-treatment schemes. Uncommon water installations like greenhouse misting systems should also be part of such schemes (Zietz et al., 2006).

Although many facts regarding biofilm communities in drinking water systems and influencing factors have been revealed, many questions remain open. It is essential to better understand the complex interactions to adequately manage microbial water quality in piped distribution systems. Appling new methods of molecular biology may be an interesting tool to achieve this.

References

Alsam, S; Jeong, SR; Sissons, J; Dudley, R; Kim, KS; Khan, NA. *Escherichia coli* interactions with *Acanthamoeba*: a symbiosis with environmental and clinical implications. *J Med Microbiol* 2006, 55, 689-694.

Berry, D; Xi, C; Raskin, L. Microbial ecology of drinking water distribution systems. *Curr Opin Biotechnol* 2006, 17, 297-302.

Donlan, RM; Costerton, JW. Biofilms: survival mechanisms of clinically relevant microorganisms. *Clin Microbiol Rev* 2002, 15, 167-193.

Exner, M; Kramer, A; Lajoie, L; Gebel, J; Engelhart, S; Hartemann, P. Prevention and control of health care-associated waterborne infections in health care facilities. *Am J Infect Control* 2005, 33 (Suppl), S26-40.

Fields, BS; Benson, RF; Besser, RE. *Legionella* and Legionnaires' disease: 25 years of investigation. *Clin Microbiol Rev* 2002, 15, 506-526.

Fraser, DW; Tsai, TR; Orenstein, W; Parkin, WE; Beecham, HJ; Sharrar, RG; Harris, J; Mallison, GF; Martin, SM; McDade, JE; Shepard, CC; Brachman, PS. Field Investigation team. Legionnaires' disease: description of an epidemic of pneumonia. *N Engl J Med* 1977, 297, 1189-97.

Garau, J; Gomez, L. *Pseudomonas aeruginosa* pneumonia. *Curr Opin Infect Dis* 2003, 16, 135-143.

Hall-Stoodley, L; Costerton, JW; Stoodley, P. Bacterial biofilms: from the natural environment to infectious diseases. *Nature Rev Microbiol* 2004, 2, 95-108.

Kalmbach, S; Manz, W; Wecke, J; Szewzyk, U. *Aquabacterium* gen. nov., with description of *Aquabacterium citratiphilum* sp. nov., *Aquabacterium parvum* sp. nov. and *Aquabacterium commune* sp. nov., three in situ dominant bacterial species from the Berlin drinking water system. *Int J Syst Bacteriol* 1999, 49, 769-777.

Marciano-Cabral, F; Cabral, G. *Acanthamoeba spp.* as agents of disease in humans. *Clin Microbiol Rev* 2003, 16, 273-307.

Martiny, AC; Jorgensen, TM; Albrechtsen, HJ; Arvin, E; Molin S. Long-term succession of structure and diversity of a biofilm formed in a model drinking water distribution system. *Appl Environ Microbiol* 2003, 69, 6899-6907.

McDade, JE; Shepard, CC; Fraser, DW; Tsai, TR; Redus, MA; Dowdle, WR. Legionnaires' disease: isolation of a bacterium and demonstration of its role in other respiratory disease. *N Engl J Med* 1977, 297, 1197-1203.

Rickard, AH; Stead, AT; O'May, GA; Lindsay, S; Banner, M; Handley, PS; Gilbert, P. *Adhaeribacter aquaticus* gen. nov., sp. nov., a Gram-negative isolate from a potable water biofilm. *Int J Syst Evol Microbiol* 2005, 55, 821-829.

Rohr, U; Weber, S; Michel, R; Selenka, F; Wilhelm, M. Comparison of free-living amoebae in hot water systems of hospitals with isolates from moist sanitary areas by identifying genera and determining temperature tolerance. *Appl Environ Microbiol* 1998, 64, 1822-1824.

Sack, DA; Sack, RB; Nair, GB; Siddique, AK. Cholera. *Lancet* 2004, 363, 223-233.

Schuster, FL; Visvesvara, GS. Free-living amoebae as opportunistic and non-opportunistic pathogens of humans and animals. *Int J Parasitol* 2004, 34, 1001-1027.

Sibille, I; Sime-Ngando, T; Mathieu, L; Block, JC. Protozoan bacterivory and *Escherichia coli* survival in drinking water distribution systems. *Appl Environ Microbiol* 1998, 64, 197-202.

Stover, CK; Pham, XQ; Erwin, AL; Mizoguchi, SD; Warrener, P; Hickey, MJ; Brinkman, FS; Hufnagle, WO; Kowalik, DJ; Lagrou, M; Garber, RL; Goltry, L; Tolentino, E; Westbrock-Wadman, S; Yuan, Y; Brody, LL; Coulter, SN; Folger, KR; Kas, A; Larbig, K; Lim, R; Smith, K; Spencer, D; Wong, GK; Wu, Z; Paulsen, IT; Reizer, J; Saier, MH; Hancock, RE; Lory, S; Olson, MV. Complete genome sequence of *Pseudomonas aeruginosa* PA01, an opportunistic pathogen. *Nature* 2000, 406, 959-964.

Thompson, FL; Iida, T; Swings, J. Biodiversity of vibrios. *Microbiol Mol Biol Rev* 2004, 68, 403-431.

Trautmann, M; Bauer, C; Schumann, C; Hahn, P; Hoher, M; Haller, M; Lepper, PM. Common RAPD pattern of *Pseudomonas aeruginosa* from patients and tap water in a medical intensive care unit. *Int J Hyg Environ Health* 2006, 209, 325-331. Erratum in: *Int J Hyg Environ Health* 2006, 209, 585-586.

Volk, CJ; LeChevallier, MW. Impacts of the reduction of nutrient levels on bacterial water quality in distribution systems. *Appl Environ Microbiol* 1999, 65, 4957-4966.

von Pettenkofer, MJ. Untersuchungen und Beobachtungen über die Verbreitungsart der Cholera nebst Betrachtungen über Maßregeln, derselben Einhalt zu thun. Munich: Literarisch-artistische Anstalt; 1855.

WHO (World Health Organization). Pathogenic mycobacteria in water: A guide to public health consequences, monitoring and management. Ed. Bartram, J; Cotruvo, JA; Dufour, A; Rees, G; Pedley, S. Geneva: WHO; 2004a.

WHO (World Health Organization). Safe piped water: Managing microbial water quality in piped distribution systems. Ed. Ainsworth, R. Geneva: WHO; 2004b.

WHO (World Health Organization). Guidelines for drinking-water quality. Third edition, incorporating first addendum. Volume 1 – Recommendations. Geneva: WHO; 2006.

WHO (World Health Organization). *Legionella* and the prevention of legionellosis. Ed. Bartram, J; Chartier, Y; Lee, JV; Pond, K; Surman-Lee, S. Geneva: WHO; 2007.

Zietz, BP; Dunkelberg, H; Ebert, J; Narbe, M. Isolation and characterization of *Legionella spp.* and *Pseudomonas spp.* from greenhouse misting systems. *J Appl Microbiol* 2006, 100, 1239-1250. Erratum in: *J Appl Microbiol* 2006, 101, 976. (Please also refer to our chapter in this book).

In: Progress in Environmental Microbiology
Editor: Myung-Bo Kim, pp. 7-10
ISBN: 978-60021-940-5

Expert Commentary B

Network Study of Interspecies Relationships Will Open New Aspects of Microbial Ecology

Shin Haruta and Yasuo Igarashi
Graduate School of Agricultural and Life Sciences, the University of Tokyo, Japan

Abstract

In nature, microorganisms exist by interacting with each other. Microbiology of pure culture is not enough to describe their behaviors. Several microbial interactions have been recognized to affect the growth or metabolism of others; e.g., syntrophic co-metabolism, competition for substrate or space, production of inhibitors or activators, and predation. A harmonious balance of negative and positive interactions is found in stable microbial communities. In addition, third-party organisms easily affect the two-species relationships. For example, a third-party will cancel an inhibitory interaction by removing an inhibitory chemical. Therefore, a network of these relationships as a total system should be depicted for a complete understanding. We should begin the challenging study by detecting all members of a microbial community. However, unlike flora and fauna, we can now detect only a part of a dominant species and can characterize limited species by isolation. Thus, a defined mixed culture system stably comprising four to five strains will serve as a model community. The relationships obtained by comprehensive experiments will be evaluated with the help of bioinformatics and systems biology.

Interaction between Bacterial Species or Cells

When a bacterial species lives with another species, they will affect each other through their metabolism (e.g., symbiosis, competition), signaling molecules (e.g., antibiotics, growth promoting factor, pheromones), or direct contact (e.g., biofilm formation, predation). These two species interactions have been studied widely.

Recently, molecular or micro-(nano-)scale studies addressed new insights into interspecies interactions. For example, direct communications in symbiotic relationships are found in several anaerobic environments. Gorby et al. observed a cell-to-cell connection with nanowires in syntrophic methanogenic cocultures of *Pelotomaculum thermopropionicum* and *Methanothermobacter thermautotrophicus* [Gorby *et al.*, 2006]. There have been numerous studies of small-molecule signals as reviewed by Camilli and Bassler [Camilli & Bassler 2006]. Antibiotics also work as a signaling molecule to affect biofilm formation of others at low concentration, although they are traditionally believed to work as weapons to remove others [Linares *et al.*, 2006]. Moreover, direct contact will change bacterial behaviors, such as biofilm formation. *Bdellovibrio* sp. predates gram-negative bacteria such as *Escherichia coli* by direct attachment on the prey cells and replication within the cells [Koval & Bayer 1997], and certain *E. coli* strain inhibits the growth of other *E. coli* strains in a contact-dependent manner [Aoki *et al.*, 2005].

In addition to these interactions, space occupied by one species will reduce the population of the other. We can easily imagine the competition for space under biofilm conditions. It may also occur in homogenous liquid cultures, since our research group indicates that the maximum total cell-density in liquid medium is not affected by the number of species in mixed cultures [Kato *et al.*, unpublished data].

Effect of Third-party Organisms Developing Network Relationship

Interspecies interactions seem to be easily affected by third-party organisms. However, few studies have focused on the importance of third-party organisms. A simple example was reported for the predation by *Bdellovibrio* sp. [Hobley *et al.*, 2006]. The predation efficiency was reduced by the existence of a decoy such as *Bacillus* sp. which was a nonsusceptible bacterium. Moreover, protease production by the decoy increased in an amino acids concentration to promote the growth of prey.

Narisawa et al. found that an antibiotics producing bacterium and a sensitive bacterium against the antibiotics co-existed in the presence of a tolerant species in biofilm conditions (unpublished data).

Lethal interaction was detected between the members in a stable mixed culture [Kato *et al.*, unpublished data]. Third-party bacteria in the culture supported the growth of the weak by a promoting effect on the weak and by a suppressive effect on the strong. They indicate that a combination of several interactions compensated for unfavorable effects to maintain the microflora by comprehensive analyses of microbial interactions.

When we focus on third-party organisms in various environments, more interesting observations will be obtained. However, it is difficult to evaluate these effects on a whole microbial community in environments. Whilst we should keep in mind the fact that global reservoirs of diversity are an important feature of natural ecosystems [Curtis & Sloan, 2004], firstly a model community will be useful to look for the way to analyze (e.g., to find out hub in a complex web of microflora) and to find out any primitive theories in microbial society. A valuable model may require the following properties; 1) all the members are isolated, 2) the members stably co-exist under certain conditions, 3) the community has a simple metabolic function, 4) cultivation conditions are homogenous (no specific niche) and chemostat culture is favorable. Unfortunately, we are not aware of the microflora which satisfies all these properties, although the first three are defined for a stable mixed culture reported by Kato et al. [Kato *et al.*, 2005].

Integration of Microbial Behaviors in Community

Environmental microbiologists have described their environment by the component species defined by phylogenetic information. However, ecophysiology has been focused in environmental microbiology, since metabolic or growth properties under laboratory pure culture conditions are not always detected in environments or microflora due to environmental complex physicochemical factors or interspecies interactions. In addition, behaviors of microorganisms are changeable by metabolic adaptation or genetic evolution [e.g., Ibara *et al.*, 2002, Kirkelund Hansen *et al.*, 2007]. Moreover, fluctuation of gene expression among cells in pure culture suggested the importance of single cell analyses [e.g., Suel *et al.*, 2006]. Microbial ecology should move toward physiological diversity of cell from phylogenetic diversity of species or strain. To build this new generation, we need advanced analytical tools, i.e., time lapse and single cell analysis.

Then we will obtain a huge number of and high quality of data about microbial behaviors. After that, mathematical data analyses or novel theory will be required as Buckley, M. R. proposed in Systems Microbiology -Beyond Microbial Genomics, a report from the American Academy of Microbiology 2004- that "...systems microbiology seeks to treat the organism or COMMUNITY as a whole, integrating fundamental biological knowledge with...to create an integrated picture of how a microbial cell or COMMUNITY operates". Complex systems research will help to develop new microbial ecology. Becks et al. successfully observed chaotic behavior of the population in competitive relationships through collaboration with mathematician [Becks *et al.*, 2005]. As well as continued progress of ecophysiological studies on each strain or cell, for the present we will begin to communicate well with research groups in the field of mathematical biology, systems biology or computer science developing novel analytical approaches and theoretical interpretation of experimental observations.

References

Aoki, S. K., Pamma, R., Hernday, A. D., Bickham, J. E., Braaten, B. A., and Low, D. A. (2005) Contact-dependent inhibition of growth in *Escherichia coli.* Science 309, 1245-1248.

Becks, L., Hilker, F. M., Malchow, H., Jurgens, K., and Arndt, H. (2005) Experimental demonstration of chaos in a microbial food web. *Nature* 435, 1226-1229.

Camilli, A. and Bassler, B. L. (2006) Bacterial small-molecule signaling pathways. *Science* 311, 1113-1116.

Curtis, T. P. and Sloan, W. T. (2004) Prokaryotic diversity and its limits: microbial community structure in nature and implications for microbial ecology. *Current Opinion in Microbiology* 7, 221-228.

Gorby, Y. A., Yanina, S., McLean, J. S., Rosso, K. M., Moyles, D., Dohnalkova, A., Beveridge, T. J., Chang, I. S., Kim, B. H., Kim, K. S., Culley, D. E., Reed, S. B., Romine, M. F., Saffarini, D. A., Hill, E. A., Shi, L., Elias, D. A., Kennedy, D. W., Pinchuk, G., Watanabe, K., Ishii, S., Logan, B., Nealson, K. H., and Fredrickson, J. K. (2006) Electrically conductive bacterial nanowires produced by *Shewanella oneidensis* strain MR-1 and other microorganisms. *Proceedings of the National Academy of Science of the United States of America* 103, 11358-11363.

Hobley, L., King, J. R., and Elizabeth Sockett, R. (2006) *Bdellovibrio* predation in the presence of decoys: Three-way bacterial interactions revealed by mathematical and experimental analyses. *Applied and Environmental Microbiology* 72, 6757-6765.

Ibara, R. U., Edwards, J. S., and Palsson, B. O. (2002) *Escherichia coli* K-12 undergoes adaptive evolution to achieve *in silico* predicted optimal growth. *Nature* 420, 186-189.

Kato, S., Haruta, S., Cui, Z. J., Ishii, M., and Igarashi, Y. (2005) Stable coexistence of five bacterial strains as a cellulose-degrading community. *Applied and Environmental Microbiology* 71, 7099-7106.

Kirkelund Hansen, S., Rainey, P. B., Haagensen, J. A. J., and Molin S. (2007) Evolution of species interactions in a biofilm community. *Nature* 445, 533-536.

Koval, S. F. and Bayer, M. E. (1997) Bacterial capsules: no barrier against *Bdellovibrio. Microbiology* 143, 749-753.

Linares, J. F., Gustafsson, I., Baquero, F., and Martinez, J. L. (2006) Antibiotics as intermicrobial signaling agents instead of weapons. *Proceedings of the National Academy of Science of the United States of America* 103, 19484-19489.

Suel, G. M., Garcia-Ojalvo, J., Liberman, L. M., and Elowitz, M. B. (2006) An excitable gene regulatory circuit induces transient cellular differentiation. *Nature* 440, 545-550.

In: Progress in Environmental Microbiology
Editor: Myung-Bo Kim, pp. 11-67
ISBN: 978-60021-940-5

Chapter I

Evolution of Symbiotic Bacteria in "Plant-Soil" Systems: Interplay of Molecular and Population Mechanisms

Nikolai A. Provorov and Nikolai I. Vorobyov
All-Russia Research Institute for Agricultural Microbiology, Podbelsky Sh. 3, St.-Petersburg, Pushkin-8, 196608, Russia

Abstract

The molecular and population mechanisms involved in the evolution of bacteria, beneficial in symbioses with plants are reviewed. These bacteria possess the complicated genomes characterized by enormous plasticity (due to saturation with mobile DNA elements) and usually differentiated into several large replicons. In root nodule bacteria (rhizobia), evolution of symbiotic (*sym*) gene networks occurs under the impacts of partners' feedbacks, which may be either negative (at the early nodulation stages controlled by *nod*/*nol*/*noe* genes encoding for signaling and host penetration) or positive (at the late stages controlled by *nif* genes encoding for N_2 fixation). At the population level, these feedbacks are organized into "Infection and Release" Cycle (IRC), which includes: competition among virulent strains for host infection; *in planta* propagation of the winners; release of endosymbiotic bacteria into soil and their interactions with resident strains. The resulted microevolutionary factors include Darwinian selection (at all stages of IRC), frequency dependent selection and genetic drift (during inter-strain competition for plant nodulation), group selection (due to preferential *in planta* propagation of the host-beneficial clones) and population waves (after the release of bacteria into environment). Co-ordinated operation of these factors is responsible for the enormous polymorphism and for the panmictic structures in bacterial populations reflecting the crucial roles of intra-genome rearrangements and of horizontal gene transfer in their evolution. Circulation of rhizobia in host-environment systems results in the "Gain-and-Loss" dynamics of *sym* genes leading to: (i) their rapid evolution due to gene recruiting from different metabolic/regulatory pathways; (ii) expansion of *sym* genes within plant-associated bacterial communities, which may be stimulated greatly by plant invasions into novel environments. Evolution of mutualistic (host-beneficial) traits in bacteria is discussed in the terms of inter-species (reciprocal) altruism. Inter-deme and kin selection pressures are addressed as the major forces, which ensure the evolution of *nif* genes responsible for mutualistic (late) interactions, while *nod/nol/noe* genes responsible for pathogenic-like (early) interactions may evolve under the impacts of individual selection.

Introduction

Symbiotic interactions between prokaryotes and eukaryotes attract an increasing attention as the mechanisms generating the novel types of biological organization and providing the partners with the greatly extended adaptive facilities [Seckbach, 2002]. The enormous creative potential of symbioses evolution was realized by plants, which colonized the land due to a tight co-operation with endo-mycorrhizal fungi and later evolved the genetic systems for symbioses with various beneficial microorganisms [Parniske, 2000]. In many groups of plants, successful adaptations to the environmental stresses are relied on the bacterial symbionts (Table 1) some of which provide their hosts with biologically fixed nitrogen or facilitate the phosphate assimilation (nodule bacteria or rhizobia, actinomycetes *Frankia*, cyanobacteria *Nostoc* and *Anabaena*, various endophytic and rhizospheric microbes [Tikhonovich & Provorov, 2007]), while the others protect the hosts from pathogens (plant growth promoting rhizobacteria [Lugtenberg et al., 2001]) or herbivores (*Clavibacter* endophytes [Metzler et al., 1997]).

At present, the bacterial symbionts of plants are in focus of genetic, molecular and ecological research aimed to reveal the fine mechanisms of symbiotic adaptations. The available data allow us to correlate the reorganizations observed in the bacterial genomes and in symbiotic (*sym*) gene networks to the microevolutionary factors operating in the bacterial populations. The output of this research is not restricted to the theoretical biology since it is required to: (i) integrate the diverse empirical data and facilitate the experimental research of symbiotic systems; (ii) analyse and regulate the environmental dynamics of microbial populations which depends greatly on their success in the habitats provided by eukaryotic hosts; (iii) create the realistic strategies to improve the beneficial microbe-plant associations and to engineer the novel types of symbiotic organisms for the sustainable crop production.

Table 1. Bacteria forming beneficial symbioses with plants

Types of symbiosis	Beneficial impacts on plants	Mechanisms	Best studied symbionts
Nutritional	Supply with nitrogen	Fixation of atmospheric N_2	Nodule bacteria (rhizobia)*; actinomycetes *Frankia***; filamentous cyanobacteria (*Nostoc, Anabaena*), endophytic (*Azoarcus, Acetobacter, Herbaspirillum*) and rhizospheric (*Azospirillum, Flavobacterium, Enterobacter*) bacteria
	Supply with phosphorous	Extraction of P from insoluble soil compounds	Phosphate dissolving bacteria (*Bacillus*)
Defensive	Suppression of pathogens	Exclusion of pathogens from the root vicinities (e.g., synthesis of antibiotics), induction of systemic resistance in plants	Plant growth promoting rhizobacteria (*Pseudomonas fluorescens, P. chlororaphis, P. putida, Serratia marcescens, Bacillus cereus*)
	Suppression of phytophagans	Synthesis of toxins	Gram-positive coryneform bacteria (*Clavibacter*)

* associated with legumes (Fabaceae) and some non-legumes (*Parasponia*: Ulmaceae); **associated with plants from eight dicot families (Betulaceae, Casuarinaceae, Myricaceae, Elaegnaceae, Rhamnaceae, Rosaceae, Coriariaceae, Datisticaceae) grouped into the clade Rosid I (together with Fabaceae and Ulmaceae); the other symbionts are not restricted to particular plant taxa.

1. Molecular Evolution of Symbiotic Bacteria

The most comprehensively studied symbiotic microbes are the root nodule bacteria (rhizobia), the N_2-fixing symbionts of leguminous plants (Fabaceae). These bacteria may be cultivated readily on the simple laboratory media and are susceptible to diverse genetic and molecular manipulations [Spaink et al., 1998]. A potential for using rhizobia as the models for the evolutionary genetics is based on their enormous biodiversity expressed with respect to systematics and symbiotic specificity (Table 2). Taxonomically, rhizobia include several groups of α-proteobacteria from the Rhizobiaceae family (*Rhizobium*, *Allorhizobium*, *Mesorhizobium* and *Sinorhizobium* are related to the tumor-forming plant pathogens *Agrobacterium*, *Azorhizobium* – to *Xanthobacter*, *Bradyrhizobium* – to *Rhodopseudomonas*) and Brucellaceae family (*Ochrobactrum*) as well as some β-proteobacteria (close to *Burkholderia* and *Ralstonia*) [Young & Haukka, 1996; Moulin et al., 2001; Chen et at., 2001, 2005; Rasolomampianina et al., 2005; Trujillo et al., 2005].

Table 2. The principle taxonomic groups of rhizobia

Genera	Species	Major host plants*	Location of *sym* genes	N_2-fixing activity *ex planta*
Rhizobium (fast-growing)	*R. leguminosarum*		On *Sym* plasmids (200-400 kb)	Absent
	bv. *viceae*	*Pisum, Vicia, Lathyrus, Lens*		
	bv. *trifolii*	*Trifolium*		
	bv. *phaseoli*	*Phaseolus*		
	R. galegae	*Galega*		
Sinorhizobium (fast-growing)	*S. meliloti, S. medicae*	*Medicago, Melilotus, Trigonella*	On *Sym* plasmids (up to 1400 kb)	Absent
	S. fredii	Different tropical and subtropical legumes; some strains have a very broad host range		
Mesorhizobium (fast- and medium-growing)	*M. loti*	*Lotus, Lupinus*	In chromosomes (symbiotic islands), sometimes on *Sym* plasmids	Absent
	M. huakuii	*Astragalus*		
	M. amorphae	*Amorpha*		
Bradyrhizobium (slow-growing)	*B. japonicum*, *B. elkanii*	Different tropical and subtropical legumes; some strains have a very broad host range	In chromosomes	Present but not sufficient for diazotrophic growth
Azorhizobium (fast-growing)	*A. caulinodans*	*Sesbania, Aeshynomene* (display the root and stem nodulation)	In chromosomes	Sufficient for diazotrophic growth

* For many tropical (e.g., *Vigna, Dalbergia, Acacia, Mimosa*) and some temperate (*Lotus, Astragalus*) legumes an ability for symbiosis with rhizobia from distant genera was reported. No congruence between the rhizobia and legume phylogenies was revealed [Doyle, 1998; Provorov, 1998].

In their host affinities, rhizobia may be classified into: (i) symbiotic specialists nodulating single plant genus (e.g., *R. leguminosarum* bv. *trifolii* is the symbiont for *Trifolium* while *R. galegae* – for *Galega* species) or to several genera from the same tribe (*S. meliloti* inoculate *Medicago*, *Melilotus* and *Trigonella* species from the Trifolieae); (ii) symbiotic generalists nodulating genera from different tribes/subfamilies of the legumes. A recognized champion in the symbiotic promiscuity is the *S. fredii* isolate NGR234 which nodulates more than 110 genera from all three legume subfamilies (Papilionoideae, Mimosoideae, Caesalpinioideae) as well as *Parasponia* – a unique non-legume (from the Ulmaceae) capable of symbiosis with rhizobia [Broughton & Perret, 1999].

1.1. Genome Architectures and Functions

Due to application of genetic and genomic approaches the extended molecular data are now available for comparing the diverse rhizobia species in organization of their genomes and of *sym* gene networks. In these bacteria the complicated genomes were revealed the architectures of which reflect the molecular evolution directed by symbiotic interactions.

1.1.1. Composition and Plasticity

The rhizobia genomes are characterized by the huge sizes (6.8 Mb in *S. meliloti*, 7.7 Mb in *R. leguminosarum*, 9.1 Mb in *B. japonicum*) reflecting a necessity to maintain the gene networks expressed at the alternative (symbiotic or free-living) stages of living cycles [Kahn et al., 2004; Young et al., 2006]. In the fast-growing *(Sino)Rhizobium* species, the genomes are separated into chromosome and large plasmids, which may reach 1400 kb in size. In some *Agrobacterium* strains close to fast-growing rhizobia the genomes are separated into circular and linear chromosomes accompanied by large plasmids [Farrand, 1998].

In fast-growing (*Sino*) *Rhizobium species,* the majority of *sym* genes are located on the special *Sym* plasmids (p*Sym*s) [Mercado-Blanko & Toro, 1996; Garcia-de los Santos et al., 1996]. In the pea and clover rhizobia (*R. leguminosarum* biovars *viceae* and *trifolii*) p*Sym*s sizes usually range within 200 and 400 kb, while in the alfalfa rhizobia (*S. meliloti*) these sizes exceed 1200 kb. However, the plasmid location of *sym* genes is not universal: for example, in *Brady-* and *Azorhizobium* strains the symbiotic functions are encoded by chromosomes while in *Mesorhizobium* they can be either chromosomal or plasmid encoded [Qian & Parker, 2002]. Interestingly, *Sinorhizobium*, *Mesorhizobium* and *Rhizobium* strains do not express N_2-fixing activity *ex planta*, while for *Brady-* and *Azorhizobium* this expression was revealed (*Azorhizobium* can grow diazotrophically using N_2 as a sole nitrogen source). Therefore, segregation of *sym* genes into the separate replicons may reflect the specialization of N_2-fixing machinery for the symbiotic purposes.

At present the total genomic sequences are available for *S. meliloti, M. loti, B. japonicum* and *R. leguminosarum* bv. *viceae* (reviewed in [Young et al., 2006]). They suggest diversification of the genome into core (conservative) and accessory (variable) parts, roughly corresponding to free-living and symbiotic functions. The core elements are characterized by relatively high G+C contents (62-64%) and are mostly chromosomal while the accessory elements are scattered at all replicons (if chromosomal, they are usually located within the mobile "islands") and are lower in G+C contents (≤60%) [Downie & Young, 2001]. The genomic analysis suggest [van Sluys et al., 2002] that rhizobia and the related agrobacteria when compared to "typical" plant pathogens (*Ralstonia, Xanthomonas, Xylella*) possess an increased number of genes responsible for intermediary metabolism, regulatory and transport functions.

A universal feature of the rhizobia genomes is represented by their enormous plasticity [Martinez et al., 1990; Romero & Palacios, 1997]. It was studied comprehensively in common bean (*Phaseolus vulraris*) symbionts (*R. etli, R. tropici*) that harbor a range of amplicons for which the copy numbers may be increased several-fold due to a *recA*-based recombination at the flanking repeats [Romero & Palacios, 1997; Mavingui et al., 1998]. The 120 kb amplicon located on p*Sym* of *R. etli* CFN42 includes genes required for host nodulation and N_2 fixation, and their amplification may result in an increased symbiotic activity. In symbiotically promiscuous *S. fredii* NGR234, genomic plasticity is expressed as the frequent (10^{-2}–10^{-3}) fusions and dissociations of three replicons: chromosome (>3.7 Mb), megaplasmid (> 2.0 Mb) and p*Sym* (536 kb) [Mavingui et al., 2002].

The elevated plasticity of rhizobia genomes is due to their saturation with IS elements, transposons and reiterated DNA sequences: in *B. japonicum* strains, some of these recombinationally active elements have 86-175 copies [Minimisawa et al., 1998, 1989]. The highest concentration of the mobile genetic was revealed in the vicinities of *sym* genes [Hahn & Hennecke, 1987; Krishnan & Pueppke, 1991; Rodrigues-Quinones et al., 1992; Romero et al., 1995] suggesting a causal relationship between the plasticity of genomes and their symbiotic functions. In p*Sym*s of NGR234 and CFN42 strains, these elements comprise 18-20% of DNA. For many rhizobia, some *sym* genes (e.g., *nodD, nifH*) are reiterated being the hot spots for recombination and gene conversion [Flores et al., 1998; Minimisawa et al., 1998; Rodriguez & Romero, 1998]. In *S. meliloti* genome, the whole megaplasmid p*Sym*A represents a major hot spot for rearrangements driving the intra-species diversification [Giuntini et al., 2005].

1.1.2. Symbiotic Functions

In order to address a relationship between elevated genome plasticity and its symbiotic functions we should summarize briefly the molecular organization of the rhizobial *sym* genes. In co-operation with complementary host genes, they control the multi-step developmental programs, which are started from penetration of rhizobia into the root hairs wherein the special tunnel structures, infection threads (ITs) are initiated [Brewin, 2004]. In *Pisum*, *Medicago* and *Lotus* species, ITs grow inter- and intra-cellularly into the root cortex wherein the nodule primordium develops and differentiates into several tissue types some of which are infected by intensively branching ITs. In evolutionary advanced papilionoid and mimosoid legumes, their development is culminated in the release of rhizobia from ITs into the plant cell cytoplasm where they are surrounded by the host membranes to form the organelle-like symbiosomes [Roth & Stacey, 1989]. Inside them, rhizobia differentiate into the bacteroids in which the nitrogenase enzyme is synthesized and N_2 fixation occurs.

The whole-genome microarray analyses suggest that 800-1000 genes from each partner do change (increase or decrease) their expression significantly during the symbiosis between *S. meliloti* and alfalfa [de Bruijn et al., 2004; Udvardi et al., 2004]. Therefore, a portion of genome involved in symbiosis is 15-20% for rhizobia and about 1% for the legumes suggesting that the bacterial genomes are much more specialized for symbiosis that the plant genomes.

The induction of symbiosis is implemented by lipo-chito-oligosaccharide (LCO) Nod factors, which represent the unique signals not known outside rhizobia [Ovtsyna & Staehelin, 2005]. These chitin-like molecules consist of 3-6 N-acetylglucosamine residues and of a fatty acid chain (16-20 C atoms). Synthesis of Nod factors is encoded by nodulation (*nod*/*nol*/*noe*) genes, which are transcriptionally activated by signals released from roots, mainly by flavonoids [Denarie et al., 1992]. After entering rhizobia into root vicinities, this activation is implemented by NodD protein (sensor of plant signals) interacting with *nod*-box (47 bp consensus) sequences located in the promoters of nodulation genes [Schlamann et al., 1992].

Only some nodulation genes (e.g., *nodABC*) are universal for all rhizobia encoding for common (core) Nod factor structures, an acylated oligochitin backbone. The majority of nodulation genes are host specific: they are responsible for affinities of different rhizobia species to particular cross-inoculation groups. The host specificity is dependent mainly on the chemical modifications in Nod factor core structures. For example, sulfation of R6 position at the non-reducing terminus encoded by *nodP*, *nodQ*, *nodH* genes is required for interaction between *S. meliloti* and alfalfa (*Medicago* spp.) while the acetylation encoded by *nodX* is necessary for *R. leguminosarum* bv. *viceae* to inoculate the "Afghan" pea (*Pisum sativum*) genotypes [Ovtsyna et al., 2000]. Genes *nodEF* found in different *Sino*(*Rhizobium*) species are responsible for synthesis and attachment to Nod factors of highly α,β-unsaturated fatty acids required for interactions with "galegoid" legumes (tribes Vicieae, Trifolieae, Galegae) growing in temperate areas [Terefework et al., 2000]. Nod factors from other rhizobia possess the saturated fatty acids derived from the common lipid metabolism [Yang et al., 1999; Debelle et al., 2001].

Due to symbiosis between legumes and rhizobia, two fundamental processes are combined: fixation of N_2 and of CO_2. A tightly integrated network of the partners' C and N metabolic pathways formed in nodules is responsible for acquiring of a novel adaptive property in plants – symbiotrophic nitrogen nutrition [Kaminski et al., 1998; Provorov & Tikhonovich, 2003]. The central role in the biochemical machinery of nodule is implemented by nitrogenase eliciting the N_2 reduction to NH_4^+ [Fisher, 1994]. This function is encoded by bacterial genes some of which are common for all N_2-fixers (e.g., *nifHDK* genes encoding for nitrogenase proteins and *nifBEN* genes encoding for MoFe-cofactor). However, many genes responsible for transcriptional regulation (*fixLJ*, *fixK*) or for energy supply of N_2 fixation (*fixGHIS*, *fixNOPQ*) are specific for symbiotic N_2 fixation.

In order to fuel the energy consuming machinery in nodules, the host spends 20-30% products of its photosynthesis [Vance & Heichel, 1991]. Being supplied to the infected nodule cells, these products are fermented to malate, which is consumed by bacteroids *via* DctA permease [Jording et al., 1994]. Among C-catabolic pathways, tricarboxylic acid cycle (TCA) dominates in bacteroids and the produced ATP is used to supply energy to nitrogenase. The immediate product of its activity, NH_4^+ is excreted into the host cytoplasm although a part of nitrogen may be exported in the form of alanine [Waters et al., 1998; Allaway et al., 2000].

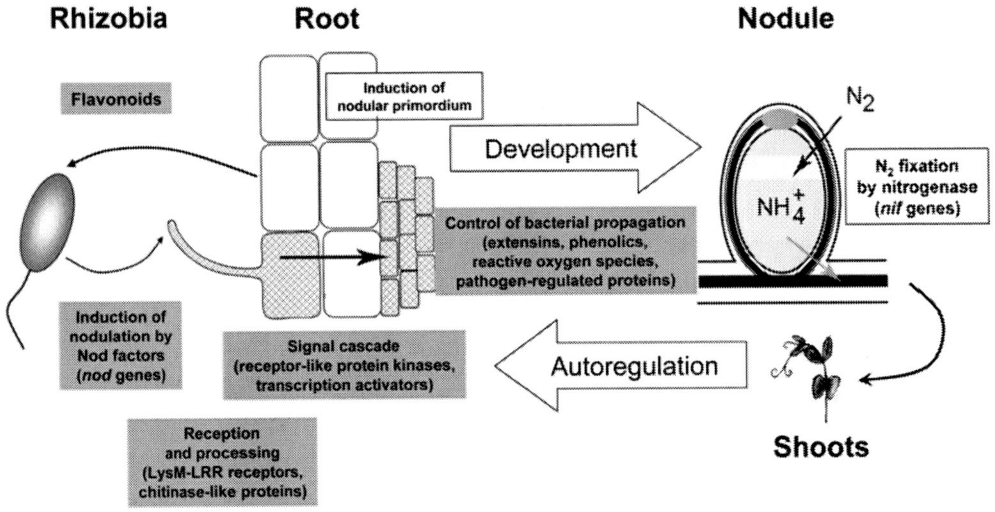

Figure 1. Regulation of the legume nodulation by rhizobia.
Components common to pathogenic interactions are given in gray boxes; components specific for mutualism are in white arrows/boxes (comments are in Section 1.2).

1.2. Natural Histories of *Sym* Genes

The available data allow us to dissect the nodule development into: (1) pathogenic-like early interactions related to the operation of Nod factors, which elicit the bacteria penetration into roots and the nodule development; (2) mutualistic (pathogenic-unlike) late interactions related to operation of the nitrogenase which implements N_2 fixation (Figure 1). These differences are reflected in the strategies used for regulation of early and late symbiotic functions and in phylogenies of the involved bacterial genes.

1.2.1. Regulation of Early and Late Nodulation Steps

For Nod factors, a pronounced similarity is evident to the chitin-like elicitors of fungal phytopathogens inducing the plant hypersensitive responses [Ovtsyna & Staehelin, 2005]. Nod factors are percepted by the plant receptor-like kinases (RLK) containing the LysM domains and transmitted the signal information to internal messengers which regulate the symbiosis-specific cellular and tissue responses [Madsen et al., 2003; Radutoiu et al., 2003]. Among these messengers, the other RLK operate which contain the leucine rich repeats (LRR) and are similar to the products of plant *R*-genes responsible for pathogen resistances [Stahl & Bishop, 2000].

At present, nearly all stages of symbiosis developmental are dissected using the mutations in plant genes controlling IT growth and nodule histogenesis; more than 50 *sym* genes were identified in the crop (*Pisum sativum, Glycine max, Vicia faba, Cicer arietinum, Arachis hypogaea*) and model (*Medicago truncatula, Lotus japonicus*) legumes [LaRue, 1980; Provorov et al., 2002; Borisov et al., 2004]. The mutational analysis suggests that the plant genes responsible for hosting bacteria in the nodule tissues and cells usually do not influence the resistance to pathogens. However, many of these genes are involved in the other mutualistic symbiosis, arbuscular mycorrhiza [Gianinazzi-Pearson, 1996].

At the beginning of nodule ontogeny, the autoregulation of nodulation (AON) is induced to balance the symbiosis with the nutritional and energy status of the plant organism [Kinkema et al., 2006]. AON is activated by a hypothetical signal transmitted from the emerging nodule into shoots and eliciting the feedback response which prevents formation of the excessive nodules. The products of some plant genes involved in this feedback (*NARK*) are close to *CLAVATA* genes that regulate the meristimatic activities and have no homologues among the known *R*-genes. However, after inducing AON, some processes common to Systemic Acquired Resistance are detected in legumes, including salicilate and jasmonate syntheses.

Dissection of symbiosis ontogeny into the pathogenic-like and pathogenic-unlike processes may be confirmed using the data on composition of plant nodulin genes induced specifically in nodules [Sanchez et al., 1991]. In infected root, the signaling cascade induces the set of early nodulin (*ENOD*) genes for some of which the products are similar to PR (pathogen-regulated) proteins and extensins [Gamas et al., 1998; Brewin, 2004]. The other defense-like reactions induced during IT growth in the epidermis and cortex includes syntheses of phenolics, flavonoids, lytic enzymes and reactive oxygen species [Spaink, 1995; Santos et al., 2001]. Due to these reactions, nodule development that is quite different from the known defence reactions at the organ/tissue levels looks very similar to them at the molecular level. The defence reactions may be enhanced greatly if rhizobia are mutated in genes controlling exopolysaccharide (*exo*, *exp*), lipopolysaccharide (*lps*) or cyclic β-glucan (*ndv*) syntheses responsible for a dialogue with host immune systems [Breedveld & Miller, 1994; Kannenberg et al., 1998; Becker & Pühler, 1998].

When the nodules are fully developed and ready for N_2 fixation, plant induces the range of late nodulin genes which do not have homologues in the defence systems. Late nodulins are represented mainly by the nodule-specific enzymes for C metabolism (sucrose synthase, malate dehydrogenase, phosphoenol pyruvatcarboxylase) and N metabolism (glutamine synthetase, glutamate synthase, asparagine synthetase, aspartate amino transferase) involved in the nodule supply with energy and in assimilation of fixed nitrogen [Fedorova et al., 1999]. Among late nodulins, the most abandoned is leghaemoglobin (Lb) which comprise more than a half of total proteins in rhizobia-infected plant cells. The major function of Lb is to protect nitrogenase from O_2 which arrests N_2 fixation at very low concentrations (> 50 nM).

At all stages of symbiosis, the feedback interactions between partners are implemented due to their signal exchange. Pathogenic-like stages involve the negative regulatory feedbacks, e.g.,: (i) Nod factors induce their own degradation by the plant chitinase-like enzymes [Ovtsyna et al., 2000]; (ii) rhizobia penetration induces defense-like responses that restrict their distribution to the specially organized inter- and intra-cellular compartments; (iii) nodule initiation induces the AON response.

However, the pathogenic-unlike (mutualistic) stages mostly involve the positive regulatory feedbacks. For example, the onset of N_2 fixation may stimulate the plant photosynthesis and the allocation of its products into N_2-fixing nodules resulting in enhancement of nitrogen fixation [LaRue, 1980]. The host can stimulate bacteroids to uptake the principle C sources, C4-dicarboxylates, using the symbiosis-specific signals [Jording et al., 1994]. A coordinated expression of late nodulins may be controlled by some signals coming from bacterial cells and interacting with the nodulin gene promoters. In *Sesbania rostrata* the promoter of *Lb3* contains two sites specific for the bacterial gene regulators. In *Azorhizobium caulinodans* (*Sesbania* symbiont) two proteins were identified which bind these sequences and the bacterial mutations that knockout the regulatory proteins lead to a decrease in Lb synthesis and in N_2 fixation [de Bruijn et al., 1994].

Deep differences between the early and late symbiotic processes are emphasized by the data demonstrating that later involve the marked features of a programmed altruism expressed by bacteria towards their hosts. Being provided with the excess of carbon, bacteroids fix much more N_2 than is required for the inter-nodule clone and the excessive nitrogen is donated to host [Udvardi & Kahn, 1992]. In rhizobia, *nif* genes are activated in nodules under the excess of combined nitrogen that inhibits N_2 fixation by free-living diazotrophs [Kaminsky et al., 1998]. Special genes are required for the irreversible bacteroid differentiation, e.g., *bacA* which has close homologues in *Brucella* – intracellular pathogen of vertebrates [Ichige & Walker, 1997; Oke & Long, 1999] and *minCDE* having homologues involved in control of cell morphology in *E. coli* [Cheng et al., 2007]. After a period of functioning, the bacteroids may be consumed by plant cells [Brewin, 1991]: the bacteria display a genetically controlled ability to sacrifice themselves to feed the host with nitrogen! However, not all bacteria within N_2-fixing nodule are converted into bacteroids: some cells retain the reproductive activity and enter the soil population after nodule death. These bacteria do not fix N_2 and consume the host metabolites behaving as the plant pathogens. In alfalfa nodules, this consumption occurs in the basal senescence zone where bacteria leave the infection threads and propagate in cytoplasm of the decaying plant cells [Timmers et al., 2000].

1.2.2. Combinative Evolution of Sym Gene Networks

The early genetic research [Beringer et al., 1980] suggested that since nodulation represents an unusual developmental program, it might be controlled by the unique rhizobial genes. However, it was demonstrated later that the *sym* genes are not unique for symbiosis. Specifically, Nod factors are synthesized by enzymes the homologues of which fulfill diverse metabolic and regulatory functions in non-symbiotic bacteria [Ovtsyna & Staehelin, 2005]. For example, NodD represents the LysM-AraC family of transcriptional regulators involved in a variety of metabolic processes [Schlamann et al., 1992], while the homologues of NodHPQ are involved in synthesis of sulfur containing amino acids. An exception may be represented by NodA responsible for attachment of the fatty acid to oligochitin chain: this enzyme have no homologues in gram-negative bacteria and may be originated by a horizontal gene transfer (HGT) from gram-positive bacteria or even from some fungi [Hirsch et al., 2001].

In contrast to Nod factors eliciting the early symbiotic steps, the nitrogenase synthesized at the late steps is not unique for rhizobia. However, in organization of the involved genes rhizobia display many peculiarities with respect to free-living diazotrophs reflecting the gene adaptation for *in planta* expression. No more than a half of 24-25 genes from *nif* cluster of the model free-living N_2-fixer *Klebsiella pneumoniae* [Pühler & Klipp, 1981] were found in rhizobia. For example, they lack the negative regulator *nifL* which downregulates the

Klebsiella nif gene transcription by combined N and O_2. The absence of *nifL* allows rhizobia to fix N_2 within plant cytoplasm rich in combined N (in symbiosis the co-ordination between N_2 fixation and N assimilation from the environment is regulated by the host *via* systemic mechanisms [Kinkema et al., 2006]) while the oxygen control of *nifHDK* in rhizobia is implemented by NifA (which in contrast to *Klebsiella* NifA possesses an oxygen-binding domain) and FixLJ (represent the two-component transcriptional regulators widespread in gram-negative bacteria but not involved in regulation of free-living N_2 fixation) [Hennecke, 2004; Bobik et al., 2006]. In rhizobia, *nif* genes are included into the extended regulatory network which co-ordinates many processes expressed *in planta* (uptake of C4-dicarboxylates, electron transport, oxygen sensing, N metabolism) and is possibly managed by the symbiosis-specific signals [Hauser et al., 2006].

These data suggest that evolution of the major symbiotic traits in rhizobia (signaling, host penetration and N_2 fixation) did not involve generation of novel enzymatic functions or of novel genes. This evolution was dominated by intra- and inter-genome recombination resulted in the reorganizations (recruiting) of the pre-existing genes into the novel host-controlled operons and regulons [Provorov, 1998]. The elevated sizes and plasticity of rhizobia genomes as well as an increased redundancy of many metabolic pathways and regulatory elements [van Slyus et al., 2002; Kahn et al., 2004] reflect this evolution.

Table 3. Evidence for different natural histories in rhizobia genes encoding for the early and late symbiotic interactions

Comparisons for	Genes	
	nod/nol/noe	*nif*
Stages/functions	Pathogenic-like early interactions (signaling and root penetration)	Mutualistic late interactions (symbiotic N_2 fixation)
Encoded products	Nod factors	Nitrogenase
Feedback with the host	Negative	Positive
Involvement in "gene-for-gene" interactions with the host	Evident	Absent
Correlation with bacterial divergence:		
• at the intra-species level	Absent	Present or variable
• at the inter-species level	Absent	Present
Correlation with host divergence	Present	Absent

Being subjected to the common (combinative) evolutionary strategy, *nod* and *nif* genes have quite different natural histories related to the peculiarities of their regulation (Table 3). This difference is illustrated by the molecular phylogeny data suggesting that the polymorphism of nodulation genes does not correlate to the divergence of core bacterial genomes (studied using 16S rDNA polymorphism at the inter- and intra-species levels) but correlates to the taxonomy of hosts, the interactions with which follow the "gene-for-gene" schemes [Dobert et al., 1994; Paffetti et al., 1996; Laguerre et al., 1996; Wernegreen & Riley, 1999; Zhang et al., 2000; Ba et al., 2002; Jarabo-Lorenzo et al., 2003]. The similar regularities were found in genes responsible for host infection in pathogenic bacteria [Guttman & Sarkar, 2004; Shan et al., 2007]. However, polymorphism of *nif* genes usually correlates to the divergence of bacterial core genome, not to the host divergence [Dobert et al., 1994; Brunel et al., 1996; Laguerre et al., 1996; Chen et al., 2000; Saleena et al., 2001; Jebara et al., 2001]. Discordant phylogenies of *nod* and *nif* genes were demonstrated for *R. leguminosarum* biovars *viceae, trifolii* and *phaseoli* [Schofield et al., 1987; Laguerre et al., 1996], *R. galegae* [Vlades & Pinero, 1992] and *S. meliloti* [Roumiantseva et al., 2002; Bena et al., 2005] suggesting different pathways of evolution in these genes.

2. Bacterial Populations in "Plant-Soil" Systems

The majority of symbiotic bacteria combine the abilities for genetically controlled interactions with hosts and for effective exploitation of the *ex planta* (soil) environments. The impacts of different biotic and abiotic stresses on these bacteria are well studied [Bottomley, 1992; Martinez-Romero & Caballero-Mellado, 1996; Martin, 2002; Kent & Triplett, 2002; Horner-Devine et al., 2004] while operation of microevolutionary factors induced by these stresses are understood poorly and their impacts on the molecular evolution are obscure. In order to analyse these impacts we suggest that the differences in natural histories between *nod* and *nif* genes (Section 1.2.2) reflect the variable selective pressures operating in the rhizobia populations at the successive stages of symbiosis. These pressures may reflect the partners' evolutionary conflict concerning the major parameters of symbiotic interactions: their specificity and efficiency [Frank, 1996; Douglas, 1998]. It is generally assumed that natural selection operating from the host side tends to narrow the symbiotic specificity (controlled by rhizobial *nod/nol/noe* genes) and to improve the beneficial effects (dependent on *nif* genes). From the microbial side selection tends to overcome the restrictions in specificity (resulting in "gene-for-gene" co-evolution) but may hamper the beneficial functions expressed for the hosts' sake. In order to trace the interplay of microevolutionary factors resulting from this conflict we should address the processes occurring to rhizobia populations circulating in "plant-soil" systems.

2.1. Population Diversity Induced by Hosts

During interaction with hosts, rhizobia surpass a series of habitats in "plant-soil" system the occupation of which results in reorganizations of spatial and genetic population structures. The suggested "Infection and Release" Cycle (IRC) involves competition among the virulent strains for host infection, propagation of the winners *in planta*, release of endosymbiotic bacteria into environment and their interactions with the resident soil strains (Figure 2, Table 4). The microevolutionary pressures induced in ICR results from the symbiosis ontogeny suggesting that the rhizobia evolution represents a programmed process controlled genetically by both partners.

Table 4. Reorganizations in the rhizobia population structures during the symbiosis development (for indeterminate nodules [Provorov et al., 2002])

Symbiosis stages	The bacterial functions involved (the relevant genes are given in parentheses)	Bacteria performance in the "plant-soil" system	Microevolutionary factors operating in the bacteria populations
Pre-infection	Motility (*mot, fla*), chemotaxis (*che*), adsorption on roots	Colonization of rhizosphere and rhizoplane	Darwinian selection in favor of strains with the high root-colonizing activities
Root infection and nodule development	Nod factor synthesis (*nod*/*nol*/*noe*), competition for nodulation (*cmp*)	Inoculation of roots	Frequency dependent selection in favor of strains with the high competition-for-nodulation activities; genetic drift related to restriced number of infection sites
Nodule functioning	Modifications in surface components (*exo, exp, lps*) required to adapt the *in planta* habitats, development of bacteroids (*bacA*), N_2 fixation (*nif*), import of dicarboxylic acids (*dct*) and construction of electron transport chains (*fix*) for energy supply of nitrogenase	Colonization of inter-cellular (infection threads) and intra-cellular (symbiosomes) compartments	Clonal propagation of winners *in planta*; Darwinian and group (inter-deme, kin) selection in favor of strains capable of using the host-provided carbon for *in planta* propagation
Exit from symbiosis	Degradation of bacteroids and propagation of non-differentiated cells in the senescent nodules	Release into the soil	Darwinian selection in favor of strains with the high competitiveness for *ex planta* survival enhanced due to population waves

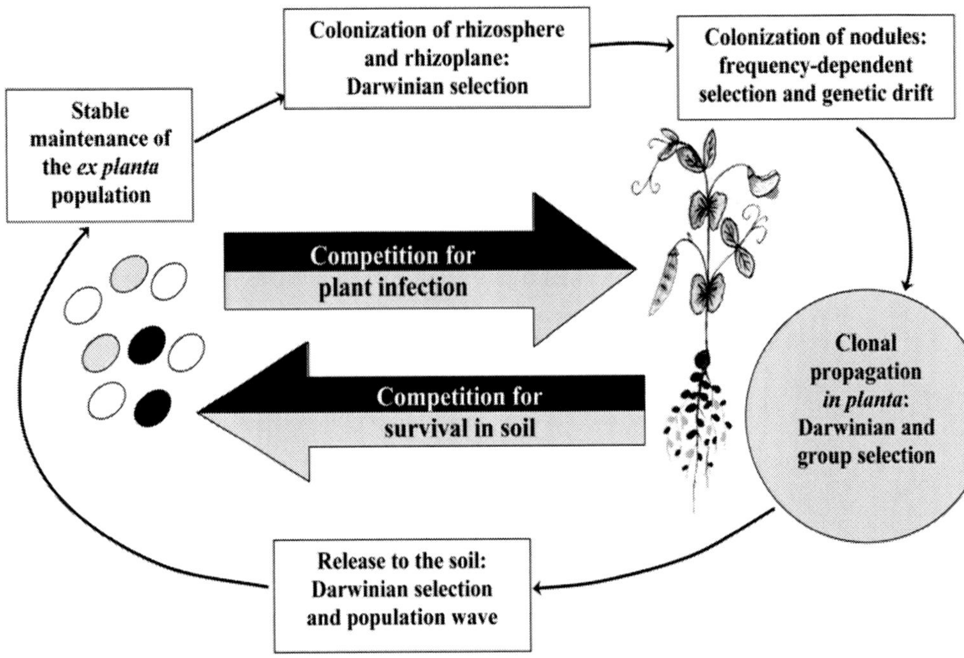

Figure 2. Infection and Release Cycle (IRC) involved in reorganizations of the rhizobia populations in the "plant-soil" systems. Mathematical expressions for the microevolutionary factors are given in [Provorov & Vorobyov, 2006]; relationships between the involved molecular and population processes are in Table 4.

2.1.1. Balanced Polymorphism

An immediate goal for analyzing IRC mechanisms is to correlate them to the peculiarities revealed in rhizobia populations using a broad range of methods including analyses of serological properties [Broughton et al., 1987a; Brockman & Bezdicek, 1989; Olsen et al., 1994], intrinsic antibiotic resistance [Broughton et al., 1987a; Shishido & Pepper, 1990; Nour et al., 1994], responces to phage infections [Bromfield et al., 1986, 2001] multi-locus enzyme electrophoresis (MLEE) profiles [Young et al., 1987; Harrison et al., 1989; Nour et al., 1994], multi-locus sequence typing [van Berkum et al., 2006], plasmid profiles [Broughton et al., 1987a; Brockman & Bezdicek, 1989; Shishido & Pepper, 1990; Kuykendall et al., 1996; Corich et al., 2001], IS (insertion sequences) content [Kosier et al., 1993; Deng et al., 1995; Laberge et al., 1995; Mazurier et al., 1996] and various genetic markers analysed using a variety of PCR-based techniques [Dye et al., 1995; Perret & Broughton, 1998; Doignon-Bourcier et al., 2000; Sanchez-Contreras et al., 2000; Corich et al., 2001].

The coincidence of population structures revealed using different methods vary from good [Hartmann & Amarger, 1991; Strain et al., 1994; Dye et al., 1995; Nour et al., 1994; Corich et al., 2001] to zero [Noel & Brill, 1980; Kuykendall et al., 1996] making it difficult to compare the data obtained in various bacteria. The most suitable for such comparisons are the MLEE data for which the Nei's [1978] index of diversity may be calculated as $H = (1-\Sigma p_i^2)\cdot[n/(n-1)]$, where p_i is a frequency of i-th genotype, n – number of analyzed strains (H values vary from 0 if population consists from a single genotype, up to 1 if each isolate represents a separate genotype [Selander et al., 1986]). These calculations suggest [Martinez-Romero & Caballero-Mellado, 1996; Provorov, 2000] that the average values of H are: 0.59

for rhizobia (*Rhizobium, Sinorhizobium, Bradyrhizobium*), 0.45 for the specialized animal pathogens (*Bordetella, Borrelia, Eryispelotrix, Haemophilus, Helicobacter, Listeria, Mycobacterium, Neisseria, Staphylococcus*) and 0.37 for enterobacteria (*Escherichia, Salmonella, Shigella*). The enormous polymorphism found in rhizobia populations may be finely balanced because under edaphic stresses the sensitive strains are often retained at high frequencies in soil and host-associated populations. For example, in soybean [Sadowsky & Graham, 1998] and alfalfa [del Papa et al., 1999] rhizobia isolated from acid soils, pH-sensitive strains may comprise more than a half of population.

For rhizobia populations, the legume hosts are more important for establishment and diversification than the soil/climatic factors [Spoerke et al., 1996; Paffetti et al., 1998; Carelli et al., 2000; Andronov et al., 2003]. In presence of hosts, the rhizobial densities may be increased by several orders of magnitude and reach 10^7–10^8 cells per g of soil or 20-25% of total bacterial counts [Bottomley, 1992; Hirsch, 1996]. The major reasons for this increase are the rhizobia multiplications in legume rhizosphere (which are usually more intensive than in rhizospheres of non-legumes) and in nodules the decay of which results in release of bacterial cells into soil.

High population densities reached by rhizobia due to symbiosis are usually accompanied by the elevated diversities [Harrison et al., 1989; Dye et al., 1995; Andrade et al., 2002] suggesting the crucial role of hosts in promoting the bacteria evolution. Specifically, the genetic differences in hosts may be more important in shaping the microsymbiont population than the soil type as it was demonstrated in *S. meliloti* nodulating different alfalfa cultivars [Bromfield et al., 1986; Paffetti et al., 1998; Hartmann et al., 1998; Carelli et al., 2000] and in *Bradyrhizobium* spp. nodulating the wild-growing legume *Amphicarpaea* [Spoerke et al., 1996; Wilkinson et al., 1996]. The absence of host is correlated to a decreased diversity in *S. meliloti* [Kosier et al., 1993] and *B. japonicum* [Minimisawa et al., 1999; Ferreira et al., 2000] populations. The mechanisms of host influence may involve the direct induction of bacterial mutations as indicated by an increased diversity in colony types [Krassilinikov, 1941], plasmid profiles [Roumiantseva et al., 2004] or N_2-fixing activity [Weaver & Wright, 1987] in the clones isolated from nodules. In *Agrobacterium,* the frequencies of point mutations and micro-deletions in the clones from tumors may be increased with respect to free-living clones [Belanger et al., 1995].

A key mechanism for maintaining the balanced rhizobia polymorphism in "plant-soil" system may be the interplay of Darwinian selection pressures operating during root penetration (which requires the *nod/nol/noe* gene activities) and during survival in soil (wherein the strains devoid of these genes may be successful). An importance of this interplay is illustrated by the stable co-existence of symbiotic and asymbiotic strains (referred further as an "ecotypic polymorphism" [Provorov & Vorobyov, 2000a]). In rhizobia, the asymbiotic strains arise from the symbiotic ones (by loss of *sym* genes or p*Sym*s which may constitute an excessive burden on the cell during saprophytic survival) and may comprise more than 90% of the soil rhizobia populations being similar to endosymbiotic populations in a genetic structure [Pinero et al., 1988; Segovia et al., 1991]. The avirulent strains are numerous in populations of pathogenic bacteria in which the virulence genes may be lost at the saprotrophic stages of living cycles [Schuster & Coyne, 1974; Frank, 1992; Thompson & Burdon, 1992]. We suggest that the ecotypic polymorphism in symbiotic bacteria ensures maximal average fitness of their populations that circulate between alternative habitats in host-environment ecosystems.

2.1.2. Frequency Dependent Selection

A key role in supporting the population diversity in organisms evolving under the strong impacts of biotic factors is implemented by Frequency-Dependent Selection (FDS), which favors the rare genotypes [Ayala & Campbell, 1974]. In the "gene-for-gene" controlled plant-pathogen interactions, FDS is mediated by the lack of host resistance against the rare virulence genes [Frank, 1992; Thompson & Burdon, 1992]. With increasing their frequencies, pathogens often lose the frequent virulence genes (due to the "costs" required to support them [Rausher, 2001]) that in turn become the objects for FDS. This type of selection (negative FDS) was revealed also in viruses [Elena et al., 1997] and in bacterial pathogens of invertebrates [Carius et al., 2001] while its operation in mutualistic symbionts is studied poorly.

Previously we suggested [Provorov & Vorobyov, 2000b] that FDS is operating in rhizobia during competition for the host nodulation that involves an empirically revealed non-linear transformation of cell numbers [Amarger & Lobreau, 1982]:

$$N_1:N_2 = c\cdot(I_1:I_2)^a, \qquad (1)$$

where I_1 are I_2 are the initial cell numbers of two strains in the inoculum; N_1 and N_2 are their resulted numbers in nodules; a and c are constants which depend on the genotypes of competing strains (and of the inoculated host [Beattie et al., 1989]) but do not depend on the strains' ratio. It was demonstrated that constant c may be either more or less than 1 reflecting the relative nodulation competitiveness of the inoculated strains at $I_1=I_2$. However, constant a appeared to be permanently less than 1 (usually $0.2<a<0.8$). One can see that if strain №1 is rare in inoculum ($I_1 < I_2$), its frequency in nodules is increased ($N_1:N_2 > I_1:I_2$ at $c=1$) leading to the negative FDS resulted from the transition of bacteria from soil (rhizospheric) to nodular habitats.

Operation of FDS in rhizobia at the early nodulation stages may be caused by their similarities to "gene-for-gene" controlled pathogenic interactions (Section 1.2). The molecular mechanisms responsible for FDS can involve activation of some nodulation genes with decreasing the cell density in environment suggesting that the rare strains become more virulent than the dominating strains. In pea [Rodelas et al., 1999; Lithgow et al., 2000], pole bean [Rosemeyer et al., 1998] and soybean [Loh et al., 2001, 2002] rhizobia this activation is mediated by the quorum sensing (density dependent) regulatory circuits. The non-linear (S-like) dependencies between concentrations of Nod factors and the molecular responses induced in root hairs [Oldroyd et al., 2001] may contribute sufficiently into the frequency dependent dynamics of competitive process.

Moreover, the competition for nodulation involves the rhizobia adsorption on root hairs [Matthysse & Kijne, 1998] that may follow the nonlinear Freundlich equation similar to dependency (1) between the bacterial ratios in nodules and inoculum [Amarger & Lobreau, 1982]. This similarity is consistent with the data demonstrating that coefficient a responsible for FDS pressure is dependent sufficiently on the host genotype [Amarger & Lobreau, 1982;

Beattie et al., 1989], which controls the dynamics for adsorption of microbial cells and for perception of their signals [Matthysse & Kijne, 1998; Oldroyd et al., 2001]. A casual relation of FDS to the endosymbiotic interactions is evident from the decreased diversity of rhizospheric bacteria in the root vicinities due to multiplication of a narrow range of genotypes consuming effectively the root exudates [Schlöter et al., 2000].

It is logical to suggest that the evolutionary impacts of the revealed FDS pressure involve the establishment of high polymorphism typical for rhizobia populations. This polymorphism may be due to the balance of FDS with the other microevolutionary factors, firstly with Darwinian selection which operates at all stages of symbiotic interaction due to differential multiplication/extinction of competing strains in soil, rhizospheral and nodular habitats [Provorov & Vorobyov, 2006]. This effect of FDS may be enhanced by the genetic drift which occurs due to root infection by a restricted number of cells which propagate clonally in nodules.

The other potential impact of FDS on symbiosis evolution includes the support of its mutualistic characters since the rhizobia strains with a high N_2-fixing potential are often the poor soil survivors. For example, in clover rhizobia populations, highly effective N_2 fixers were revealed in the minor genetic classes (identified by probing different chromosomal and plasmid genes) not in the major classes [Leung et al., 1994a, 1994b]. In this rhizobia species, poor soil survivors are usually good nodule colonizers and *vice versa* [Duodu et al., 2005]. In alfalfa rhizobia populations, active N_2-fixers are more frequent among pH-sensitive strains than among pH-tolerant strains [del Papa et al., 1999].

One can suggest that FDS saving the active N_2 fixers within *ex planta* habitats may be responsible for maintenance of mutualistic interactions under the stress conditions. However, FDS-based mechanism is not related functionally to efficiency of mutualism (measured as the improvements of fitness obtained by partners due to symbiosis [Douglas, 1994]) that may evolve under the impacts of factors, which differ considerably from the individual (Darwinian, frequency-dependent) selective pressures (Section 3.2).

2.2. Gene Transfer in Evolution of Symbiotic Bacteria

Since bacteria are devoid of sexual reproduction, their typical population constitutes a commonwealth of clones [Young, 1989] or meroclones [Feil & Spratt, 2001] in which the vertical gene transmission dominates over the horizontal one. However, distant bacteria that occupy the same environment are usually not isolated genetically due to a DNA exchange *via* transformation or conjugation [Thomas & Nielsen, 2005]. The resulting "non-Mendelian" organization of bacterial populations (lack of free gene combination within populations and of genetic isolation from populations of the other species) should be considered to estimate a role of HGT in bacterial evolution [Ochman & Bergthorsson, 1998; Ochman et al., 2000; Gogarten & Townsend, 2005]. For deleterious bacteria, the mobility of "pathogenicity islands" in populations is well documented that results in the expansion of virulence genes in microbial communities leading to generation of the novel pathogens [Groisman & Ochman, 1996; Stephens & Murray, 2001; Dorrell et al., 2005; O'Connell, 2006]. However, in mutualistic systems the population dynamics of symbiotic genes is not studied in details. In this section, we use the model of legume-rhizobia symbiosis to trace the role of HGT in the microevolution of symbiotic microbes under the ecological impacts induced by plant hosts.

2.2.1. Population Structures

A key property of rhizobia populations is represented by their tendency to form the panmictic structures due to symbiotic interactions [Maynard Smith et al., 1993; Gordon et al., 1995; Martinez-Romero & Caballero-Mellado, 1996; Vinuesa et al., 2005]. By using the index of allele association I_A ($I_A=V_o/V_e$, where V_o is the observed variance for the number of allele mismatches in the pair-wise strain comparisons, V_e is the same expected under the random allele recombination) a lack of linkage disequilibrium between alleles of izozyme loci was revealed in many populations [Gordon et al., 1995] although this disequilibrium is typical for free-living bacteria [Lenski, 1993; Maynard Smith et al., 1993]. The other approach demonstrating high panmixia in rhizobia populations is the analysis of associations between plasmid and chromosomal (e.g., 16S rRNA or MLEE) markers [Schofield et al., 1987; Kinkle & Schmidt, 1991; Laguerre et al., 1993, 1996; Gordon et al., 1995; Louvrier et al., 1996].

Mechanisms for development of panmictic structures in rhizobia populations are essentially symbiosis-specific because for p*Sym*s the distribution with respect to chromosomal markers is more random than for "non-symbiotic" plasmids [Laguerre et al., 1992, 1993, 1996]. The "phylogeny congruency test" suggested that in *R. etli* and *R. galegae* populations 16-30% of strains have been the donors or recipients of p*Sym*s in their recent past [Vlades & Pinero, 1992]. *R. etli* populations isolated from individual host plants are more panmictic than populations from different plants in the same field or from separate fields [Souza et al., 1992]. The random associations with 16S rRNA genes were revealed for *glnII*, not for *glnI* gene encoding for glutamine synthetases II and I [Wernegreen & Riley, 1999; Turner & Young, 2000] consistent with involvement of *glnII*, not of *glnI* in co-regulation with *nif* genes [Hauser et al., 2006].

An important factor influencing the degree of panmixia in rhizobia populations may be the variable stability of transferred genes in recombinants depending on genetic relatedness of donor and recipient strains. Due to this variability, p*Sym*s in *Rhizobium leguminosarum* populations associate randomly only with the particular chromosomes [Young & Wexler, 1988; Louvrier et al., 1996]. Different authors [Eardly et al., 1990; Bailly et al., 2006] demonstrated that the plasmid/chromosomal associations are random inside two related species of alfalfa rhizobia (*S. meliloti*, *S. medicae* [Rome et al., 1996; Biondi et al., 2003]) while they are not random across these genetically isolated species. The same isolation was demonstrated for two biovars (subspecies) of *R. galegae* effective in symbiosis with the goats' rue species, *Galega orientalis* and *G. officianalis* [Andronov et al., 2003]. Due to a limited plasmid-chromosomal compatibility, high levels of panmixia are characteristic mostly for homogeneous, not for heterogeneous bacterial populations. For 18 rhizobia populations [Gordon et al., 1995] a coefficient of correlation between I_A and H values is +0.60 ($P_0<0.01$) while for 21 populations of free-living and symbiotic bacteria [Maynard Smith et al., 1993] it is +0.47 ($P_0<0.05$).

Another mechanism restricting the p*Sym*-chromosome combination in rhizobia may be a decreased competitivenss in some recombinants. Non-random p*Sym*-chromosome associations were found in *R. leguminosarum* bv. *viceae* strains isolated from nodules while for soil isolates these associations were random [Harrison et al., 1988; Engvild et al., 1990; Louvrier et al., 1996]. Variation of the nodulation competitiveness in various p*Sym*-chromosome associations can result in the epidemic *in planta* propagation of only some strains which shade the initially panmictic structure in soil populations [Maynard Smith et al., 2000]. The same conclusion may be deduced from comparison of symbiotic and non-symbiotic *R. etli* subpopulations (isolated from *Phaseolus vulgaris* nodules and from bulk soil) using MLEE analysis: the first one was clonal, the second one – panmictic [Gordon et al., 1995].

A crucial role of HGT in rhizobia microevolution is supported by the data on intra-species transfer of p*Sym*s within *(Sino)Rhizobium* species under laboratory conditions. The mating experiments suggest that: (i) at least some p*Sym*s may be transferred due to their own *tra* functions with the frequencies up to 10^{-5} per recipient within or between different biovars of *R. leguminosarum*; (ii) an ability to form N_2-fixing nodules with the donor's hosts is often inherited after transfer of p*Sym*s between *R. leguminosarum* biovars [Beringer, 1982; Martinez et al., 1990]. However, the degree of panmixia in natural rhizobia populations is higher than it would be expected from the data on gene transfer since the p*Sym* conjugation was revealed only in some *Rhizobium* and *Sinorhizobium* strains while in many strains p*Sym*s are non-transmissible. Some rhizobia possess the intrinsic systems for genetic transformation and transduction [Beringer et al., 1980; Hynes & Finan, 1998] but their efficiencies are usually low. With this respect, rhizobia differ greatly from enterobacteria (*Escherichia*, *Salmonella*) which possess the highly effective systems for gene transfer (e.g., *F*- and *R*-factors) but usually display the clonal population structures [Selander & Levin, 1980; Caugant et al., 1981; Maynard Smith et al., 1993; Ochman et al., 1999]. The genetic structures are usually more panmictic in *Rhizobium* populations isolated from limited soil plots than in *E. coli* populations isolated from different human bodies [Souza et al., 1992]. Also it is important to note that the panmictic structures are formed in rhizobia populations in spite of a regular clonal propagation occurring within nodules.

2.2.2. "Gain-and-Loss" Model

The input of HGT into the bacterial population structures may be modified greatly by selective pressures which determine the fate of recombinants in the environment. For the human pathogens (*Neisseria*, *Streptococcus*) J. Maynard Smith and co-workers [2000] suggested that the background population structure may be transformed from panmictic to the clonal/epidemic due to multiplication of a narrow range of virulent clones under the impacts of Darwinian selection (Figure 3A). In rhizobia, controversy between the poor *sym* gene transfer and high panmixia in natural populations may be explained by the frequency dependent dynamics of competition for nodulation which ensures multiplication of rare recombinants (Figure 3B). Using the specially constructed mathematical simulation technique, we demonstrated that under realistic parameters of rhizobia-legume system, FDS might ensure multiplication of recombinants generated with frequencies less than 10^{-15} per recipient [Provorov & Vorobyov, 2000b]. Therefore, impacts of HGT on adaptive evolution of symbiotic bacteria may be less dependent on the gene transfer frequencies than on the availability of selective pressures which pick up the rare recombinants and anchor them in populations.

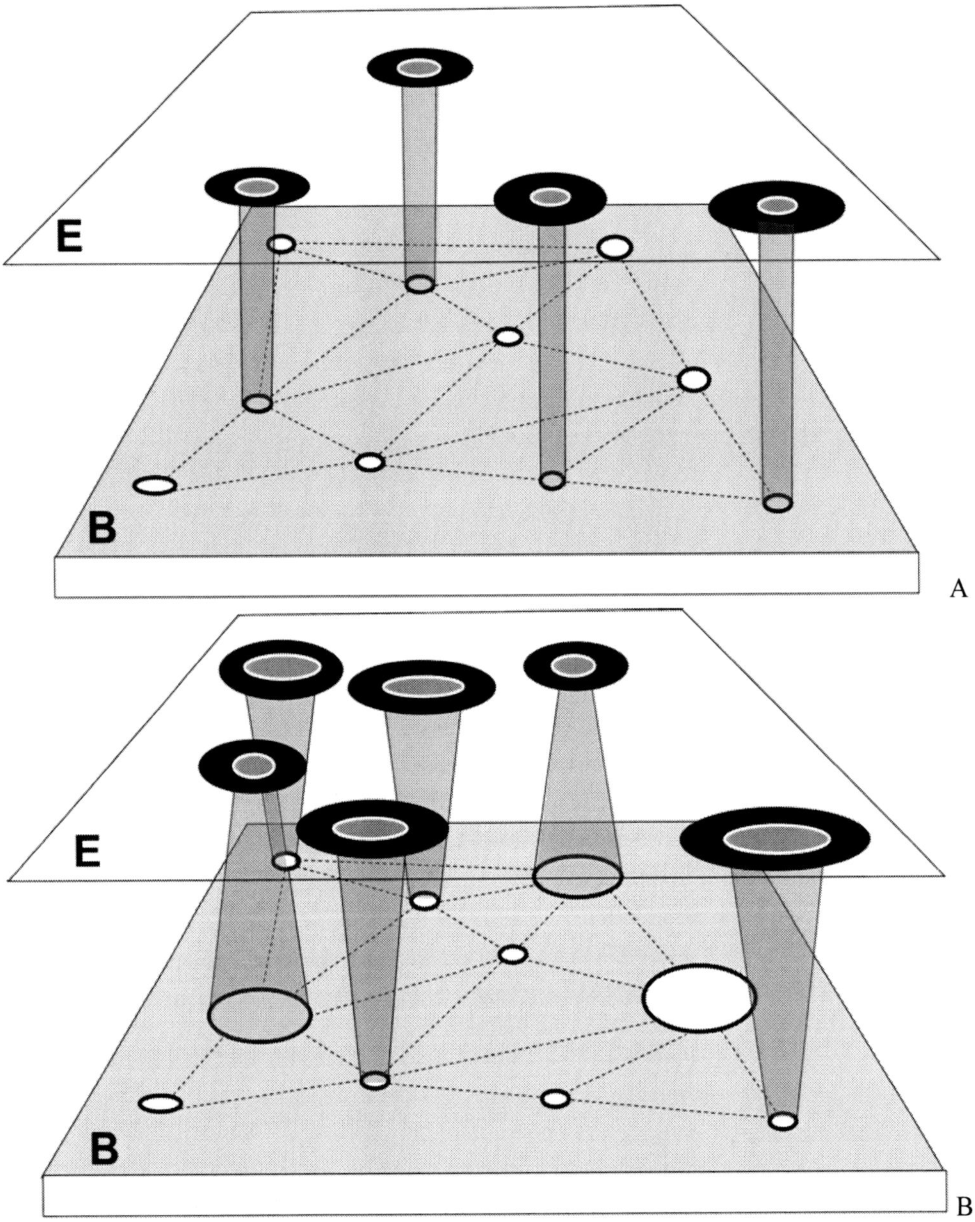

Figure 3. Transformation of the background (B) to empirically revealed (E) population structures in symbiotic bacteria under impacts of the host-induced selective pressures.
A – transformation of the panmictic into the epidemic structure die to multiplication of infective clones *via* Darwinian selection (modified from [Maynard Smith et al., 2000]; linkage equilibrium in B-population is shown by the equivalent sizes of circles representing different gene combinations);
B – transformation of clonal (epidemic) into the panmictic structure due to propagation of rare infective recombinants *via* frequency dependent selection (Section 2.1.2; linkage disequilibrium in B-population is shown by the variable sizes of circles representing different gene combinations).

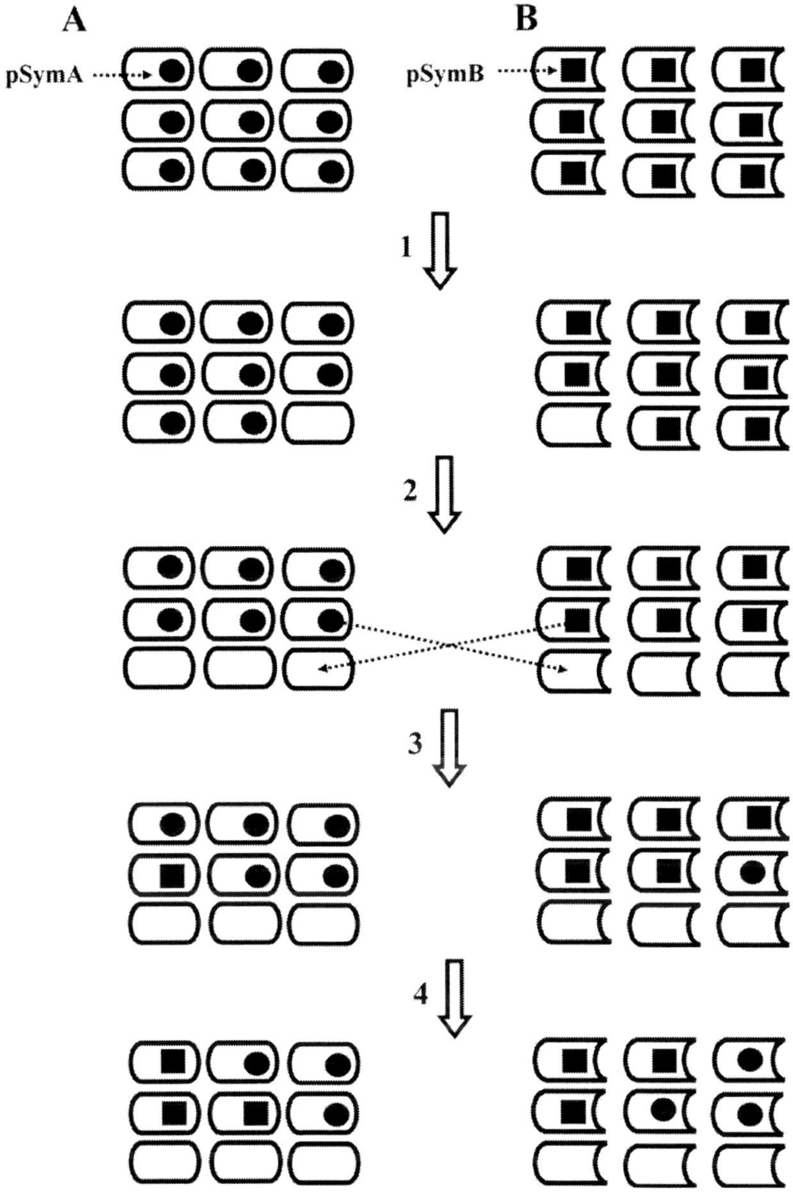

Figure 4. "Gain-and-Loss" strategy in formation of a typical (panmictic, ecotypically polymorphic) population in rhizobia due to combination of *sym* gene transfer and natural selection pressures.
The initial populations consists of two virulent strains (A and B) harboring the *Sym* plasmids p*Sym*A and p*Sym*B.
1 – generation of avirulent mutants *via* the p*Sym* losses;
2 – formation of the ecotypically polymorphic population due to propagation of avirulent mutants having an improved saprophytic competence;
3 – generation of virulent recombinants due to p*Sym* transfer;
4 – multiplication of virulent recombinants (A-p*Sym*B and B-p*Sym*A) due to frequency dependent selection during competition for nodulation. The resulted populations acquire a panmictic structure based on broad distribution of recombinant genotypes.

In order to compose an integral picture for the formation of rhizobia populations we propose a scheme for arrangement of their typical (panmictic, ecotypically polymorphic) structures based on the "Gain-and-Loss" dynamics of *sym* genes (Figure 4). We suggest that a combination of the events generating novel genotypes (gene losses and gains) with the selective pressures ensuring their anchoring/fixation in the host-environment system results in formation of the bacterial population structure. The "Gain-and-Loss" scheme is based on broad distribution of asymbiotic (lacking p*Sym*s) strains in the "ecotypically polymorphic" populations [Soberon-Chavez & Najera, 1989; Segovia et al., 1991] which can be optimal recipients for the *sym* genes. The avirulent strains may regain the symbiotic properties or even improve the N_2-fixing activities after acquisition of p*Sym*s from the genetically related strains [Soberon-Chavez et al., 1991; Jarvis et al., 1989, 1995]. However, after p*Sym* transfer into symbiotically active strains their N_2-fixing activities are often reduced due to functional incompatibility between the incoming and resident plasmids [Wang et al., 1986; Provorov et al., 2004].

The "Gain-and-Loss" strategy may be followed also in the evolution of pathogens, for instance, in salmonellas whose uniform population clonality was demonstrated *via* MLEE analysis while the *fim* genes responsible for virulence combine randomly with chromosomal markers [Beltran et al., 1988]. The evolutionary scenario including deletions of virulence genes, their transfer to the resulted avirulent mutants from virulent strains and the host-induced multiplication of the novel recombinants have been suggested for pathogenic *Salmonella* strains [Baumler et al., 1997].

We suggest that "Gain-and-Loss" model represents a general symbiosis-specific mechanism ensuring high levels of panmixia in host-associated bacteria which may be accompanied by an induction of gene transfer during symbiosis as it was revealed for *S. meliloti* in alfalfa nodules or rhizospheres [Broughton et al., 1987b; Pretorius-Guth et al., 1990]. The *Ti*-plasmid transfer in *Agrobacterium* is controlled *via* the "quorum-sensing" mechanism switched on by plant-released opines [Faqua & Vinans, 1994; Miller & Bassler, 2001]. This mechanism enables agrobacteria populations to retain a high virulence potential in spite of maintaining the huge *Ti* (*Ri*) plasmid in the minority (<1% [Mougel et al., 2001]) of cells: symbiosis-specific gene transfer balances the accumulation of avirulaent mutants in soil habitats [Finan, 2002].

The stimulation of gene transfer by host may be observed also in ecto-symbiotic bacteria, since the phytosphere represents a favorable environment for implementing the parasexual processes due to high cell densities, or to cell attachment on plant surfaces facilitating the bacterial mating [Nielsen et al., 2001]. For free-living and root-associated bacteria (*Pseudomonas, Serratia, Escherichia*), plasmid transfer is more intensive in rhizosphere/spermasphere than in bulk soil [Van Elsas et al., 1988; Pukall et al., 1996].

The gene transfer in ecto- and endo-symbiotic bacterial communities may be facilitated by the broad host range plasmids conjugated with high frequencies among taxonomically diverse bacteria [Tauch et al., 2002]. In different rhizobia species, migration of such plasmids may be ensured by *repC* or *traABCDEFG* loci that are distributed randomly in populations with respect to chromosomal markers (Turner et al., 1996; Perez-Mendosa et al., 2004). A small (22 kb) plasmid identified in *S. meliloti* can migrate to *R. leguminosarum, R. tropici* and *A. tumefaciens* strains and facilitate the transfer of other replicons [Pistorio et al., 2003].

2.3. Co-Evolution with Plants and Speciation in Symbiotic Microbes

The performance of symbiotic microbes to "host-environment" systems is usually described in terms of partners' co-evolution which was initially defined as a constellation of selective processes in the interacting populations leading to their mutual adaptations [Mode, 1958; Janzen, 1980]. Later, a more broad concept of co-evolution was developed to include the parallel gene changes which reflect the macroevolutionary processes expressed as the congruence of partners' phylogenies [Doyle, 1998]. However, relationships between the co-evolutionary processes occurring at the molecular and population levels remain obscure.

2.3.1. Nodulation and Origins of Rhizobia

The clearest evidence on the co-evolution of legumes and rhizobia was obtained by analyzing the diversities in nodulation (*nod/nol/noe*) genes encoding for Nod factor signals. In spite of different chromosomal backgrounds, the sets of nodulation genes are quite similar in *Sino*- and *Mesorhizobium* species nodulating *Acacia tortilis* [Ba et al., 2002] and *Astragalus sinicus* [Zhang et al., 2000] or in *Azorhizobium* and *Sinorhizobium* species nodulating *Sesbania rostrata* [Terefework et al., 2000]. The most probable explanation for these similarities is the expansion of nodulation genes (responsible for triggering the same host receptors) in the diverse bacterial communities associated with the promiscuous hosts.

The data on molecular diversities in nodulation gene (Nod factor) structures suggest that the macroevolution of extant rhizobia diversity involved: (1) origination of the abilities for synthesis of Nod factor core structures ("minimal" Nod factors) in the ancient rhizobia; (2) broad distribution of nodulation genes within soil and plant-associated bacterial communities [Perret et al., 2000]. Similarity of Nod factors to the chitin oligomeres suggest that the first stage might be implemented *via* molecular mimicry of some ancient fungal symbionts by the bacteria. At the second stage, partners' co-evolution resulted in adaptation of rhizobia to the variable host reception/signaling systems the operation of which relies on chemical modifications in Nod factors. A key step in the evolution of nodulation was represented by generation of the ability to synthesize and attach to Nod factor the highly unsaturated fatty acids that was originated in the rhizobia coevolved with "galegoid" legumes [Yang et al., 1999; Debelle et al., 2001].

The presented data allow us to enlighten the origins of rhizobia which diverged from a hypothetical "common ancestor" much earlier than the flowering plants originated and therefore, this ancestor could not be a N_2-fixing legume symbiont [Young, 1992; Sprent & Raven, 1992; Young et al., 2006]. In spite of polyphyletic origins, rhizobia harbor a range of common *sym* genes involved in synthesis the Nod factor core (*nodABC*), in symbiosis-specific regulation (*fixLJ, fixK*) and energy supply of nitrogen fixation (*fixABCX, fixGHIS, fixNOPQ*). These genes were apparently acquired by an ancestral symbiotic bacterium and later distributed to the other bacteria *via* HGT.

Sometimes rhizobia are addressed as the refined (N_2-fixing) plant pathogens [Djordjevic et al., 1987] while the nodulation process is considered as "sympathogenesis" [Spaink, 1995]. These suggestions were primarily based on the close relatedness of rhizobia to agrobacteria, however, the molecular analyses demonstrated that mechanisms for relevant interactions are quite different (Table 5) and the direct origination (filiation) of nodulation from tumor-genesis (or *vice versa*) is impossible. Much more probable looks the origin of nodulation from root associations with N_2-fixing PGPR (plant growth promoting rhizobacteria) similar to *Azospirillum* which have the homologues of some *nod* genes [Kennedy, 1996]. In structures of their *nif* genes, *Azospirillum* is close to the *Bradyrhizobium* which may represent the ancestral forms of nodule bacteria [Norris, 1956; Rivas et al., 2002]. In *Azospirillum* the mutants were generated which excrete ammonium from the N_2-fixing cells as it is typical for rhizobial bacteroids. The rhizobia strains can function as PGPR in rhizospheres of non-legumes responsible for a root growth promotion or for an improved nutrient assimilation [Chabot et al., 1996; Tan et al., 2001].

The presented data allow us to suggest that the first stages of nodulation gene evolution had been occurred in the free-living N_2 fixers similar to *Azo-* or *Bradyrhizobium* which can be referred as "primary rhizobia" filiated directly form their ancestral forms. In these bacteria, a range of features was found which are considered as the evolutionary primitive [Sprent & Raven, 1992]: N_2 fixation *ex planta*, absence of special systems for transfer of *sym* genes, an ability for photosynthesis (used as energy source for N_2-fixation) combined with nodulation of legume stems (e.g., in *Sesbania* and *Aeshynomene*). However, in *Rhizobium*, *Mesorhizobium* and *Sinorhizobium* the *sym* gene systems are located on special plasmids or chromosomal islands that move actively in populations. These bacteria cannot fix N_2 *ex planta*, therefore representing the genuine symbiotic N_2-fixers, which can be referred as the "secondary rhizobia" originated from the "primary" ones by acquisition of *sym* genes *via* HGT.

Table 5. Comparative analysis of plant interactions with rhizobia and agrobacteria

Functions	Rhizobia	Agrobacteria
Induction of virulence genes by plant signals	By flavonoids (via one-component regulator NodD)*	By phenolics (via two-component regulator VirA/VirG)
Transfer of bacterial genes into plant cells	Not involved	Obligatory (integration of T-DNA into host chromosome)
Proliferation of plant cells is induced by:	Lipo-chito-oligosaccharide Nod factors produced by bacteria	Phytohormones produced by plant cells transformed with T-DNA
Development of novel structures in plants	Histologically differentiated nodules	Non-differentiated tumors or hairy roots
Transfer of nutrients:		
plant→bacteria	Dicarboxylic acids (sources of C)	Opines (sources of C and N)
bacteria→plant	NH4+ (amino acids?)	Not involved
Maintenance of bacteria in the host tissues	Due to cyclic β-glucans encoded by homologous genes: ndv (rhizobia) and chv (agrobacteria)	

*In *Bradyrhizobium japonicum* the two-component system *nodV/nodW* is involved in sensing the flavonoid signals, in addition to *nodD* [Stacey, 1995].

The hypothesis on the broad involvement of HGT in the rhizobia speciation is consistent with the results of laboratory experiments suggesting that acquisition of *sym* genes can impart an ability to infect and even nodulate the donor's hosts to: a) the rhizobia species which are taxonomically distant from the donors and do not overlap their host specificities; b) various bacteria which are either related to rhizobia (*Agrobacterium, Phyllobacterium, Ensifer*) or are not related to them (*Pseudomonas, Escherichia, Lignobacter, Sphingobacterium*) [Hirsch et al., 1984; Plazinski & Rolfe, 1985; Van Veen et al., 1988; Novikova & Safronova, 1992; Fenton & Jarvis, 1994; Rogel et al., 2001]. The occurrence of these processes in natural populations is suggested by the field isolation of tumor-forming clover rhizobia [Skotnicki & Rolfe, 1978] or of the pole bean rhizobia harboring the nearly full set of agrobacterial *vir* genes [Bittinger et al., 2000]. Some avirulent agrobacteria can be converted into N_2-fixing legume endosymbionts by substitution of p*Ti*s for p*Sym*s as it was suggested for fast-growing rhizobia which share many genetic markers (including the replication/transfer origins and some IS elements) with agrobacteria [Deng et al., 1995; Eardly et al., 2005]. For example, p*Sym* of *S. fredii* NGR234 possesses the replication and transfer systems (*oriV*/*rep*, *tra*, *trb*) very close to p*Ti* (*A. tumefaciens*) and p*Ri* (*A. rhizogenes*) suggesting the p*Sym* origin by replacement of *vir* for *nod*/*nif* genes [Freiberg et al., 1997; Moriguchi et al., 2001].

2.3.2. Role of the Host Plant Invasions

The evolutionary processes leading to origination of the novel rhizobia species may be intensified greatly by the host plant movements (invasions) into novel environments which are facilitated greatly if the plants are accompanied by their beneficial symbionts [Richardson et al., 2000; Bena et al., 2005]. These movements occurred many times in the natural evolution of legumes (which originated in tropics but were later distributed all over the world) and in their artificial evolution caused by domestication and allocation from the centres of origins into agricultural areas [Allen & Allen, 1981]. Although the legumes have no special mechanism for the "vertical transmission" of root symbionts, they can be distributed *via* seeds [Perez-Ramirez et al., 1998] providing a pseudo-vertical transmission of symbiotic microbes [Wilkinson, 1997]. The growing body of evidence suggests that several evolutionary scenarios may be implemented by moving the plant-microbe systems into novel environments (Figure 5).

The most simple scenario is implemented if the genetic processes in microsymbionts which accompany the host are restricted to their adaptations to the novel edaphic conditions. This scenario is illustrated by migration of lupine and seradella bradyrhizobia to South Africa and Western Australia where bacteria retained the basic genetic markers typical for the source European populations [Stepkowski et al., 2005]. Under such evolutionary scenario the initial diversity of microsymbionts may be decreased: e.g., Kura clover, *Trifolium ambiguum* symbionts in North America are characterized by a low value (3.5) of Shannon-Weawer diversity index (calculated from PCR data) while in population from Caucasian centre of origin this value was 10.76 [Seguin et al., 2001] (Figure 5A).

A more complex scenario (Figure 5B) may involve adaptations of the invasive plants to the local bacteria that may be caused by a sufficiently changed composition of host symbiotic genes. For example, pea (*Pisum sativum*) in the center of its origin (Middle East) is characterized by high polymorphism concerning the specificity for *Rhizobium leguminosarum* bv. *viceae* strains [Lie et al., 1987]. This diversity is controlled by the plant genes restricting nodulation only to some effective bacterial genotypes, for example, *Sym2* recessive alleles (e.g., *sym2*^{A} found in "Afghan" pea lines) encoding a specific resistance to *R. l.* bv. *viceae* strains (surmounted by the bacterial *nodX* gene; Section 1.1.2).

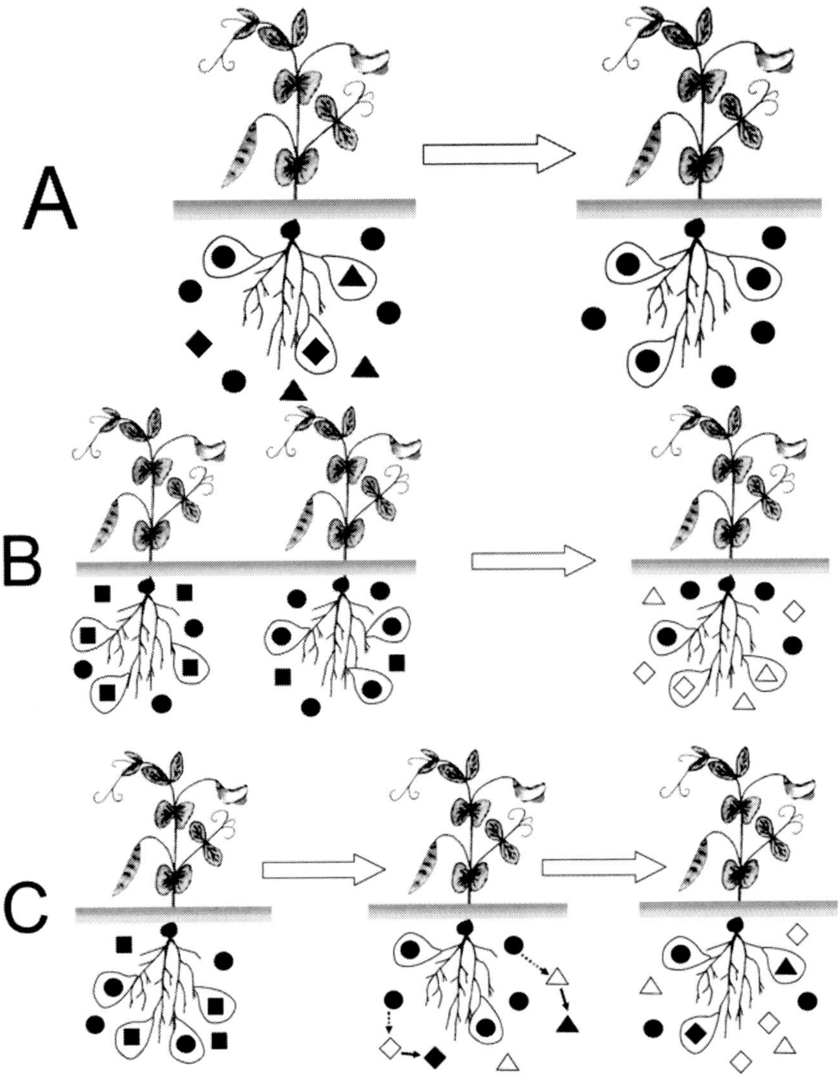

Figure 5. Different scenarios of symbiosis evolution induced by invasions of legume hosts into the novel environments (details are in Section 2.3.2).
A – adaptation of initial microsymbiont population to the novel environment (followed by a decrease in microbial diversity).
B – adaptation of the legumes to local rhizobia (followed by a decrease of specificity expressed by host towards its microsymbionts).
C – generation of novel symbionts *via* horizontal *sym* gene transfer (represented by dotted arrows) from the initial symbionts to local bacteria (see Table 6).

Table 6. Evolution of the novel rhizobia *via* horizontal *sym* gene transfer from introduced rhizobia to the local bacteria as related to the host invasions into novel environments

Host plant	Direction (age) of the host invasion	Rhizobia introduced together with the host (presumable *sym* gene donors)	Local bacteria (presumable *sym* gene recipients)	Novel symbionts found in the area of invasion	Transferred genetic material and mode of transfer	References
Lotus corniculatus	Europe → New Zealand (7 years)	*Mesorhizobium loti*	Non-symbiotic local mesorhizobia	*Mesorhizobium loti* strains which combine core genome markers from local strains and symbiotic markers from introduced ones	Chromosomal symbiotic islands (up to 611 kb) integrated into phe-tRNA genes	Sullivan et al., 1995, 2002
Biserrula pelecinus	Europe (Mediterranean region) → Western Australia (12 years)	*Mesorhizobium* sp.	The same	The same	Chromosomal symbiotic island integrated into phe-tRNA genes *via* P4-like inegrase (IntS)	Nandesena et al., 2006
Robinia pseudoacacia	Northern America → Europe (≈ 300 years)	*Mesorhizobium huakuii*	Local rhizobia or agrobacteria	*M. loti, M. huakuii, M. amorphae, R. leguminosarum, R. tropici*	Presumably, symbiotic islands	Ulrich & Zaspel, 2000
Amorpha fruticosa	Northern America → China (≈ 50 years)	*Mesorhizobium amorphae*	Local rhizobia	*M. spp., R. leguminosarum, B. elkanii*	Presumably, symbiotic islands	Wang et al., 1999
Phaseolus vulgaris	South America, Mesoamerica → Europe (≈ 500 years)	*Rhizobium tropici, R. etli*	Local rhizobia or agrobacteria	*R. tropici, R. leguminosarum* bv. *phaseoli, R. gallicum, R. giardinii*	Presumably, *Sym* plasmids	Amarger et al., 1994; Laguerre et al., 1993, 2001; Brom et al., 2002
Glycine max	North America → South America (≈ 80 years)	*Bradyrhizobium japonicum*	Local rhizobia or agrobacteria	*Rhizobium* spp., *B. japonicum, B. elkanii*	Not known	Chen et al., 2000

In the European and American pea cultivars the restrictive *sym2*^{A} alleles are absent and the plants may be inoculated by bacteria lacking *nodX*, e.g., by local *R. l.* bv. *viceae* populations that are adapted to wild-growing *Vicia* and *Lathyrus* hosts. These data suggest that the pea invasions from centre of origin to European agricultural areas (which occurred in Roman times) resulted in narrowing the partners' symbiotic polymorphism, and the plants become susceptible to interactions with non-specific and low effective local strains which can even block the nodulation by specific (*nodX* harboring) strains [Winarno & Lie, 1979].

The similar tendencies were followed by soybeans introduced to North America from the Chinese centre of origin where the wild-growing *Glycine max* and *G. soja* lines are polymorphic for nodulation restricting genes: *Rj2, Rj3* and *Rj4* control interactions with different *B. japonicum* serotypes, *Rj5* - with *S. fredii* strains [Triplett & Sadowsky, 1992]. In Chinese soils, indigenous rhizobia populations are taxonomically diverse including *B. japonicum*, *B. elkanii* and *S. fredii* strains while in USA soils the populations are composed of *B. japonicum* strains correlating to the low host polymorphism for *Rj* genes [Devine & Kuykendall, 1996].

The most complex evolutionary scenario (Figure 5C; Table 6) is implemented when the invasions of legumes into the novel areas lead to a greatly extended diversity in microbial partners based on transfer of *sym* genes from initial mirosymbionts. This scenario was demonstrated for the symbionts of pole bean (*Phaseolus vulgaris*) the nodulation of which in South American and Mesoamerican centers of origin is implemented by *Rhizobium tropici* and *R. etli* [Eardly et al., 1995; Bernal & Graham, 2001; Aguilar et al., 2004]. In European agricultural areas where pole bean was introduced about 500 years ago, much broader spectrum of symbionts was found with includes *R. gallicum*, *R. giardinii* and *R. leguminosarum* bv. *phaseoli* along with the initial symbionts, *R. tropici* [Amarger et al., 1994; Laguerre et al., 2001]. In spite of broad chromosomal diversity, these species have homologous sets of nodulation genes and the similar Nod factor profiles.

These data suggest that pole been was introduced into Europe in combination with some of the initial symbionts which donated their symbiotic genes to local microbes. One possibility is that co-migrated donor was *R. tropici* in which the ecological plasticity may be due to the resistance for soil acidity [Glenn & Dilworth, 1994]. The recipients were most probably, local *R. leguminosarum* biovars *trifolii* and *viceae*, or the *Agrobacterium* strains transformed into the regular pole bean co-habitants [Laguerre et al., 1993]. The other possibility is that initial source of *sym* genes found in European pole bean rhizobia may be *R. etli* as it was suggested from the data on distribution of p*Sym* markers from *R. etli* in *R. gallicum*, *R. giardinii*, *R. leguminosarum* bv. *phaseoli* and in some *S. fredii* strains [Brom et al., 2002]. However, *R. etli* co-migration with *R. tropici* was documented only when pole bean was introduced to some areas in Africa, e.g., to Senegal and Gambia [Diouf et al., 2000].

The similar processes of generating the novel symbionts *via* HGT proceed very quickly in the *Mesorhizobium* species harboring *nod* and *nif* genes in chromosomal islands that may be transferred horizontally as the conjugative transposons *via* type IV secretion systems [Sullivan et al., 2002]. The most convincing data were obtained for the trefoil (*Lotus corniculatus*) symbionts introduced from Europe into New Zealand where a very rapid formation of novel symbionts was observed having markers of local avirulent mesorhizobia and the *sym* genes (symbiotic islands) from the introduced *M. loti* inoculant strains. Such *in situ* evolution occurred in 7 years after which the stable populations of novel *L. corniculatus* symbionts were established in the New Zealand soils. Similar processes accompanied the invasions of *Robinia pseudoacacia* from North America to Europe, of *Amorpha fruticosa* from North America to China and of *Biserrula pelecinus* from Europe to Australia (Table 6). Co-transfer of *nod* and *nif* genes *via* the symbiotic islands may be responsible for their phylogenetic congruence observed in some *M. loti* populations [Qian & Parker, 2002].

The enormous diversification of symbionts occurred also after introduction of soybeans from Northern to Southern America (Paraguay) accompanied by *Bradyrhizobium* strains. Due to donation of their *sym* genes into the diverse local broad-host-range rhizobia (which occurred in spite of absence of p*Sym*s or motile symbiotic islands in *Bradyrhizobium*), the variety of novel symbionts was evolved in 80 years including different *(Brady)Rhizobium*- and *Agrobacterium*-like genotypes [Chen et al., 2000]. One can suggest that the involvement of selective pressures in favor of novel genotypes is more important for their rapid *in situ* evolution than a high-frequency HGT. The previously discussed FDS pressures (Section 2.2.2) which occur during rhizobia competition for the host nodulation may be a potent factor driving this evolutionary scenario.

3. Evolution of Symbiotic Mutualism

The population processes discussed in Section 2 do not pertain specifically the evolution of symbiotic mutualism. For a long time it remained enigmatic for the researchers who followed the idea that natural selection can not support an organisms' trait which sole effect is to benefit another organism [Person et al., 1962]. Discussing this enigma from the viewpoint of microbial genetics, J. Maynard Smith [1989] proposed the question: "Why should a symbiont benefit its host if it gains an immediate advantage by injuring it?" Obviously, it would be difficult to answer this question until the natural selection is reduced to its most simple (Darwinian) form which improves the fitness of individual organisms. Although mechanisms for this improvement may be diverse (good nutrition, consumption of additional energy, adequate behavior in the changing environment), the universal expression of Darwinian fitness was considered as an individual reproductive success of the organisms [Mayr, 1970; Timofeeff-Ressovsky et al., 1977]. In symbiotic systems, improvement of fitness is considered in the terms of benefit (symbiotic efficiency) which should be determined for each partner to differentiate mutualistic (+/+) from antagonistic (+/–) symbioses [de Bary, 1879; Douglas, 1994].

The legume-rhizobia coevolution for nodulation (Section 2.3.1) gives an immediate benefit to microbes, but for hosts the advantage is conditional, dependent on the activity of N_2 fixation in nodules. The involved adaptive processes may be quite different from those operating in the partners' coevolution for nitrogen fixation: nitrogenase activity in bacteria is of immediate use for the host, while for bacteria the benefit is conditional, depending on the "return" of costs paid to implement N_2 fixation and to export its products. The individual adaptations are hardly compatible with the evolution of mutualism until it is correlated to an altruistically reduced reproduction in microbial cells (Section 1.2.1). However, many years ago the evolutionary ideas were coined which could enlighten the mechanisms responsible for the symbiotic mutualism.

3.1. Symbiosis and Biological Altruism

J.B.S. Haldane [1932] was among the first why attempted to reveal the genetic and evolutionary backgrounds for altruistic interactions. He considered biological altruism as an intra-species interaction that improves fitness in one individual at the expense of the other one. These interactions were firstly illustrated by parental care that may be accompanied by a decrease of parents' fitness required for their progenitors to survive and to save the "altruistic genes" for next generations. Using these instructive ideas, W. Hamilton [1964] and J. Maynard Smith [1964] suggested a mathematical description for the relevant evolutionary processes speculating that altruism may be supported in a population if the net benefit (b) obtained by recipients of altruism, the cost (c) paid by donors of altruism and the genetic relatedness between the donors and recipients (k) satisfy the inequality:

$$k \cdot b - c > 0. \tag{2}$$

This simple formula assumes that altruism is to be supported in a group of organisms wherein the total balance of fitness is improved due to its decrease in some individuals (including their programmed death). The involved mechanism was named "kin selection" [Maynard Smith, 1964] to stress that it operates within the close relatives representing a restricted gene pool of the same family (deme).

Developing the ideas about operation of natural selection at the group level, some authors considered a possibility of inter-group (inter-deme) selection resulted in evolution of the altruistic traits and of some other types of social behavior [Williams & Williams, 1957; Wynne-Edwards, 1963]. The sophisticated mathematical models were suggested to simulate fixation of "altruistic adaptations" within a group of individuals, which permit some of them to survive by using the "inclusive fitness" obtained from the others [Hamilton, 1970; Michod & Abugov, 1980; Graften, 2007]. However, these speculations were criticized heavily since under the real ecological conditions the competition occurs mainly among individuals, not among their teams [Mayr, 1970; Timofeeff-Ressovsky et al., 1977]. The discussion lead to a compromise view that even if inter-deme selection is involved in the adaptive evolution, it is less effective than Darwinian selection due to: (i) limited number of demes; (ii) relatively inter-deme variation (with respect to individual variation); (iii) absence of solid inter-deme barriers leading to migration of individuals between demes and restricting their competition [Wilson, 1977; Wade, 1977].

Even more problematic was the possibility for an inter-species altruism: if the non-related organisms follow the self-sacrificing strategy in their interactions, who will save an "altruistic gene" after death of its owner (donor of altruism)? In order to extend the notion of altruism to mutualistic symbioses, one can suggest them as a type of interaction where for both partners the benefits from co-existence overcome its costs [Frank, 1994; Herre et al., 1999]. In a simplified form, this suggestion may be expressed as a system of two inequalities:

$$\begin{cases} r \cdot b_1 - c_1 > 0 \\ r \cdot b_2 - c_2 > 0 \end{cases}, \tag{3}$$

where $r>0$ – is the coefficient of correlation between the values of benefit obtained by partners. In its biological meaning, r is similar to k from formula (2): both coefficients characterize the maintenance of the altruistic traits in the super-organism systems the stabilities of which are due to interactions between their components. This similarity means that the inter-species altruism may evolve if the partner returns its costs either to the immediate donor of symbiotic altruism or to its kin relatives. Therefore, mutualistic symbiosis is possible if both partners possess the tightly co-regulated genes for which the primary effects are to improve the Darwinian fitness not in their owners, but in its co-habitants. In order to simulate the evolution of such genes we should address the selective pressures, which support the inter-species altruism in symbiotic systems.

3.2. Group Selection in the Beneficial Symbionts

It was repeatedly attempted to represent the mutualism as a reciprocal exploitation of two interacting organisms assuming that each partner behaves as a pathogen towards the other and the co-evolutionary process may be described in terms of host-pathogen "arms race" [Preston et al., 1998; Herre et al., 1999; Steinert et al., 2000]. If rhizobia are the refined pathogens of legumes [Djordjevic et al., 1987] which use N_2 fixation to avoid (suppress) the host defense [Udvardi & Kahn, 1992], *nif* genes should be subjected to the same evolutionary mechanisms as virulence factors in typical pathogens and the rapid inter-conversions (direct filiations) between nodulation and pathogenic interactions should occur. However, the molecular data suggest that pathogenic agrobacteria being genetically close to fast-growing rhizobia differ from them completely in organization of genes required for symbiotic interactions (Table 5). In rhizobia, the phylogenetic discordance of genes controlling mutualistic and pathogenic-like interactions with hosts was revealed (Section 1.2.2). N_2-fixing cyanobacteria (*Nostoc, Anabaena*), arbuscular mycorrhizal and ectomycorrhizal fungi [Hibbett et al., 2000; Meeks & Elhai, 2002; Schußler, 2002] represent the broadly distributed groups of plant symbionts, which are not related to pathogens. The direct filiations may occur in the defensive symbionts *Clavibacter* [Metzler et al., 1997], which represent the narrow group of plant co-habitants. These data suggest that quite different natural selection pressures may be involved in evolution of mutualistic and antagonistic microbe-plant interactions.

3.2.1. Selection Based on Positive Genetic Feedbacks

Proceeding from a qualitative difference between mutualism and antagonism, we can try to differentiate the enigmatic population mechanisms of mutualism evolution from the mechanisms of antagonism evolution. It is generally assumed that the evolution of pathogenesis is based on the "negative genetic feedback" among the interacting organisms [Pimentel, 1968; Frank, 1992]: in symbiotic system, the genetic changes improving fitness in one partner usually lead to a decreased fitness is the other one (e.g., high virulence is beneficial for pathogens, but deleterious for the hosts and *vice versa*). In this case, co-evolutionary process is based on the operation of Darwinian selection in both partners creating the "gene-for-gene" systems. For example, acquisition of new virulence gene(s) by pathogen leads to a decreased host fitness and elicits its "arms race" towards new complementary resistance(s) which in turn stimulate pathogen to acquire a novel virulence, *etc.* [Frank, 1992; Mitchell-Olds & Bergelson, 2000].

However, models of the host-pathogen (gene-for-gene) coevolution are generally not valid for mutualistic symbioses [Person et al., 1962] where the selective processes should support the "positive genetic feedback": due to mutualistic interaction, an increase of fitness in one partner leads to its parallel increase in the other one. This type of feedback assumes that the mutualistic symbiosis may be represented as a product of inter-species (reciprocal) altruism that can differ greatly in its evolutionary mechanisms from antagonistic interactions (reciprocal exploitation). Specifically, evolution of mutualism may occur *via* the same forms of group selection that supports the intra-species altruism. Considering rearrangements which occur during IRC (Figure 2) we suggest that the inter-group (inter-deme) selection may be induced in rhizobia due to the clonal propagation of N_2-fixing strains in nodules (Figure 6A). This propagation may be based on allocation of the plant photosynthates into N_2-fixing nodules where the excess of carbon is consumed by bacteria to be released into soil after the nodule death [Jimenez & Casadesus, 1989]. Moreover, impairment of N_2 fixation in some nodules may result in "sanctions" from host which restricts the bacteria nutrition and can suppress them using the defense reactions [Denison, 2000].

Therefore, rhizobia may represent a convincing example of inter-deme selection that is based on competition between the nodular clones for the host nutrients. In supporting the mutualism, the inter-deme selection from the rhizobial side may be combined with the individual selection from the host side improving the symbiotic efficacy (Figure 6B).

The additional strategy of natural selection for mutualism may result from separation of the microsymbiont clones into two types of subclones: those, which benefit the host due to a decreased own fitness (altruists), and those, which consume the resources provided by hosts (egoists) (Figure 6C). In the legume-rhizobia system, this separation occurs *via* transformation of bacteria into non-reproducible bacteroids and leads to the kin selection represented by inequality (2). It is important to note that the pronounced differentiation of non-viable bacteroids is typical to the evolutionary advanced "galegoid" legumes characterized by very high symbiotic efficiencies: e.g., in *Galega orientalis* and *Medicago sativa* up to 500-600 kg/ha/season of N_2 may be fixed [Provorov & Tikhonovich, 2003].

An attempt was made to extend the altruistic model on the interactions between legumes and rhizospheric rhizobia populations [Olivieri & Frank, 1994]. A preferential *ex planta* multiplication of Fix^+ clones was suggested due to consumption of some metabolites excreted specifically from N_2-fixing nodules. The rhizospheral altruism may contribute significantly to the evolution of rhizobia which possess the genes for metabolism of inositol-like rhizopines: *mos* genes for their synthesis induced in bacteroids and *moc* genes for catabolism induced in free-living cells. The rhizopine synthesis is activated *via* the same (NifA/NtrA) circuit as the nitrogenase (*nif*) genes and *mos*/*moc* genes are linked to *nif* genes [Murphy et al., 1988, 1993; Saint et al., 1993]. In Mos^+Fix^+ strains, *moc* genes are activated *ex planta* being responsible for high rhizospheral competence of rhizobia [Murphy et al., 1995; Heinrich et al., 1999; Jiang et al., 2001]. Similarly to *nod*/*nol*/*noe*, *mos*/*moc* genes are actively transferred in rhizobia populations [Rao et al., 1995] and undergo a rapid molecular evolution [Murphy et al., 1993].

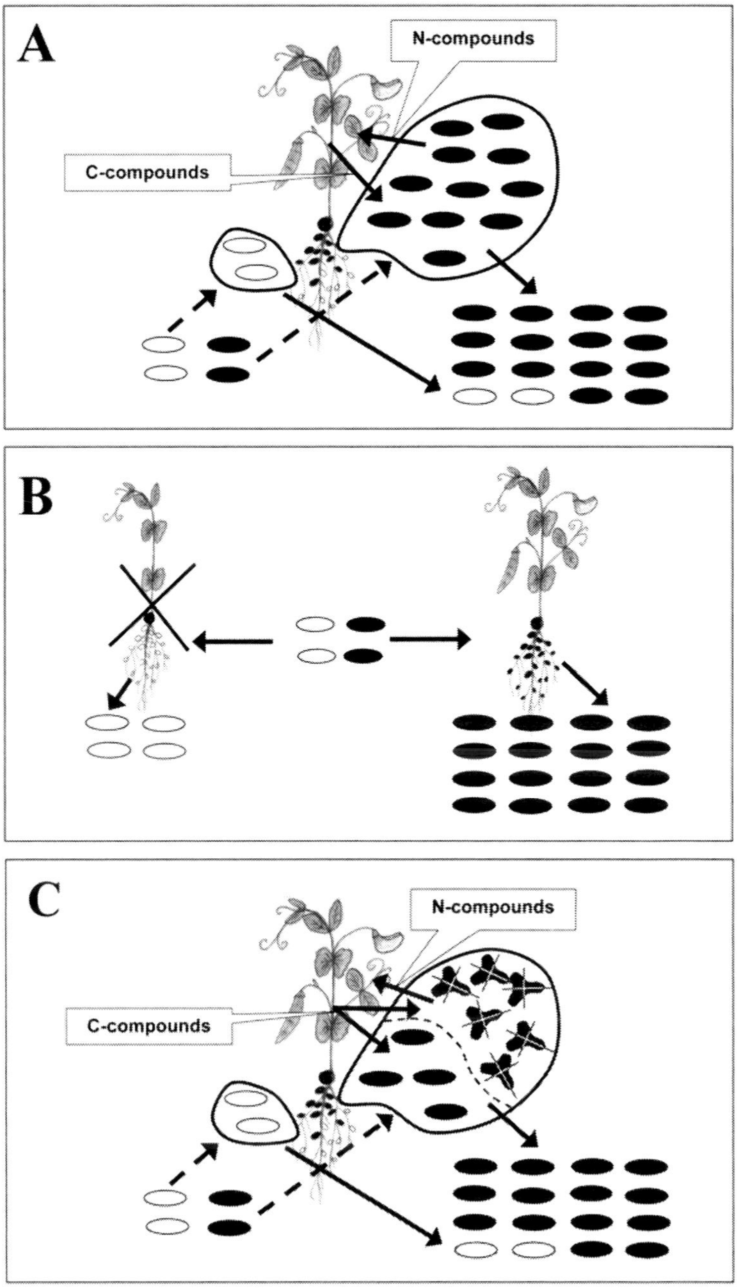

Figure 6. Group selection in the rhizobia population which occurs due to their positive feedback with the hosts.
Nitrogen-fixing bacteria are shown in black, non-fixing bacteria – in white color.
A – inter-deme selection caused by allocation of fixed carbon to the N_2-fixing nodules;
B – combination of inter-deme selection in bacteria with the Darwinian selection in the plant population leading to the improvement of symbiotic efficiency;
C – kin selection in bacterial populations due to a programmed cell death within the endosymbiotic clones (non-reproducing bacteroids are shown by the crossed Y-like cells).

However, *mos*/*moc* genes occur only in some (10-14%) *S. meliloti* and *R. leguminosarum* bv. *viceae* strains and are not found in the other rhizobia species [Murphy et al., 1995; Gordon et al., 1996]. The consumption of nodule-excreted rhizopines may be effective if the kin relatives of intra-nodule clone are structured properly in the *ex planta* habitats [Bever & Simms, 2000] (e.g., are located in the nodule surface), but no experimental data suggesting such spatial structures are available. Obviously, the kin selection in rhizobia populations is based mainly on the intra-nodule altruism wherein the return of costs for N_2 fixation is facilitated by the regular localization of donors and recipients of altruism inside the nodule structures.

3.2.2. Group Selection in Symbiotic vs Free-living Organisms

The mechanisms of group selection in symbiotic microbes may be clarified by their comparisons to free-living organisms. Among them are the microbes which adapt the environment being the members of structurally and functionally differentiated colonies or biofilms having some attributes of multi-cellular organisms [Shapiro, 1998]. An example of highly specialized altruistic adaptations is represented by the programmed cell death in the colonies of enteric bacteria. These adaptations are expressed due to apoptosis under starvation, when the nutrients released from decaying cells are consumed by their clonal relatives. This apoptosis is controlled by *mazFE* operon encoding for a stable toxin inducing cell autolysis (MazF) and an non-stable antitoxin (MazE). Under good nutrition, these genes are expressed actively and cell viability is ensured by blocking the MazF lytic activity by MazE. Under starvation, *maz* expression is hampered (*via* action of 3',5'-bispyrophosphate), non-stable antitoxin MazE degrades rapidly permitting the cell lysis by MazF [Aizenman et al., 1996].

A well-studied bacterial property which apparently can evolve *via* group selection is the "quorum-sensing" gene activation which is frequently involved in the symbiotic interactions. It was firstly revealed for the *lux* operon of *Vibrio fischeri*, the luminescent symbiont of marine animals [Fuqua & Winans, 1994; Hardman et al., 1998]. The quorum-sensing circuits regulate virulence genes in various plant and animal pathogens, possibly because infection is successful when inoculum size is high enough to overcome the defense barriers [Hardman et al., 1998].

The adaptive evolution of altruistic interactions in symbioses may be clarified also by their comparison to the social systems supported by group selection in non-symbiotic organisms [Wade, 1977; Wilson, 1977]. In sexually reproducing species (e.g., in animals which display a parental care) kin selection is operating under the restricted relatedness of donors and recipients of altruism ($k \leq 0.5$). However, in rhizobia the kin selection efficiency may be markedly higher since each nodule usually harbors a clone originated form a single bacterial cell that initially infected the root ($k \rightarrow 1$). Surprisingly, the "primitive" bacteria when entering in symbiosis with plants gain more opportunities to implement the biological altruism in their populations than the more "advanced" but free-living animals!

Using these examples we can suggest that in mutualistic symbioses the efficiency of evolution for altruistic traits may be higher than in free-living organisms due to a specific hierarchy of selective pressures: individual selection in hosts lead to the inter-deme selection in microbes (Figure 6A, 6B) which may be enhanced by the kin selection (Figure 6C). The highly effective kin selection in rhizobia is possible because the legume host serves as an effective intermediate for transmitting the altruistic effects within the intra-nodule clone while the fixed nitrogen may be considered as a salary for this service (Figure 7A). Moreover, in the effectiveness of kin selection, symbiotic microbes may exceed the free-living microbes wherein the donors and recipients of altruism interact directly (Figure 7B) and the efficiency of altruistic cooperation may be suppressed by the intra-species competition.

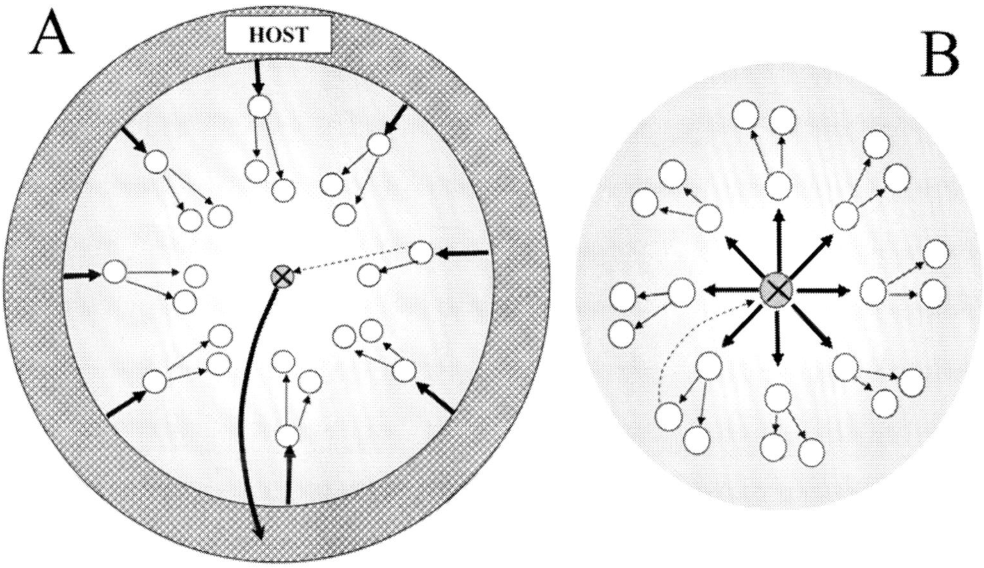

Figure 7. Altruistic interactions in the bacterial clones evolved *via* kin selection.
A – endosymbionts of higher organisms (interactions between the donors and recipients of altruism are mediated by the host);
B – free-living bacteria (immediate interactions between donors and recipients of altruism).
Solid arrows represent the altruistic interactions; thin arrows – multiplication of the recipients of altruism; dotted arrows – differentiation of the donors; the crossed circles represent the decreased fitness (extinction) in the donors of altruism.

Conclusion and Prospects

The microbe-plant systems provide us a unique opportunity to analyze the interplay between molecular and population mechanisms in the bacterial evolution. To address their relationships, we used the model of rhizobia-legume symbiosis the development of which may be dissected into the pathogenic-like early stages (controlled by *nod*/*nol*/*noe* genes encoding for signaling and root penetration) and the mutualistic late stages (controlled by *nif* genes encoding for symbiotic N_2 fixation). Evolution of both gene groups follows the combinative strategy (recruiting from the non-symbiotic genes) which is implemented due to high genomic plasticity in rhizobia. This plasticity results in the complications of rhizobia genomes that are characterized by elevated sizes, separation into several large replicons, and by saturation with the mobile DNA elements.

Molecular approaches suggest different natural histories for the rhizobial genes responsible for early (*nod*/*nol*/*noe*) and late (*nif*) symbiotic functions expressed as a discordance in the gene phylogenies. This discordance may be due to contrast types of "genetic feedback": negative for pathogenic-like (early) interactions but positive for mutualistic (late) interactions. One can suggest that in mutualistic symbioses the positive feedback among partners is expressed at two levels: at the development of symbiotic systems *via* cross-regulation of the partners' genes (Section 1.2) and at the interaction of partners' populations *via* balancing the natural selection pressures operating from the plant and bacterial sides (Sections 2.1.2, 2.3 and 3.2).

In order to reveal the relationships between molecular and adaptive evolution of rhizobia, their genomic changes may be correlated to the peculiarities of population dynamics described using the IRC model (Figure 2) which assumes the host as a principle organizer of microbial evolution. The symbiosis-specific constellation of microevolutionary factors elicited by the bacteria circulation in plant-soil systems results in the "Gain-and-Loss" dynamics of *sym* genes in rhizobia populations (Figure 4) leading to the combinative evolution of *sym* gene networks *via* inter- and intra-genome recombinations. Specifically, the frequent reorganizations in the spatial/genetic population structures in cooperation with HGT may be responsible for inducing plasticity in the bacterial genomes that facilitate evolution of *sym* gene networks.

The role of "Gain-and-Loss" strategy in rhizobia evolution is not only to stimulate *sym* gene exchange within populations of genetically related strains but also to expand these genes among taxonomically distant bacteria leading to generation of novel symbionts. The evolutionary potential of HGT is increased due to the host invasions (natural or artificial) into the novel environment which provide: (1) parallel migrations of symbiotic microbes resulted in their interactions with local bacteria; (2) selective pressures in favour of rare recombinants which arise from hybridization between introduced and local bacteria [Provorov & Vorobyov, 2000b]. These pressures ensure a rapid *in situ* evolution of novel rhizobia species which is well documented for the pole bean rhizobia and for mesorhizobia interacting with different legumes (Table 6).

An important goal of the microbial evolutionary genetics is to correlate the organization and expression in different groups of genes to their functions and to the types of involved selective pressures. For example, comparison of synonymous Vs non-synonymous substitution rates in the proteins is used to measure a balance between the neutral and directed gene evolution [Tibayrenc, 1996]. Different types of selection were proposed to control evolution in *Sinorhizobium meliloti nod* genes (purifying selection) and *exo* genes (balancing selection) [Bailly et al., 2006] which may be correlated in their symbiotic functions: induction the nodule development and dialogue with host defense systems, respectively.

Analysis of the natural selection pressures is required to address the evolution of mutualism, which may be considered either as a refined pathogenesis (reciprocal exploitation) or as a reciprocal altruism of partners. If the first approach is valid, the phylogenies of genes responsible for mutualistic and pathogenic interactions should be essentially similar and the direct evolutionary inter-conversions (filiations) should be common among the microbes implementing pathogenic and mutualistic interactions with the plants. However, molecular and phylogenetic data suggest quite different phylogenies for beneficial and pathogenic microbes (as well as for genes which control the contrasting types of interactions), considering that the reciprocal altruism would be a more adequate representation of mutualism than the reciprocal exploitation.

In order to analyze the involvement of natural selection in evolution of mutualistic interactions we suggest to extend the notion of biological altruism from intra-species (social) interactions towards the inter-species (symbiotic) ones. This approach enables us to suggest the operation of group (inter-deme, kin) selection in evolution of beneficial plant-microbe interactions. In rhizobia, group selection pressures may control evolution of *nif* genes creating the sufficient differences in their phylogenies with respect to *nod*/*nol*/*noe* genes which presumably evolve under impacts of individual (Darwinian, frequency dependent) selection.

Evolutionary research is required not only for reconstructing the natural histories of symbiotic systems but also for their improvement and application. Agricultural potential of plant-microbe symbioses is very high since many legume and non-legume crops have decreased symbiotic activities with respect to their wild-growing relatives [Provorov & Tikhonovich, 2003]. This decrease is due to dis-balancing the natural co-evolutionary processes in symbiotic systems, which are directed mainly towards an increased efficiency of mutualism [Sprent & Raven, 1992]. Therefore, in our attempts to improve and engineer the agronomically attractive plant-microbe systems we should proceed from the knowledge on their natural evolution and, possibly, simulate its population and molecular mechanisms.

Acknowledgements

The research presented in this paper is supported by grants from Russian Foundation of Basic Research (06-04-48800, 06-04-89000NWO); NWO Centre of Excellence: 047.018.001; CRDF and RF Ministry for Higher Education and Science (Annex BP2M12, Award RWXO-012-ST-05, Y2-B-12-05); State Contract with RF Ministry for Higher Education and Science (N02.445.11.74.92).

References

Aguilar, O.M., Riva, O. & Peltzer, E. (2004). Analysis of *Rhizobium etli* and its symbiosis with *Phaseolus vulgaris* supports coevolution in centers of host diversification. *Proc. Natl. Acad. Sci. USA, 101*, 13548-13553.

Aizenman, E., Engelberg-Kulka, H. & Glaser, G. (1996). An *Escherichia coli* chromosomal "addiction module" regulated by 3',5'-bispyrophosphate: a model for programmed bacterial cell death. *Proc. Natl. Acad. Sci. USA, 93*, 6059-6063.

Allaway, D., Lowdig, E.M., Crompton, L.A., Wood, M., Parsons, R., Wheeler, T.R. & Poole, P.S. (2000). Identification of alanine dehydrogenase and its role in mixed secretion of ammonia and alanine by pea bacteroids. *Molec. Microbiol., 36*, 508-515.

Allen, O.N. & Allen, E.K. (1981). *The Leguminosae. A Source Book of Characteristics, Uses and Nodulation*. Madison: The University of Wisconsin Press.

Amarger, N., Bours, M., Revoy, F., Allard, M.R. & Laguerre, G. (1994). *Rhizobium tropici* nodulates field-growing *Phaseolus vulgaris* in France. *Plant and Soil, 161*, 147-156.

Amarger, N. & Lobreau, J.P. (1982). Quantitative study of nodulation competitiveness in *Rhizobium* strains. *Appl. Environ. Microbiol., 44*, 583-588.

Andrade, D.S., Murphy, P.J. & Giller, K.E. (2002). The diversity of *Phaseolus*-nodulating rhizobial populations is altered by liming of acid soils planted with *Phaseolus vulgaris* L. in Brasil. *Appl. Environ. Microbiol., 68*, 4025-4034.

Andronov, E.E., Terefework, Z., Roumiantseva, M.L., Dzyubenko, N.I., Onichtchouk, O.P., Kurchak, O.N., Dresler Nurmi, A., Young, J.P.W., Simarov, B.V. & Lindström, K. (2003). Symbiotic and genetic diversity of *Rhizobium galegae* isolates collected from the *Galega orientalis* gene centre in the Caucasus. *Appl. Environ. Microbiol., 69*, 1067-1074.

Ayala, F.J. & Campbell, C.A. (1974). Frequency-dependent selection. *Annu. Rev. Ecol. Syst., 5*, 115-138.

Ba, S., Willems, A., de Lajudie, P., Roche, P., Jeder, H., Quatrini, P., Neura, M., Ferro, M., Prome, J.-C., Gillis, M., Boivin-Masson, C. & Lorquin, J. (2002). Symbiotic and taxonomic diversity of rhizobia isolated from *Acacia tortilis* subsp. *raddiana* in Africa. *System. Appl. Microbiol., 25*, 130-145.

Bailly, X., Olivieri, I., de Mita, S., Cleyet-Marel, J.-C. & Bena, G. (2006). Recombination and selection shape the molecular diversity pattern of nitrogen-fixing *Sinorhizobium* sp. associated to *Medicago. Molec. Ecol., 15*, 2719-2734.

Baumler, A.J., Gilde, A.J., Tsolis, R.M., van der Velden, A.W.M., Ahmer, B.M.M. & Heffron, F. (1997). Contribution of horizontal gene transfer and deletion events to development of distinctive patterns of fimbrial operons during evolution of *Salmonella* serotypes. *J. Bacteriol., 179*, 317-322.

Beattie, G.A., Clayton, M.C. & Handelsman, J. (1989). Quantitative comparison of the laboratory and field competitiveness of *Rhizobium leguminosarum* bv. *phaseoli. Appl. Environ. Microbiol., 55*, 2755-2761.

Becker, A. & Pühler, A. (1998). Production of exopolysaccharides. In H.P. Spaink, A. Kondorosi, & P.J.J. Hooykaas (Eds.), *The Rhizobiaceae. Molecular Biology of Model Plant-Associated Bacteria* (pp. 87-118). Dordrecht, Boston, London: Kluwer Acad. Publ.

Belanger, C., Canfield, M.L., Moore, L.W. & Dion, P. (1995). Genetic analysis of non-pathogenic *Agrobacterium tumefaciens* mutants arising in crown gall tumors. *J. Bacteriol., 177*, 3752-3757.

Beltran, P., Musser, J.M., Helmuth, R., Farmer, J.J., Frerichs, W.M., Wachsmuth, I.K., Ferris, K., McWhorter, A.C., Wells, J.G., Cravioto, A. & Selander, R.K. (1988). Towards a population genetic analysis of *Salmonella* genetic diversity and relationships among strains of serotypes *S. choleraesuis, S. derby, S. dublin, S. enterotidis, S. heidelberg, S. infantis, S. newport, S. typhimurium. Proc. Natl. Acad. Sci. USA, 85*, 7753-7757.

Bena, G., Lyet, A., Huguet, T. & Olivieri, I. (2005). *Medicago – Sinorhizobium* symbiotic specificity evolution and the geographic expansion of *Medicago. J. Evol. Biol., 18*, 1547-1558.

Beringer, J.E. (1982). The genetic determination of host range in the Rhizobiaceae. *Israel J. Botany*, *31*, 89-93.

Beringer, J.E., Brewin, N.J. & Johnston, A.W.B. (1980). The genetic analysis of *Rhizobium* in relation to symbiotic nitrogen fixation. *Heredity*, *45*, 161-186.

Bernal, G. & Graham, P.H. (2001). Diversity on the rhizobia associated with *Phaseolus vulraris* L. in Ecuador, the comparison with Mexican bean rhizobia. *Canad. J. Microbiol.*, *47*, 526-534.

Bever, J.D. & Simms, E.L. (2000). Evolution of nitrogen fixation in spatially structured populations of *Rhizobium*. *Heredity*, *85*, 366-372.

Biondi, E.G., Pilli, E., Giuntini, E., Roumiantseva, M.L., Andronov, E.E., Onichtchouk, O.P., Kurchak, O.N., Simarov, B.V., Dzyubenko, N.I., Mengoni, A. & Bazzicalupo, M. (2003). Genetic relationship of *Sinorhizobium meliloti* and *Sinorhizobium medicae* strains isolated from Caucasian region. *FEMS Microbiol. Lett.*, *220*, 207-213.

Bittinger, M.A., Gross, J.A., Widom, J., Clardy, J. & Handelsman, J. (2000). *Rhizobium etli* CE3 carries *vir* gene homologs on a self-transmissible plasmid. *Molec. Plant-Microbe Interact.*, *13*, 1019-1021.

Bobik, C., Meilhoc, E. & Batut, J. (2006). FixJ: a major regulator of the oxygen limitation response and late symbotic functions of *Sinorhizobium meliloti*. *J. Bacteriol.*, *188*, 4890-4902.

Borisov, A.Y., Danilova, T.N., Koroleva, T.A., Naumkina, T.S., Pavlova, Z.B., Pinaev, A.G., Shtark, O.Y., Tsyganov, V.E., Voroshilova, V.A., Zhernakov, A.I., Zhukov, V.A. & Tikhonovich, I.A. (2004). Pea (*Pisum sativum* L.) regulatory genes controlling development of nitrogen-fixing nodule and arbuscular mycorrhiza: fundamentals and application. *Biologia*, *59/Suppl. 13*, 137-144.

Bottomley, P.J. (1992). Ecology of *Bradyrhizobium* and *Rhizobium*. In G. Stacey, R.H. Burris, & H.J. Evans (Eds.), *Biological Nitrogen Fixation* (pp. 293-384). London: Chapman and Hall.

Breedveld, M.J. & Miller, K.J. (1994). Cyclic β-glucans of members of the family Rhizobiaceae. *Microbiol. Rev.*, *58*, 145-161.

Brewin, N.J. (1991). Development of the legume root nodule. *Annu. Rev. Cell Biology*, *7*, 191-226.

Brewin, N.J. (2004). Novel symbiotic organelles in the *Rhizobium*-legume interaction. In I.A. Tikhonovich, B.J.J. Lugtenberg, & N.A. Provorov (Eds.), *Biology of Plant-Microbe Interactions* (V. 4, pp. 476-482). St.-Petersburg: Biont.

Brockman, F.J. & Bezdicek, D.F. (1989). Diversity within serogroups of *Rhizobium leguminosarum* biovar *viceae* in the Palouse region of Eastern Washington as indicated by plasmid profiles, intrinsic antibiotic resistance and topography. *Appl. Environ. Microbiol.*, *55*, 109-115.

Brom, S., Girard, L., Garcia-de los Santos, A., Sanjuan-Pinilla, J.M., Olivares, J. & Sanjuan, J. (2002). Conservation of plasmid-encoded traits among bean-nodulating *Rhizobium* species. *Appl. Environ. Microbiol.*, *68*, 2555-2561.

Bromfield, E.S.P., Butler, G. & Barran, L.R. (2001). Temporal effects on the composition of a population of *Sinorhizobium meliloti* associated with *Medicago sativa* and *Melilotus alba*. *Canad. J. Microbiol.*, *47*, 567-573.

Bromfield, E.S.P., Sinha, I.B. & Wolynetz, M.S. (1986). Influence of location, host cultivar and inoculation on the composition of naturalized populations of *Rhizobium meliloti* in *Medicago sativa* nodules. *Appl. Environ. Microbiol., 51*, 1077-1084.

Broughton, W.J., Heycke, N., Priefer, U., Schneider, G.-M. & Stanley, J. (1987a). Ecological genetics of *Rhizobium meliloti*: diversity and competitive dominance. *FEMS Microbiol. Lett., 40*, 245-249.

Broughton, W.J. & Perret, X. (1999). Genealogy of legume-*Rhizobium* symbiosis. *Curr. Opinion in Plant Biology, 2*, 305-311.

Broughton, W.J., Samrey, U. & Stanley, J. (1987b). Ecological genetics of *Rhizobium meliloti*: symbiotic plasmid transfer in the *Medicago sativa* rhizosphere. *FEMS Microbiol. Lett., 40*, 251-255.

Brunel, B., Rome, S., Ziani, R. & Cleyet-Marel, J.C. (1996). Comparison of nucleotide diversity and symbiotic properties of *Rhizobium meliloti* populations from annual *Medicago* species. *FEMS Microbiol. Ecol., 19*, 71-82.

Carelli, M., Gnocchi, S., Fancelli, S., Mengoni, A., Paffetti, D., Scotti, C. & Bazzicalupo, M. (2000). Genetic diversity and dynamics of *Sinorhizobium meliloti* populations nodulating different alfalfa cultivars in Italian soils. *Appl. Environ. Microbiol., 66*, 4785-4789.

Carius, H.J., Little, T.J. & Ebert, D. (2001). Genetic variation in a host-parasite association: potential for co-evolution and frequency-dependent selection. *Evolution, 55*, 1136-1145.

Caugant, D.A., Levin, B.B. & Selander, R.K. (1981). Genetic diversity and temporial variation in the *E. coli* population of a human host. *Genetics, 98*, 467-490.

Chabot, R., Antoun, H., Kloepper, J.W. & Beauchamp, C.J. (1996). Root colonization of maize and lettuce by bioluminescent *Rhizobium leguminosarum* biovar *phaseoli*. *Appl. Environ. Microbiol., 62*, 2767-2772.

Chen, L.A., Figueredo, A., Pedrosa, F.O. & Hungria, M. (2000). Genetic characterization of soybean rhizobia in Paraguay. *Appl. Environ. Microbiol., 66*, 5099-5103.

Chen, W.-M., de Faria, S.M., Straliotto, R., Pitard, R.M., Simoes-Araujo, J.L., Chou, J.-H., Chou, Y.-J., Barrios, E., Prescott, A.R., Elliott, G.N., Sprent, J.I., Young, J.P.W. & James, E.K. (2005). Proof that *Burkholderia* strains form effective symbioses with legumes: a study of novel *Mimosa*-nodulating strains from South America. *Appl. Environ. Microbiol., 71*, 7461-7471.

Chen, W.-M., Laevens, S., Lee, T.M., Coenye, T., de Vos, P., Mergeay, M. & Vandamme, P. (2001). *Ralstonia taiwanensis* sp. nov. isolated from root nodules of *Mimosa* species and sputum of a cystic fibrosis patient. *Intern. J. Syst. Evol. Microbiol., 51*, 1719-1735.

Cheng, J., Sibley, C.D., Zaheer, R. & Finan, T.M. (2007). A *Sinorhizobium meliloti minE* mutant has an altered morphology and exhibits defects in legume symbiosis. *Microbiology, 153*, 375-387.

Corich, V., Giacomini, A., Carlot, M., Simon, R., Tichy, H.-V., Squartini, A. & Nuti, M.P. (2001). Comparative strain typing of *Rhizobium leguminosarum* bv. *viceae* natural populations. *Canad. J. Microbiol., 47*, 580-584.

de Bary, A. (1879). Die Erscheinung der Symbiose. Strassburg: Verlag Von Karl J. Trübner.

de Bruijn, F., Chen, R., Fujimoto, S.Y., Pinaev, A., Silver, D. & Szczyglowski, K. (1994). Regulation of nodulin gene expression. *Plant and Soil, 161*, 59-68.

de Bruijn, F.J., Davey, M.E., Berges, H., Krol, E., Bruand, C., Rüberg, S., Capela D., Lauber, E., Meilhoc, E., Ampe, F., Fourment, J., Francez-Charlot, A., Kahn, D., Küster, H., Liebe, C., Pühler, A., Weidner, S., Batut, J. & Becker, A. (2004). Genetics and genomics of nutrient deprivation-induced, microaerobic and symbiotic gene expression in *Sinorhizobium meliloti*. In I.A. Tikhonovich, B.J.J. Lugtenberg, & N.A. Provorov (Eds.), *Biology of Plant-Microbe Interactions* (V. 4, pp. 405-410). St.-Petersburg: Biont.

Debelle, F., Moulin, L., Mangin, B., Denarie, J. & Boivin, C. (2001). *nod* genes and Nod signals and the evolution of *Rhizobium*-legume symbiosis. *Acta Biochim. Polon., 48*, 359-365.

del Papa, M.F., Balague, L.J., Sowinski, S.C., Wegener, C., Segundo, E., Abarca, F.M., Toro, N., Niehaus, K., Pühler, A., Aguilar, O.M., Martinetz-Drets, G. & Lagares, A. (1999). Isolation and characterization of alfalfa-nodulating rhizobia in acid soils of Central Argentina and Uruguay. *Appl. Environ. Microbiol., 65*, 1420-1427.

Denarie, J., Debelle, F. & Rosenberg, C. (1992). Signaling and host range variation in nodulation. *Annu. Rev. Microbiol., 46*, 497-531.

Deng, W., Gordon, M.P. & Nester, E.W. (1995). Sequence and distribution of IS*1312*: evidence for horizontal DNA transfer from *Rhizobium meliloti* to *Agrobacterium tumefaciens*. *J. Bacteriol., 177*, 2554-2559.

Denison, R.F. (2000). Legume sanctions and the evolution of symbiotic cooperation by rhizobia. *The Amer. Naturalist, 156*, 567-576.

Devine, T.E. & Kuykendall, L.D. (1996). Host genetic control of symbiosis in soybean (*Glycine max* L.). *Plant and Soil, 186*, 173-187.

Diouf, A., de Lajudie, P., Neyra, M., Kersters, K., Gillis, M., Martinez-Romero, E. & Gueye, M. (2000). Polyphasic characterization of rhizobia that nodulate *Phaseolus vulgaris* in West Africa (Senegal and Gambia). *Intern. J. System. Evol. Microbiol., 50*, 159-170.

Djordjevic, M.A., Gabriel, D.W. & Rolfe, B.G. (1987). *Rhizobium* - the refined parasite of legumes. *Annu. Rev. Phytopathol., 25*, 145-168.

Dobert, R.C., Breil, B.T. & Triplett, E.W. (1994). DNA sequence of the common nodulation genes of *Bradyrhizobium elkanii* and their phylogenetic relationship to those of other nodulating bacteria. *Molec. Plant-Microbe Interact., 7*, 564-572.

Doignon-Bourcier, F., Willems, A., Coopman, R., Laguerre, G., Gillis, M. & de Lajudie, P. (2000). Genotypic characterization of *Bradyrhizobium* strains nodulating small Senegalese legumes by 16S-23S rRNA intergenic gene spacers and amplified fragment length polymorphism fingerprint analyses. *Appl. Environ. Microbiol., 66*, 3987-3997.

Dorrell, N., Hinchliffe, S.J. & Wren, B.W. (2005). Comparative phylogenomics of pathogenic bacteria by microarray analysis. *Curr. Opin. Microbiol., 8*, 620-626.

Douglas, A.E. (1994). *Symbiotic Interactions*. Oxford; New York; Toronto: Oxford Univ. Press.

Douglas, A.E. (1998). Host benefit and the evolution of specialization in symbiosis. *Heredity, 81*, 599-603.

Downie, J.A. & Young, J.P.W. (2001). The ABC of symbiosis. *Nature., 412*, 597-598.

Doyle, J.J. (1998). Phylogenetic perspectives on nodulation: evolving views of plants and symbiotic bacteria. *Trends in Plant Sci., 2*, 473-478.

Duodu, S., Bhuvaneswari, T.V., Gudmundsson, J. & Svenning, M.M. (2005). Symbiotic and saprophytic survival of three unmarked *Rhizobium leguminosarum* biovar *trifolii* strains introduced into the field. *Environ. Microbiol., 7*, 1049-1058.

Dye, M., Skot, L., Mytton, L.R., Harrison, S.P., Dooley, J.J. & Cresswell, A. (1995). A study of *Rhizobium leguminosarum* biovar *trifolii* populations from soil extracts using randomly amplified polymorphic DNA profiles. *Canad. J. Microbiol., 41*, 336-344.

Eardly, B.D., Materon, L.A., Smith, N.H., Johnson, D.A., Rumbaugh, M.D. & Selander, R.K. (1990). Genetic structure of natural populations of nitrogen-fixing bacterium *Rhizobium meliloti. Appl. Environ. Microbiol., 56*, 187-194.

Eardly, B.D., Nour, S.M., van Berkum, P. & Selander, R.K. (2005). Rhizobial 16S rRNA and *dnaK* genes: mosaicism and the uncertain phylogenetic placement of *Rhizobium galegae. Appl. Environ. Microbiol., 71*, 1328-1335.

Eardly, B.D., Wang, F.-S., Whittam, T.S. & Selander, R.K. (1995). Species limits in *Rhizobium* populations that nodulate the common bean (*Phaseolus vulgaris*). *Appl. Environ. Microbiol., 61*, 507-512.

Elena, S.F., Miralles, R. & Moya, A. (1997). Frequency-dependent selection in a mammalian RNA virus. *Evolution, 51*, 984-987.

Engvild, K.C., Jensen, E.S. & Skot, L. (1990). Parallel variation in izozyme and nitrogen fixation markers in a *Rhizobium* population. *Plant and Soil, 128*, 283-286.

Farrand, S.K. (1998). Conjugal plasmids and their transfer. In H.P. Spaink, A. Kondorosi, & P.J.J. Hooykaas (Eds.), *The Rhizobiaceae. Molecular Biology of Model Plant-Associated Bacteria* (pp. 199-233). Dordrecht: Kluwer Acad. Publ.

Fedorova, M., Tikhonovich, I.A. & Vance, C.P. (1999). Expression of C-assimilating enzymes in pea (*Pisum sativum* L.) root nodules. *In situ* localization in effective nodules. *Plant, Cell and Environment, 22*, 1249-1262.

Feil, E.J. & Spratt, B.G. (2001). Recombination and the population structures of bacterial pathogens. *Annu. Rev. Microbiol., 55*, 561-590.

Fenton, M. & Jarvis, B.D.W. (1994). Expression of the symbiotic plasmid from *Rhizobium leguminosarum* biovar *trifolii* in *Sphingobacterium multivorum. Canad. J. Microbiol., 40*, 873-879.

Ferreira, M.C., Andrade, D.S., Chueire, L.M., Takemura, S.M. & Hungria, M. (2000). Tillage method and crop rotation effects on the population sizes and diversity of bradyrhizobia nodulating soybean. *Soil Biol. Biochem., 32*, 627-637.

Finan, T.M. (2002). Evolving insights: symbiotic islands and horizontal gene transfer. *J. Bacteriol., 184*, 2855-2856.

Fisher, H.M. (1994). Genetic regulation of nitrogen fixation in rhizobia. *Microbiol. Rev., 58*, 352-386.

Flores, M., Mavingui, P., Girard, L., Perret, X., Broughton, W.J., Martinez-Romero, E., Davila, G. & Palacios, R. (1998). Three replicons of *Rhizobium* sp. strain NGR234 harbor symbiotic gene sequences. *J. Bacteriol., 180*, 6052-6053.

Frank, S.A. (1992). Models of plant-pathogen co-evolution. *Trends in Genetics, 8*, 213-219.

Frank, S.A. (1994). Genetics of mutualism: the evolution of altruism between species. *J. Theor. Biol., 170*, 393-400.

Frank, S.A. (1996). Host-symbiont conflict over the mixing of symbiotic lineages. *Proc. Roy. Soc. London. B., 263*, 339-344.

Freiberg, C., Fellay, R., Bairoch, A., Broughton, W.J., Rosenthal, A. & Perret, X. (1997). Molecular basis of symbiosis between *Rhizobium* and legumes. *Nature, 387*, 394-401.

Fuqua, W.C. & Winans, S.C. (1994). A *luxR-luxI* type regulatory system activates *Agrobacterium Ti* plasmid conjugal transfer in the presence of a plant tumor metabolite. *J. Bacteriol., 176,* 2796-2806.

Gamas, P., de Billy, F. & Truchet, G. (1998). Symbiosis-specific expression of two *Medicago truncatula* nodulin genes, *MtM1* and *MtN13* encoding products homologous to plant defense proteins. *Molec. Plant-Microbe Interact., 11*, 393-403.

Garsia-de los Santos, A., Brom, S. & Romero, D. (1996). *Rhizobium* plasmids in bacteria-legume interactions. *World J. Microbiol. Biotechnol., 12*, 119-125.

Gianinazzi-Pearson, V. (1996). Plant cell responses to arbuscular mycorrhizal fungi: getting closer to the roots of the symbiosis. *The Plant Cell, 8*, 1871-1883.

Giuntini, E., Mengoni, A., de Filippo, C., Cavalieri, D., Aubin-Horth, N., Landry, C.R., Becker, A. & Bazzicalupo, M. (2005). Large-scale genomic variation of the symbiosis-required megaplasmid p*Sym*A revealed by comparative genomic analysis of *Sinorhizobium meliloti* natural strains. *BMC Genomics, 6*, 158-168.

Glenn, A.R. & Dilworth, M.J. (1994). The life of root nodule bacteria in the acidic underground. *FEMS Microbiol. Lett., 123*, 1-10.

Gogarten, J.P. & Townsend, J.P. (2005). Horizontal gene transfer, genome innovation and evolution. *Nature Rev./Microbiol., 3*, 679-689.

Gordon, D.M., Reyder, M.H., Heinrich, K. & Murphy, P. (1996). An experimental test of the rhizopine concept in *Rhizobium meliloti*. *J. Bacteriol., 62*, 3991-3996.

Gordon, D.M., Wexler, M., Reardon, T.B. & Murphy, P.J. (1995). The genetic structure of *Rhizobium* populations. *Soil Biol. Biochem., 27*, 491-499.

Groisman, E.A. & Ochman, H. (1996). Pathogenicity islands: bacterial evolution in quantum leaps. *Cell, 87*, 791-794.

Graften, A. (2007). Detecting kin selection at work using inclusive fitness. *Proc. Roy. Soc. Lond. B., 274*, 713-719.

Guttman, D.S. & Sarkar, S.F. (2004). Evolutionary and functional genomics of host specificity in *Pseudomonas syringae* In I.A. Tikhonovich, B.J.J. Lugtenberg, & N.A. Provorov (Eds.), *Biology of Plant-Microbe Interactions* (V. 4, pp. 132-137). St.-Petersburg: Biont.

Hahn, M. & Hennecke, H. (1987). Mapping of a *Bradyrhizobium japonicum* DNA region carrying genes for symbiosis and an asymmetric accumulation of reiterated sequences. *Appl. Environ. Microbiol., 53*, 2247-2252.

Haldane, J.B.S. (1932). *The causes of evolution*. New York: Longmans, Green & Co.

Hamilton, W.D. (1964). The genetical evolution of social behavior. *J. Theor. Biol., 7*, 1-16.

Hamilton, W.D. (1970). Selfish and spiteful behaviour in an evolutionry model. *Nature, 228*, 1218-1220.

Hardman, A.M., Stewart, G.S. & Williams, P. (1998). Quorum sensing and cell-cell communication dependent regulation of gene expression in pathogenic and non-pathogenic bacteria. *Anton. Leeuwen., 74*, 199-210.

Harrison, S.P., Jones, D.G., Schunmann, P.H.D., Forster, J.W. & Young, J.P.W. (1988). Variation in *Rhizobium leguminosarum* biovar *trifolii Sym* plasmids and the association with effectiveness of nitrogen fixation. *J. Gen. Microbiol., 134*, 2721-2730.

Harrison, S.P., Jones, D.G. & Young, J.P.W. (1989). *Rhizobium* population genetics: genetic variation within and between populations from diverse locations. *J. Gen. Microbiol., 135*, 1061-1069.

Hartmann, A. & Amarger, N. (1991). Genotypic diversity of an indigenous *Rhizobium meliloti* population assessed by plasmid profiles, DNA fingerprinting and insertion sequence typing. *Canad. J. Microbiol., 37*, 600-608.

Hartmann, A., Giraud, J.J. & Amarger, N. (1998). Genotypic diversity of *Sinorhizobium* (formerly *Rhizobium*) *meliloti* strains isolated directly from a soil and from nodules of alfalfa (*Medicago sativa*) grown in the same soil. *FEMS Microbiol. Ecol., 25*, 107-116.

Hauser, F., Lindemann, A., Vuilleumier, S., Patrignani, A., Schlapbach, R., Fisher, H.M. & Hennecke, H. (2006). Design and validation of a partial genome microarray for transcriptional profiling of the *Bradyrhizobium japonicum* symbiotic gene region. *Molec. Gen. Genet., 275*, 55-67.

Heinrich, K., Gordon, D.M., Ryder, M.H. & Murphy, P.J. (1999). A rhizopine strain of *Sinorhizobium meliloti* remains at a competitive nodulation advantage after an extended period in the soil. *Soil Biol. Biochem., 31*, 1063-1065.

Hennecke, H. (2004). Regulation of rhizobial life in symbiosis In H.P. Spaink, A. Kondorosi, & P.J.J. Hooykaas (Eds.), *The Rhizobiaceae. Molecular Biology of Model Plant-Associated Bacteria* (pp. 411-415). Dordrecht, Boston, London: Kluwer Acad. Publ.

Herre, E.A., Knowlton, N., Mueller, U.G. & Rehner, S.A. (1999). The evolution of mutualisms: exploring the paths between conflict and cooperation. *Trends in Ecol. Evol., 14*, 49-53.

Hibbett, D.S., Gilbert, L.B. & Donoghue, M.J. (2000). Evolutionary instability of ectomycorrhizal symbioses in basidiomycetes. *Nature, 407*, 506-508.

Hirsch, A.M., Lum, M.R. & Downie, J.A. (2001).What makes the rhizobia-legume symbiosis so special? *Plant Physiol., 127*, 1484-1492.

Hirsch, A.M., Wilson, K.J., Jones, J.D.G., Bang, M., Walker, V.V. & Ausubel, F.M. (1984). *Rhizobium meliloti* nodulation genes allow *Agrobacterium tumefaciens* and *Escherichia coli* to form pseudonodules on alfalfa. *J. Bacteriol., 158*, 1133-1143.

Hirsch, P.R. (1996). Population dynamics of indigenous and genetically modified rhizobia in the field. *New Phytologist, 133*, 159-171.

Horner-Devine, M.C., Carney, K.M. & Bohannan, B.J.M. (2004). An ecological perspective of bacterial biodiversity. *Proc. R. Soc. Lond. B., 271*, 113-122.

Hynes, M. & Funan, T. (1998). General genetics. In H.P. Spaink, A. Kondorosi, & P.J.J. Hooykaas (Eds.), *The Rhizobiaceae. Molecular Biology of Model Plant-Associated Bacteria* (pp. 25-43). Dordrecht, Boston, London: Kluwer Acad. Publ.

Ichige, A. & Walker, G.C. (1997). Genetic analysis of the *Rhizobium meliloti bacA* gene: functional interchangeability with the *Escherichia coli sbmA* gene and phenotypes of mutants. *J. Bacteriol., 179*, 209-216.

Janzen, D.H. (1980). When is it coevolution? *Evolution, 34*, 611-612.

Jarabo-Lorenzo, A., Perez-Galdona, R., Donate-Correa, J., Rivas, R., Velazquez, E., Hernandez, M., Temprano, F., Martinez-Molina, E., Ruiz-Argüeso, T. & Leon-Barrios, M. (2003). Genetic diversity of bradyrhizobial populations from diverse geographic origins that nodulate *Lupinus* spp. and *Ornithopus* spp. *System. Appl. Microbiol., 26*, 611-623.

Jarvis, B.D.W., Sivakumaran, S. & Lockhart, P.J. (1995). Identification of soil bacteria expressing a symbiotic plasmid from *Rhizobium leguminosarum* bv. *trifolii*. In I.A. Tikhonovich, N.A. Provorov, V.I. Romanov, & W.E. Newton (Eds.), *Nitrogen Fixation: Fundamentals and Applications* (p. 405). Dordrecht; Boston; London: Kluwer Acad. Publ.

Jarvis, B.D.W., Ward, L.J.H. & Slade, E.A. (1989). Expression by soil bacteria of nodulation genes from *Rhizobium leguminosarum* bv. *trifolii*. *Appl. Environ. Microbiol., 56*, 1426-1434.

Jebara, M., Mhamdi, R., Aouani, M.E., Ghrir, R. & Mars, M. (2001). Genetic diversity of *Sinorhizobium* populations recovered from different *Medicago* varieties cultivated in Tunisian soils. *Canad. J. Microbiol., 47*, 139-147.

Jiang, G., Krishnan, A.H., Kim, Y.-W., Wacek, T.J. & Krishnan, H.B. (2001). A functional *myo*-inositol dehydroganase gene is required for efficient nitrogen fixation and competitiveness of *Sinorhizobium fredii* USDA191 to nodulate soybean (*Glycine max* [L.] Merr.). *J. Bacteriol., 183*, 2595-2604.

Jimenez, J. & Casadesus, J. (1989). An altruistic model of *Rhizobium*-legume association. *J. Heredity, 80*, 335-337.

Jording, D., Uhde, C., Schmidt, R. & Pühler, A. (1994). The C_4-dicarboxylate transport system of *Rhizobium meliloti* and its role in nitrogen fixation during symbiosis with alfalfa (*Medicago sativa*). *Experientia, 50*, 874-883.

Kahn, M.L., Schroeder, B.K., House, B.L., Mortimer, M.W., Yurgel, S.N., Maloney, S.C., Warren, K.L., Fisher, R.F., Barnett, M.J., Toman, C. & Long, S.R. (2004). Foraging for meaning - postgenome approaches to *Sinorhizobium meliloti*. In I.A. Tikhonovich, B.J.J. Lugtenberg, & N.A. Provorov (Eds.), *Biology of Plant-Microbe Interactions* (V. 4, pp. 416-422). St.-Petersburg: Biont.

Kaminski, P.A., Batut, J. & Boistard, P. (1998). A survey of symbiotic nitrogen fixation by rhizobia. In H.P. Spaink, A. Kondorosi, & P.J.J. Hooykaas (Eds.), *The Rhizobiaceae. Molecular Biology of Model Plant-Associated Bacteria* (pp. 431-460). Dordrecht, Boston, London: Kluwer Acad. Publ.

Kannenberg, E.L., Reuhs, B.L., Forsberg, L.S. & Carlson, R.W. (1998). Lipopolysaccharides and K-antigens: their structures, biosynthesis and functions. In H.P. Spaink, A. Kondorosi, & P.J.J. Hooykaas (Eds.), *The Rhizobiaceae. Molecular Biology of Model Plant-Associated Bacteria* (pp. 119-154). Dordrecht, Boston, London: Kluwer Acad. Publ.

Kennedy, I.R. (1996). Facilitating the evolution of an effective N_2-fixing association between *Azospirillum* and wheat. *Abstr. 7-th Intern. Symp. Biolog. Nitrogen Fixat.* (p. 46). Haisalabad, Pakistan.

Kent, A.D. & Triplett, E. (2002). Microbial communities and their interactions in soil and rhizosphere ecosystems. *Annu. Rev. Microbiol., 56*, 211-236.

Kinkema, M., Scott, P.L. & Gresshoff, P. (2006). Legume nodulation: successful symbiosis through short- and long-distance signaling. *Functional Plant Biology*, *33*, 707-721.

Kinkle, B.K. & Schmidt, E.L. (1991). Transfer of pea symbiotic plasmid pJB5JI in non-sterile soil. *Appl. Environ. Microbiol.*, *57*, 3254-3269.

Kosier, B., Pühler, A. & Simon, R. (1993). Monitoring and diversity of *Rhizobium meliloti* field and microcosm isolates with a novel rapid genotyping method using insertion elements. *Molec. Ecol.*, *2*, 35-46.

Krassilinikov, N.A. (1941). Variability of root nodule bacteria. *C.R. Acad. Sci. USSR (In Russian)*, *31*, 90-92.

Krishnan, H.B. & Pueppke, S.G. (1991). Repetitive seguences with homology to *Bradyrhizobium japonicum* DNA and the T-DNA of *Agrobacterium rhizogenes* are closely linked to *nodABC* of *Rhizobium fredii* USDA257. *Molec. Plant-Microbe Interact.*, *4*, 521-529.

Kuykendall, L.D., Swelim, D.M., Hashem, F.M., Abdel Wahab, S.M. & Hegazi, N.I. (1996). Symbiotic competence, genetic diversity and plasmid profiles of Egyptian isolates of a *Rhizobium* speces from *Leucaena leucocephala* (Lam) de Wit. *Lett. Appl. Microbiol.*, *22*, 347-352.

Laberge, S., Middleton, A.T. & Wheatcroft, R. (1995). Characterization, nucleotide sequence and conserved genomic locations of insertion sequence IS*Rm5* in *Rhizobium meliloti*. *J. Bacteriol.*, *177*, 3133-3142.

Laguerre, G., Geniaux, E., Mazurier, S.I., Rodrigues, C.R., Amarger, N. (1993). Conformity and diversity among field isolates of *Rhizobium leguminosarum* bv. *viciae*, bv. *trifolii* and bv. *phaseoli* revealed by DNA hybridization using chromosome and plasmid probes. *Canad. J. Microbiol.*, *39*, 412-419.

Laguerre, G., Mavingui, P., Allard, M.R., Charnay, M.P., Lauvrier, P., Mazurier, S.I., Rigottier-Gois, L. & Amarger, N. (1996). Typing of rhizobia by PCR DNA fingeprinting and PCR-restriction lengths polymorphism analysis of chromosomal and symbiotic gene regions: application to *Rhizobium leguminosarum* and its different biovars. *Appl. Environ. Microbiol.*, *62*, 2029-2036.

Laguerre, G., Mazurier, S.I. & Amarger, N. (1992). Plasmid profiles and restriction length polymorpism of *Rhizobium leguminosarum* bv. *viceae* in field populations. *FEMS Microbiol. Ecol.*, *101*, 17-26.

Laguerre, G., Nour, S.M., Macheret, V., Sanjuan, J., Drouin, P. & Amarger, N. (2001). Classification of rhizobia based on *nodC* and *nifH* gene analysis reveals a close phylogenetic relationship among *Phaseolus vulgaris* symbionts. *Microbiology*, *147*, 981-993.

LaRue, T.A. (1980). Host plant genetics and enhancing symbiotic nitrogen fixation. In W.D.R. Stewart, & J.R. Gallon (Eds.), *Nitrogen Fixation* (pp. 355-364). London, New York, Toronto: Acad. Press.

Lenski, R.E. (1993). Assessing the genetic structure of microbial populations. *Proc. Natl. Acad. Sci. USA*, *90*, 4334-4336.

Leung, K., Strain, S.R., de Bruijn, F.J. & Bottomley, P.J. (1994a). Genotypic and phenotypic comparisons of chromosomal types within an indigenous soil population of *Rhizobium leguminosarum* bv. *trifolii*. *Appl. Environ. Microbiol.*, *60*, 416-426.

Leung, K., Wanjage, F.N. & Bottomley, P.J. (1994b). Symbiotic characteristics of *Rhizobium leguminosarum* bv. *trifolii* isolates which represent major and minor nodule-occupying chromosomal types in field grown subclover (*Trifolium subterraneum* L.). *Appl. Environ. Microbiol., 60*, 427-433.

Lie, T.A., Göktan, D., Engin, M., Pijnenborg, J. & Anlarsal, E. (1987). Co-evolution of the legume-*Rhizobium* association. *Plant and Soil, 100*, 171-181.

Lithgow, J.K., Wilkinson, A., Hardman, A., Rodelas, B., Wisniewski-Dye, F., Williams, P. & Downie, J.A. (2000). The regulatory locus *cinRI* in *Rhizobium leguminosarum* controls a network of quorum-sensing loci. *Molec. Microbiol., 37*, 81-97.

Loh, J.T., Lohar, D.P., Andersen, B. & Stacey, G. (2002). A two-component regulator mediates population-density-dependent expression of the *Bradyrhizobium japonicum* nodulation genes. *J. Bacteriol., 184*, 1759-1766.

Loh, J.T., Yuen-Tsai, J.P.Y., Stacey, M.G., Lohar, D., Welborn, A. & Stacey, G. (2001). Population density-dependent regulation of the *Bradyrhizobium japonicum* nodulation genes. *Molec. Microbiol., 42*, 37-46.

Louvrier, P., Laguerre, G. & Amarger, N. (1996). Distribution of symbiotic genotypes in *Rhizobium leguminosarum* biovar *viceae* populations isolated directly from soils. *Appl. Environ. Microbiol., 62*, 4202-4205.

Lugtenberg, B.J.J., Dekkers, L. & Bloemberg, G. (2001). Molecular determinants of rhizosphere colonization by *Pseudomonas*. *Annu. Rev. Phytopathol., 39*, 461-490.

Madsen, E.B., Madsen, L.H., Radutoiu, S., Olbryt, M., Rakwalska, M., Szczyglowski, K., Sato, S., Kaneko, T., Tabata, S., Sandal, N. & Stougaard, J. (2003). A receptor kinase gene of the LysM type is involved in legume perception of rhizobial signals. *Nature, 425*, 637-640.

Martin, A.P. (2002). Phylogenetic aproaches for describing and comparing the diversity of microbial communities. *Appl. Environ. Microbiol., 68*, 3673-3682.

Martinez, E., Romero, D. & Palacios, R. (1990). The *Rhizobium* genome. *Crit. Rev. Plant Sci., 9*, 59-93.

Martinez-Romero, E. & Caballero-Mellado, J. (1996). *Rhizobium* phylogenies and bacterial genetic diversity. *Crit. Rev. Plant Sci., 15*, 113-140.

Matthysse, A.G. & Kijne, J.W. (1998). Attachment of Rhizobiaceae to plant cells. In H.P. Spaink, A. Kondorosi, & P.J.J. Hooykaas (Eds.), *The Rhizobiaceae. Molecular Biology of Model Plant-Associated Bacteria* (pp. 235-249). Dordrecht, Boston, London: Kluwer Acad. Publ.

Mavingui, P., Flores, M., Guo, X., Davila, G., Perret, X., Broughton, W. & Palacios, R. (2002). Dynamics of genome architecture in *Rhizobium* sp. strain NGR234. *J. Bacteriol., 184*, 171-176.

Mavingui, P., Laermans, T., Flores, M., Romero, D., Martinez-Romero, E. & Palacios, R. (1998). Genes essential for Nod factor production and nodulation are located on a symbiotic amplicon (AMP*Rtr*CFN299pc60) in *Rhizobium tropici*. *J. Bacteriol., 180*, 2866-2874.

Maynard Smith, J. (1964). Group selection and kin selection. *Nature, 201*, 1145-1147.

Maynard Smith, J. (1989). Generating novelty by symbiosis. *Nature, 341*, 284-285.

Maynard Smith, J., Feil, E.J. & Smith, N.H. (2000). Population structure and evolutionary dynamics of pathogenic bacteria. *BioEssays, 22*, 1115-1122.

Maynard Smith, J., Smith, N.H., O'Rourke, M. & Spratt, B.G. (1993). How clonal are bacteria? *Proc. Natl. Acad. Sci. USA, 90*, 4384-4388.

Mayr, E. (1970). *Populations, species, and evolution*. Cambridge, Massachusetts: The Belknap Press of Harvard Univ. Press.

Mazurier, S.I., Rigotter-Gois, L. & Amarger, N. (1996). Characterization, distribution and localization of ISRl2, an insertion sequence element from *Rhizobium leguminosarum* biovar *viceae*. *Appl. Environ. Microbiol., 62*, 685-693.

Meeks, J.C. & Elhai, J. (2002). Regulation of cellular differentiation in filamentous cyanobacteria in free-living and plant-associated symbiotic growth states. *Microbiol. Molec. Biol. Rev., 66*, 94-121.

Mercado-Blanco, J. & Toro, N. (1996). Plasmids in rhizobia: the role of non-symbiotic plasmids. *Molec. Plant-Microbe Interact., 9*, 535-545.

Metzler, M.C., Laine, M.J. & de Boer, S.H. (1997). The status of molecular biological research on the plant pathogenic genus *Clavibacter*. *FEMS Microbiol. Lett., 150*, 1-8.

Michod, R.E. & Abugov, R. (1980). Adaptive topography in family-structured models of kin selection. *Science, 210*, 667-669.

Miller, M.B. & Bassler, B.L. (2001). Quorum sensing in bacteria. *Annu. Rev. Microbiol., 55*, 165-199.

Minamisawa, K., Isawa, T., Nakatsuka, Y. & Ichikawa, N. (1998). New *Bradyrhizobium japonicum* strains that possess high copy numbers of the repeated sequence RSα. *Appl. Environ. Microbiol., 65*, 1845-1851.

Minamisawa, K., Nakatsuka, Y. & Isawa, T. (1999). Diversity and field site variation of indigenous populations of soybean bradyrhizobia in Japan by fingerprints with repeated sequences RSα and RSβ. *FEMS Microbiol. Ecol., 29*, 171-178.

Mitchell-Olds, T. & Bergelson, J. (2000). Biotic interactions: genomics and coevolution. *Curr. Opin. Plant Biol., 3*, 273-277.

Mode, C.J. (1958). A mathematical model for the co-evolution of obligate parasites and their hosts. *Evolution, 12*, 158-165.

Moriguchi, K., Maeda, Y., Satou, M., Hardayani, N.S.N., Kataoka, M., Tanaka, N. & Yoshida, K. (2001). The complete nucleotide sequence of a plant-inducing (*Ri*) plasmid indicates its chimeric structure and evolutionary relationship between tumor-inducing (*Ti*) and symbiotic (*Sym*) plasmids in Rhizobiaceae. *J. Molec. Biol., 307*, 771-784.

Mougel, C., Cournoyer, B. & Nesme, X. (2001). Novel tellurite-amended media and specific chromosomal and Ti plasmid probes for direct analysis of soil populations of *Agrobacterium* biovars 1 and 2. *Appl. Environ. Microbiol., 67*, 65-74.

Moulin, L., Munive, A., Dreyfus, B. & Boivin-Masson, C. (2001). Nodulation of legumes by members of the beta-subclass of Preoteobacteria. *Nature, 411*, 948-950.

Murphy, P.J., Heycke, N., Trenz, S.P., Ratet, P., de Bruijn, F.J. & Schell, J. (1988). Synthesis of the opine-like compound, rhizopine in alfalfa nodules is symbiotically regulated. *Proc. Natl. Acad. Sci. USA, 85*, 9133-9137.

Murphy, P.J., Trenz, S.P., Grzemski, W., de Bruijn, F.J. & Schell, J. (1993). The *Rhizobium meliloti* rhizopine *mos* locus is a mosaic structure facilitating its symbiotic regulation. *J. Bacteriol., 175*, 5193-5204.

Murphy, P.J., Wexler, W., Grzemski, W., Rao, J.R. & Gordon, D. (1995). Rhizopines – their role in symbiosis and competition. *Soil. Biol. Biochem., 27*, 525-529.

Nei, M. (1978). Estimation of average heterozygosity and genetic distance from a small number of individuals. *Genetics, 89*, 583-590.

Nandasena, K.G., O'Hara, G.W., Tiwari, R.P. & Howieson, J.G. (2006). Rapid *in situ* evoltution of nodulating strains for *Biserrula pelecinus* L. through lateral transfer of a symbiosis island from the original mesorhizobial inoculant. *Appl. Environ. Microbiol., 72*, 7365-7367.

Nielsen, K.M., van Elsas, J.D. & Smalla, K. (2001). Dynamics, horizontal transfer and selection of novel DNA in bacterial populations in the phytosphere of transgenic plants. *Ann. Microbiol., 51*, 79-94.

Noel, K.D. & Brill, W.J. (1980). Diversity and dynamics of indigenous *Rhizobium japonicum* populations. *Appl. Environ. Microbiol., 40*, 931-938.

Norris, D.O. (1956). Legumes and the *Rhizobium* symbiosis. *Emp. J. Exper. Agric., 24*, 247-270.

Nour, S.M., Cleyet-Marel, J.C., Beck, D., Effosse, A. & Fernandez, M.P. (1994). Genotypic and phenotypic diversity of *Rhizobium* isolated from chickpea (*Cicer arietinum* L.). *Canad. J. Microbiol., 40*, 345-354.

Novikova, N.I. & Safronova, V.I. (1992). Transconjugants of *Agrobacterium radiobacter* harbouring *sym* genes of *Rhizobium galegae* can form an effective symbiosis with *Medicago sativa*. *FEMS Microbiol. Lett., 93*, 261-268.

O'Connell, D. (2006). Evolving virulence. *Nature Rev./Microbiol., 4*, 83.

Ochman, H. & Bergthorsson, U. (1998). Rates and patterns of chromosome evolution in enteric bacteria. *Curr. Opin. Microbiol., 1*, 580-583.

Ochman, H., Elwyn, S. & Moran, N.A. (1999). Calibrating bacterial evolution. *Proc. Natl. Acad. Sci. USA, 96*, 12638-12643.

Ochman, H., Lawrence, J.G. & Groisman, E.A. (2000). Lateral gene transfer and the nature of bacterial innovation. *Nature, 405*, 299-304.

Oke, V. & Long, S.R. (1999). Bacteroid formation in the *Rhizobium*-legume symbiosis. *Curr. Opin. Microbiol., 2*, 641-646.

Oldroyd, G.E.D., Mitra, R.M., Wais, R.J. & Long, S.R. (2001). Evidence for structurally specific negative feedback in the Nod factor signal transduction pathway. *The Plant J., 28*, 191-199.

Olivieri, I. & Frank, S.A. (1994). The evolution of nodulation in *Rhizobium*: altruism in the rhizosphere. *J. Heredity, 85*, 46-47.

Olsen, P., Wright, S., Collins, M. & Rice, W. (1994). Patterns of reactivity between a panel of monoclonal antibodies and forage *Rhizobium* strains. *Appl. Environ. Microbiol., 60*, 654-661.

Ovtsyna, A.O., Schultze, M., Tikhonovich, I.A., Spaink, H.P., Kondorosi, E., Kondorosi, A. & Staehelin, C. (2000). Nod factors of *Rhizobium leguminosarum* bv. *viceae* and their fucosylated derivatives stimulate a Nod factor cleaving activity in pea roots and are hydrolysed *in vitro* by plant chitinases at different rates. *Molec. Plant-Microbe Interact., 13*, 799-807.

Ovtsyna, A.O. & Staehelin, C. (2005). Bacterial signals required for the *Rhizobium*-legume symbiosis. *Recent Res. Develop. Microbiol., 7*, 631-648.

Paffetti, D., Daguin, F., Fancelli, S., Gnocchi, S., Lippi, F., Scotti, C. & Bazzicalupo, M. (1998). Influence of plant genotype on the selection of nodulating *Sinorhizobium meliloti* strains by *Medicago sativa*. *Anton. Leeuwen., 73*, 3-8.

Paffetti, D., Scotti, C., Gnocchi, S., Fancelli, S. & Bazzicalupo, M. (1996). Genetic diversity of an Italian *Rhizobium meliloti* population from different *Medicago sativa* varieties. *Appl. Environ. Microbiol., 62*, 2279-2285.

Parniske, M. (2000). Intracellular accommodation of microbes by plants: a common developmental program for symbiosis and disease? *Curr. Opin. Plant Biol., 3*, 320-328.

Perez-Mendoza, D., Dominguez, A., Sepulveda, E., Pando, V., Munoz, S., Nogales, J., Herrera-Cervera, J.A., Soto, M.J., Olivares, J., Girard, L., Romero, D., Brom, S. & Sanjuan, J. (2004). Conjugative transfer of symbiotic DNA in rhizobia. In I.A. Tikhonovich, B.J.J. Lugtenberg, & N.A. Provorov (Eds.), *Biology of Plant-Microbe Interactions* (V. 4, pp. 582-584). St.-Petersburg: Biont.

Perez-Ramirez, N.O., Rogel, M.A., Wang, E., Castellanos, J.Z. & Martinez-Romero, E. (1998). Seeds of *Phaseolus vulgaris* bean carry *Rhizobium etli*. *FEMS Microbiol. Ecol., 26*, 289-296.

Perret, X. & Broughton, W.J. (1998). Rapid identification of *Rhizobium* strains by targeted PCR fingerprinting. *Plant and Soil, 204*, 21-34.

Perret, X., Staehelin, C. & Broughton, W.J. (2000). Molecular basis for symbiotic promiscuity. *Microbiol. Molec. Biol. Rev., 64*, 180-201.

Person, C., Samborski, D.J. & Rohringer, R. (1962). The gene-for-gene concept. *Nature, 194*, 561-562.

Pimentel, D. (1968). Population regulation and genetic feedback. *Science, 159*, 1432-1437.

Pinero, D., Martinez, E. & Selander, R.K. (1988). Genetic diversity and relationships among isolates of *Rhizobium leguminosarum* bv. *phaseoli*. *Appl. Environ. Microbiol., 54*, 2825-2832.

Pistorio, M., del Papa, M.F., Balague, L.J. & Lagares, A. (2003). Identification of a transmissible plasmid from an Argentine *Sinorhizobium meliloti* strain which can be mobilized by conjugative helper functions of the European strain *S. meliloti* GR4. *FEMS Microbiol. Lett., 225*, 15-21.

Plazinski, J. & Rolfe, B.G. (1985). Sym plasmid genes of *Rhizobium trifolii* expressed in *Lignobacter* and *Pseudomonas* strains. *J. Bacteriol., 162*, 1261-1269.

Preston, G.M., Haubold, B. & Rainey, P.B. (1998). Bacterial genomics and adaptation to life on plants: implications for the evolution of pathogenicity and symbiosis. *Curr. Opin. Microbiol., 1*, 589-597.

Pretorius-Guth, I.M., Pühler, A. & Simon, R. (1990). Conjugal transfer of megaplasmid 2 between *Rhizobium meliloti* strains in alfalfa nodules. *Appl. Environ. Microbiol., 56*, 2354-2359.

Provorov, N.A. (1998). Coevolution of rhizobia with legumes: facts and hypotheses. *Symbiosis, 24*, 337-367.

Provorov, N.A. (2000). Population genetics of nodule bacteria. *Zhurnal Obshei Biologii (in Russian), 61*, 229-257.

Provorov, N.A., Borisov, A.Y. & Tikhonovich, I.A. (2002). Developmental genetics and evolution of symbiotic structures in nitrogen-fixing nodules and arbuscular mycorrhiza. *J. Theor. Biol., 214*, 215-232.

Provorov, N.A., Fokina, I.G., Roumiantseva, M.L. & Simarov, B.V. (2004). Transfer of *Sym* plasmids into symbiotically active and asymbiotic rhizobia strains: properties of recombinants and the possible evolutionary consequencies. *Ecological Genetics (in Russian), 2*, 29-34.

Provorov, N.A. & Tikhonovich, I.A. (2003). Genetic resources for improving nitrogen fixation in legume-rhizobia symbiosis. *Genet. Res. Crop Evolut., 50*, 89-99.

Provorov, N.A. & Vorobyov, N.I. (2000a). Evolutionary genetics of nodule bacteria: molecular and population aspects. *Russian J. Genetics, 36*, 1323-1335.

Provorov, N.A. & Vorobyov, N.I. (2000b). Population genetics of rhizobia: construction and analysis of an "infection and release" model. *J. Theor. Biol., 205*, 105-119.

Provorov, N.A. & Vorobyov, N.I. (2006). Interplay of Darwinian and frequency-dependent selection in the host-associated microbial populations. *Theor. Populat. Biol., 70*, 262-272.

Pühler, A. & Klipp, W. (1981). Fine structure analysis of the gene region for N_2 fixation (*nif*) of *Klebsiella pneumoniae*. In H. Bothe, & A. Trebst (Eds.), *Biology of Inorganic Nitrogen and Sulfur* (pp. 276-286). Heidelberg, Berlin: Springer.

Pukall, R., Tschape, H. & Smalla, K. (1996). Monitoring and spread of broad host and narrow host range plasmids in soil microcosms. *FEMS Microbiol. Ecol., 20*, 53-66.

Qian, J. & Parker, M.P. (2002). Contrasting *nifD* and ribosomal gene relationships among *Mesorhizobium* from *Lotus oroboides* in Northern Mexico. *System. Appl. Microbiol., 25*, 68-73.

Radutoiu, S., Madsen, L.H., Madsen, E.B., Felle, H.H., Umehara, Y., Gronlund, M., Sato, S., Nakamura, Y., Tabata, S., Sandal, N. & Stougaard, J. (2003). Plant recognition of symbiotic bacteria requires two LysM receptor-like kinases. *Nature, 425*, 585.

Rao, J.P., Grzemski, W. & Murphy, P.J. (1995). *Rhizobium meliloti* lacking *mosA* synthesizes the rhizopine *scyllo*-inosamine in place of 3-o-methyl-*scyllo*-inosamine. *Microbiology, 141*, 1683-1690.

Rasolomampianina, R., Bailly, X., Fetiarison, R., Rabevohitra, R., Bena, G., Ramaroson, L., Raherimandimby, M., Moulin, L., de Lajudie, P., Dreyfus, B. & Avarre, J.-C. (2005). Nitrogen-fixing nodules from rose wood legume trees (*Dalbergia* spp.) endemic to Madagascar host seven different genera belonging to α- and β-Proteobacteria. *Molec. Ecol., 14*, 4135-4146.

Rausher, M.D. (2001). Co-evolution and plant resistance to natural enemies. *Nature, 411*, 857-864.

Richardson, D.M., Allsopp, N., d'Antonio, C.M., Milton, S.J. & Rejmanek, M. (2000). Plant invasions – the role of mutualisms. *Biol. Rev., 75*, 65-93.

Rivas, R., Velazquez, E., Willems, A., Vizcaino, N., Subba-Rao, N.S., Mateos, P.F., Gillis, M., Dazzo, F. & Martinez-Molina, E. (2002). A new species of *Devosia* that forms a unique nitrogen-fixing root-nodule symbiosis with the aquatic legume *Neptunia natans* (L.f.) Druce. *Appl. Environ. Microbiol., 68*, 5217-5222.

Rodelas, B., Lithgow, J.K., Wisniewski-Dye, F., Hardman, A., Wilkinson, A., Economou, A., Williams, P. & Downie, J.A. (1999). Analysis of quorum-sensing-dependent control of rhizosphere-expressed (*rhi*) genes in *Rhizobium leguminosarum* bv. *viceae*. *J. Bacteriol., 181*, 3816-3823.

Rodriguez, C. & Romero, D. (1998). Multiple recombination events maintain sequence identity among the members of nitrogenase multigene family in *Rhizobium etli*. *Genetics, 149*, 785-794.

Rodriguez-Quinones, F., Judd, A.K., Sadowsky, M.L., Liu, R.-L. & Cregan, P.B. (1992). Hyperreiterated DNA regions are conserved among *Bradyrhizobium japonicum* serocluster 123 strains. *Appl. Environ. Microbiol., 58*, 1878-1885.

Rogel, M.A., Hernandez-Lucas, I., Kuykendall, L.D., Balkwill, D.L. & Martinez-Romero, E. (2001). Nitrogen-fixing nodules with *Ensifer adhaerens* harboring *Rhizobium tropici* symbiotic plasmids. *Appl. Environ. Microbiol., 67*, 3264-3268.

Rome, S., Fernandez, M.P., Brunel, B., Normand, P. & Cleyet-Marel, J.C. (1996). *Sinorhizobium medicae* sp. nov. isolated from annal *Medicago* spp. *Int. J. Syst. Bacteriol., 46*, 972-980.

Romero, D., Martinez-Salazar, J., Lourdes Girard, M., Brom, S., Davila, G., Palacios, R., Flores, M. & Rodrigues, C. (1995). Discrete amplifiable regions (amplicons) in the symbiotic plasmid of *Rhizobium etli* CFN42. *J. Bacteriol., 177*, 973-980.

Romero, D. & Palacios, R. (1997). Gene amplification and genomic plasticity in prokaryotes. *Annu. Rev. Genet., 31*, 91-111.

Rosemeyer, V., Michiels, J., Verreth, C. & Vanderleyden, J. (1998). *luxI* and *luxR* - homologous genes of *Rhizobium etli* CNPA-F512 contribute to synthesis of autoinductor molecules and nodulation of *Phaseolus vulgaris*. *J. Bacteriol., 180*, 815-821.

Roth, L.E. & Stacey, G. (1989). Bacterium release into host cells of nitrogen-fixing soybean nodules: the symbiosome membrane comes from three sources. *Europ. J. Cell. Biol., 49*, 13-23.

Roumiantseva, M.L., Andronov, E.E., Sagulenko, V.V., Onichtchouk, O.P., Provorov, N.A. & Simarov, B.V. (2004). Instability of cryptic plasmids in strain *Sinorhizobium meliloti* P108 in the course of symbiosis with alfalfa *Medicago sativa*. *Russ. J. Genet., 40*, 454-461.

Roumiantseva, M.L., Andronov, E.E., Sharypova, L., Damman-Kalinowski, T., Keller, M., Young, J.P.W. & Simarov, B.V. (2002). Diversity of *Sinorhizobium meliloti* from the Central Asian alfalfa gene center. *Appl. Environ. Microbiol., 68*, 4694-4697.

Sadowsky, M.J. & Graham, P.H. (1998). Soil biology of the Rhizobiaceae. In H.P. Spaink, A. Kondorosi, & P.J.J. Hooykaas (Eds.), *The Rhizobiaceae. Molecular Biology of Model Plant-Associated Bacteria* (pp. 155-172). Dordrecht, Boston, London: Kluwer Acad. Publ.

Saint, C.P., Wexler, M., Murphy, P.J., Tempe, J., Tate, M.E. & Murphy, P.J. (1993). Characterization of genes for synthesis and catabolism of a new rhizopine induced in nodules by *Rhizobium meliloti* Rm220-3: extension of the rhizopine concept. *J. Bacteriol., 175*, 5205-5215.

Saleena, L.M., Loganathan, P., Rangarajan, S. & Nair, S. (2001). Genetic diversity of *Bradyrhizobium* strains isolated from *Arachis hypogaea*. *Canad. J. Microbiol., 47*, 118-122.

Sanchez, F., Padilla, J., Perez, H. & Lara, M. (1991). Control of nodulin genes in root-nodule development and metabolism. *Annu. Rev. Plant Physiol. Plant Mol. Biol., 42*, 507-528.

Sanchez-Contreras, M., Llortet, J., Martin, M., Villacieros, M., Bonilla, I. & Rivilla, R. (2000). PCR use of highly conserved DNA regions for identification of *Sinorhizobium meliloti*. *Appl. Environ. Microbiol., 66*, 3621-3623.

Santos, R., Herouart, D., Sigaud, S., Touati, D. & Puppo, A. (2001). Oxidative burst in alfalfa – *Sinorhizobium meliloti* symbiotic interaction. *Molec. Plant-Microbe Interact., 14*, 86-89.

Schlamann, H.R.M., Okker, R.J.H. & Lugtenberg, B.J.J. (1992). Regulation of nodulation gene expression by NodD in rhizobia. *J. Bacteriol., 174*, 5177-5182.

Schlöter, M., Lebuhn, M., Heulin, T. & Hartmann, A. (2000). Ecology and evolution of bacterial microdiversity. *FEMS Microbiol. Rev., 24*, 647-660.

Schofield, P.R., Gibson, A.H., Dudman, W.F. & Watson, J.M. (1987). Evidence for genetic exchange and recombination of *Rhizobium* symbiotic plasmids in a soil population. *Appl. Environ. Microbiol., 53*, 2942-2947.

Schußler, A., (2002). Molecular phylogeny, taxonomy and evolution of *Geosiphon pyriformis* and arbuscular mycorrhizal fungi. *Plant and Soil, 244*, 75-83.

Schuster, M.L. & Coyne, D.P. (1974). Survival mechanisms of phytopathogenic bacteria. *Annu. Rev. Phytopathol., 12*, 199-221.

Seckbach, J., Ed. (2002). *Symbiosis: Mechanisms and Model Systems*. Dordrecht; Boston; London: Kluwer Acad. Publ.

Segovia, L., Pinero, D., Palacios, R. & Martinez-Romero, E. (1991). Genetic structure of a soil population of non-symbiotic *Rhizobium leguminosarum*. *Appl. Environ. Microbiol., 57*, 426-433.

Seguin, P., Graham, P.H., Sheaffer, C.C., Ehlke, N.J. & Russelle, M.P. (2001). Genetic diversity of rhizobia nodulating *Trifolium ambiguum* in North America. *Canad. J. Microbiol., 47*, 81-85.

Selander, R.K., Caugant, D.A., Ochman, H., Musser, J.M., Gilmour, M.N. & Whittam, T.S. (1986). Methods of multilocus enzyme electrophoresis for bacterial population genetics and systematics. *Appl. Environ. Microbiol., 51*, 873-884.

Selander, R.K. & Levin, B.R. (1980). Genetic diversity and structure in *Escherichia coli* populations. *Science, 210*, 545-547.

Shan, L., He, P., Sheen, J. (2007). Endless hide-and-seek dynamic co-evolution in plant-bacterium warfare. *J. Integr. Plant Biol., 49*, 105-111.

Shapiro, J.A. (1998). Thinking about bacterial populations as multicellular organisms. *Annu. Rev. Microbiol., 52*, 81-104.

Shishido, M. & Pepper, I.L. (1990). Identification of the dominant indigenous *Rhizobium meliloti* by plasmid profiles and intrinsic antibiotic resistance. *Soil. Biol. Biochem., 22*, 11-16.

Skotnicki, M.L. & Rolfe, B.G. (1978). Transfer of nitrogen fixation genes from a bacterium with the characteristics of both *Rhizobium* and *Agrobacterium*. *J. Bacteriol., 133*, 518-526.

Soberon-Chavez, G. & Najera, R. (1989). Isolation from soil of *Rhizobium leguminosarum* lacking symbiotic information. *Canad. J. Microbiol., 35*, 464-468.

Soberon-Chavez, G., Najera, R., Espin, G. & Moreno, S. (1991). Formation of *Rhizobium phaseoli* symbiotic plasmids by genetic recombination. *Molec. Microbiol., 5*, 909-916.

Souza, V., Nguyen, T.T., Hudson, R.R., Pinero, D. & Lenski, R.E. (1992). Hierarchical analysis of linkage disequilibrium in *Rhizobium* populations: evidence for sex? *Proc. Natl. Acad. Sci. USA, 89*, 8389-8393.

Spaink, H.P. (1995). The molecular basis of infection and nodulation by rhizobia: the ins and outs of sympathogenesis. *Ann. Rev. Phytopathol., 33*, 345-368.

Spaink, H., Kondorosi, A. & Hooykaas, P.J.J., Eds. (1998). *The Rhizobiaceae. Molecular Biology of Model Plant-Associated Bacteria.* Dordrecht; Boston; London: Kluwer Acad. Publ.

Spoerke, J.M., Wilkinson, H.H. & Parker, M.A. (1996). Nonrandom genotypic associations in a legume-*Bradyrhizobium* mutualism. *Evolution, 50*, 146-154.

Sprent, J.I. & Raven, J.A. (1992). Evolution of nitrogen-fixing symbiosis. In G. Stacey, R.H. Burris, & H.J. Evans (Eds.), *Biological Nitrogen Fixation* (pp. 461-496). New York; London: Chapman and Hall.

Stacey, G. (1995). *Bradyrhizobium japonicum* nodulation genetics. *FEMS Microbiol. Lett., 127*, 1-9.

Stahl, E.A. & Bishop, J.G. (2000). Plant-pathogen arms races at the molecular level. *Curr. Opin. Plant Biol., 3*, 299-304.

Steinert, M., Hentschel, U. & Hacker, J. (2000). Symbiosis and pathogenesis: evolution of the microbe-host interaction. *Naturwissenschaften, 87*, 1-11.

Stephens, C. & Murray, W. (2001). Pathogen evolution: how good bacteria go bad. *Current Biology, 11*, 53-56.

Stepkowski, T., Moulin, L., Krzyianska, A., McInnes, A., Law, I.J. & Howieson, J. (2005). European origin of *Bradyrhizobium* populations infecting lupins and seradella in soils of Western Australia and South Africa. *Appl. Environ. Microbiol., 71*, 7041-7052.

Strain, S.R., Leung, K., Whittam, T.S., de Bruijn, F. & Bottomley, P.J. (1994). Genetic structure of *Rhizobium leguminosarum* biovar *trifolii* and *viceae* populations found in two Oregon soils under different plant communities. *Appl. Environ. Microbiol., 60*, 2772-2778.

Sullivan, J.T., Patrick, H.N., Lowther, W.L., Scott, D.B. & Ronson, C.W. (1995). Nodulating strains of *Rhizobium loti* arise through chromosomal symbiotic gene transfer in the environment. *Proc. Natl. Acad. Sci. USA, 92*, 8985-8989.

Sullivan, J.T., Trzebiatowski, J.R., Cruickshank, R.W., Gouzy, J., Brown, S.D., Elliot, R.M., Fleetwood, D.J., McCalum, N.G., Rossbach, U., Stuart, G.S., Weaver, J.E., Webby, R.J., de Bruijn, F. & Ronson, C. (2002). Comparative sequence analysis of the symbiosis island of *Mesorhizobium loti* strain R7A. *J. Bacteriol., 184*, 3086-3095.

Tan, Z., Hurek, T., Vinuesa, P., Müller, P., Ladha, J.K. & Reinhold-Hurek, B. (2001). Specific detection of *Bradyrhizobium* and *Rhizobium* strains colonizing rice (*Oryza sativa*) roots by 16S-23S ribosomal DNA intergenic spacer-targeted PCR. *Appl. Environ. Microbiol., 67*, 3655-3664.

Tauch, A., Schneiker, S., Selbitschka, W., Pühler, A., van Overbeek, L.S., Smalla, K., Thomas, C.M., Bailey, M.J., Forney, L.J., Weightman, A., Ceglowski, P., Pembroke, T., Tietze, E., Schroder, G., Lanka, E. & van Elsas, J.D. (2002). The complete nucleotide sequence and environmental distribution of the cryptic conjugative, broad-host-range plasmid pIP02 isolated from bacteria of the wheat rhizosphere. *Microbiology, 148*, 1637-1653.

Terefework, Z., Lortet, G., Souminen, L. & Lindström, K. (2000). Molecular evolution of interactions between rhizobia and their legume hosts. In *Procaryotic Nitrogen Fixation: a Model System for Analysis of a Biological Process* (pp. 187-206). Wymondham: Horizon Sci. Press.

Thomas, C.M. & Nielsen, K.M. (2005). Mechanisms of and barriers to horizontal gene transfer between bacteria. *Nature Rev./Microbiol., 3*, 711-721.

Thompson, J.N. & Burdon, J. (1992). Gene-for-gene co-evolution between plants and parasites. *Nature, 360*, 121-125.

Tibayrenc, M. (1996). Towards a unified evolutionary genetics of microorganisms. *Annu. Rev. Microbiol., 50*, 401-429.

Tikhonovich, I.A. & Provorov, N.A. (2007). Beneficial plant-microbe interactions. In Y.T.Dyakov, V. Dzhavakhiya, & T. Korpela (Eds.), *Comprehensive and Molecular Phytopathology* (pp. 365-420). Amsterdam: Elsevier.

Timmers, A.C.J., Soupene, E., Auriac, M.C., de Billy, F., Vasse, J., Boistard, P. & Truchet, G. (2000). Saprophytic intracellular rhizobia in alfalfa nodules. *Molec. Plant-Microbe Interact., 13*, 1204-1213.

Timofeeff-Ressovsky, N.W., Jablokov, A.V. & Glotov, N.V. (1977). *Grundriss der Populationslehre*. Jena: Gustav Fisher Verlag.

Triplett, E.V. & Sadowsky, M.E. (1992). Genetics of competition for nodulation of legumes. *Annu. Rev. Microbiol., 46*, 399-428.

Trujillo, M.E., Willems, A., Abril, A., Planchuelo, A.-M., Rivas, R., Ludena, D., Mateos, P.F., Martinez-Molina, E. & Velazquez, E. (2005). Nodulation of *Lupinus albus* by strains of *Ochrobactrum lupini* sp. nov. *Appl. Environ. Microbiol., 71*, 1318-1327.

Turner, S.L. Rigotter-Goiz, L., Power, R.S., Amarger, N. & Young, J.P.W. (1996). Diversity of *repC* plasmid-replication sequences in *Rhizobium leguminosarum*. *Microbiology, 142*, 1705-1713.

Turner, S.L. & Young, J.P.W. (2000). The glutamine synthetases of rhizobia: phylogenetics and evolutionary implications. *Molec. Biol. Evol., 17*, 309-319.

Udvardi, M., Bock, V., Colebatch, G., Desbrosses, G., Kloska, S., Kopka, J., Krause, K., Krusell, L., Ott, T., Trevaskis, B. & Wandrey, M. (2004). Genetic reorganisation of legume transport and metabolism during symbiotic nitrogen fixation. In I.A. Tikhonovich, B.J.J. Lugtenberg, & N.A. Provorov (Eds.), *Biology of Plant-Microbe Interactions* (V. 4, pp. 490-492). St.-Petersburg: Biont.

Udvardi, M.K. & Kahn, M.L. (1992). Evolution of the (*Brady*)*Rhizobium*-legume symbiosis: why do bacteroids fix nitrogen? *Symbiosis, 14*, 87-101.

Ulrich, A. & Zaspel, I. (2000). Phylogenetic diversity of rhizobial strains nodulating *Robinia pseudoacacia* L. *Microbiology, 146*, 2997-3005.

Van Berkum, P., Elia, P. & Eardly, B.D. (2006). Multilocus sequence typing as an approach for population analysis of *Medicago*-nodulating rhizobia. *J. Bacteriol., 188*, 5570-5577.

Van Elsas, J.D., Trevors, J.T. & Starodub, E.E. (1998). Bacterial conjugation between pseudomonads in the rhizosphere of wheat. *FEMS Microbiol. Ecol., 53*, 299-306.

Van Sluys, M.A., Monteiro-Vitorello, C.B., Camargo, L.E.A., Menck, C.F.M., da Silva, A.C.R., Ferro, J.A., Oliviera, M.C., Setubal, J.C., Kitajima, J.P. & Simpson, A.J. (2002). Comparative genomic analysis of plant-associated bacteria. *Annu. Rev. Phytopathol., 40*, 169-189.

Van Veen, R.J.M., Dulk-Ras, H., Bisseling, T., Schilperoort, R.A. & Hooykaas, P.J.J. (1988). Crown gall tumor and root nodule formation by the bacterium *Phyllobacterium myrsinacearum* after the introduction of an *Agrobacterium Ti* plasmid or a *Rhizobium Sym* plasmid. *Molec. Plant-Microbe Interact., 1*, 231-234.

Vance, C.P. & Heichel, G.H. (1991). Carbon in N_2 fixation: limitation or exquisive adaptation? *Annu. Rev. Plant Physiol., 42*, 373-392.

Vinuesa, P., Silva, C., Werner, D. & Martinez-Romero, E. (2005). Population genetics and phylogenetic inference in bacterial molecular systematics: the roles of migration and recombination in *Bradyrhizobium* species cohesion and delineation. *Molec. Phylogen. Evol., 34*, 29-54.

Vlades, A.M. & Pinero, D. (1992). Phylogenetic estimation of plasmid exchange in bacteria. *Evolution, 46*, 641-656.

Wade, M.J. (1977). An experimental study of group selection. *Evolution, 31*, 134-153.

Wang, C.L., Beringer, J.E. & Hirsch, P.R. (1986). Host plant effects on inter-specific hybrids of *Rhizobium leguminosarum* biovars *viceae* and *trifolii*. *J. Gen. Microbiol., 132*, 2063-2070.

Wang, E.T., van Berkum, P., Sui, X.H., Beyene, D., Chen, W. & Martinez-Romero, E. (1999). Diversity of rhizobia associated with *Amorpha fruticosa* isolated from Chinese soils and description of *Mesorhizobium amorphae* sp. nov. *Int. J. Syst. Bacteriol., 49*, 51-65.

Waters, J.K., Hughes, B.L., Purcell, L.C., Gerhardt, K.O., Mawhinney, T.P. & Emerich, D.W. (1998). Alanine, not ammonia, is excreted from N_2-fixing soybean nodule bacteroids. *Proc. Natl. Acad. Sci. USA, 95*, 12038-12042.

Weaver, R.W. & Wright, S.F. (1987). Variability in effectiveness of rhizobia during culture and in nodules. *Appl. Environ. Microbiol., 53*, 2972-2974.

Wernegreen, J.J. & Riley, M.A. (1999). Comparison of the evolutionary dynamics of symbiotic and housekeeping loci: a case for the genetic coherence of rhizobial lineages. *Molec. Biol. Evol., 16*, 98-113.

Wilkinson, D.M. (1997). The role of seed dispersal in the evolution of mycorrhizae. *Oikos, 78*, 394-396.

Wilkinson, H.H., Spoeke, J.M. & Parker, M.A. (1996). Divergence in symbiotic compartibility in a legume-*Bradyrhizobium* mutualism. *Evolution, 50*, 1470-1477.

Williams, G.C. & Williams, D.C. (1957). Natural selection of individually harmful social adaptations among sibs with special reference to social insects. *Evolution, 11*, 32-39.

Wilson, D.S. (1977). Structured demes and the evolution of group-advantageous traits. *The Amer. Naturalist, 111*, 157-185.

Winarno, R. & Lie, T.A. (1979). Competition between *Rhizobium* strains in nodule formation: interaction between nodulating and non-nodulating strains. *Plant and Soil, 51*, 135-142.

Wynne-Edwards, V.C. (1963). Intergroup selection in the evolution of social systems. *Nature, 4907*, 623-626.

Yang, G.-P., Debelle, F., Savagnac, A., Ferro, M., Schiltz, O., Maillet, F., Prome, D., Treilhou, M., Vialas, C., Lindström, K., Denarie, J. & Prome, J.-C. (1999). Structure of the *Mesorhizobium huakuii* and *Rhizobium galegae* Nod factors: a cluster of phylgenetically related legumes are nodulated by rhizobia producing Nod factors with α,β-unsaturated N-acyl substitutions. *Molec. Microbiol., 34*, 227-237.

Young, J.P.W. (1989). The population genetics of bacteria. In D.A. Hopwood, & Chater K.F. (Eds.), *Genetics of Bacterial Diversity* (pp. 417-438). London: Academic Press.

Young, J.P.W. (1992). Phylogenetic classification of nitrogen-fixing organisms. In G. Stacey, R.H. Burris, & Evans H.J. (Eds.), *Biological Nitrogen Fixation* (pp. 43-86). London: Chapman and Hall.

Young, J.P.W., Crossman, L.C., Johnston, A.W.B., Thomson, N.R., Ghazoui, Z.F., Hull, K.H., Wexler, M., Curson, A.R.J., Todd, J.D., Poole, P.S., Mauchline, T.H., East, A.K., Quail, M.A., Churcher, C., Arrowsmith, C., Cherevach, I., Chillingworth, T., Clarke, K., Cronin, A., Davis, P., Fraser, A., Hance, Z., Hauser, H., Jagels, K., Moule, S., Mungall, K., Noebertczak, H., Rabbinowitsch, E., Sanders, M., Simmonds, M., Whitehead, S. & Parkhill, J. (2006). The genome of *Rhizobium leguminosarum* has recognizable core and accessory components. *Genome Biology, 7*, R34.

Young, J.P.W., Demetriou, L. & Apte, R.G. (1987). *Rhizobium* population genetics: enzyme polymorphism in *Rhizobium leguminosarum* from plants and soil in a pea crop. *Appl. Environ. Microbiol., 53*, 397-402.

Young, J.P.W. & Haukka, K.E. (1996). Diversity and phylogeny of rhizobia. *New Phytologist, 133*, 87-94.

Young, J.P.W. & Wexler, M. (1988). Sym plasmid and chromosomal genotypes are correlated in field populations of *Rhizobium leguminosarum*. *J. Gen. Microbiol., 134*, 2731-2739.

Zhang, X.-X., Turner, S.L., Guo, X.-W., Yang, H.-J., Debelle, F., Yang, G.-P., Denarie, J., Young, J.P.W. & Li, F.-D. (2000). The common nodulation genes of *Astragalus sinicus* rhizobia are conserved despite chromosomal diversity. *Appl. Environ. Microbiol., 66*, 2988-2995.

In: Progress in Environmental Microbiology
Editor: Myung-Bo Kim, pp. 69-110
ISBN: 978-60021-940-5

Chapter II

Mixtures of Microorganisms in Biocontrol

Magdalena Szczech*
Department of Plant Protection, Research Institute of Vegetable Crops,
Konstytucji 3-Maja 1/3, 96-100 Skierniewice, Poland

Abstract

In this review, the possibility of the use of mixtures or combinations of active microorganisms as a more consistent and effective method of disease control than the application of a single biocontrol agent (BCA) is discussed. The growing pollution of the environment, the general concern of harmful residues in food, and resistance of numerous pathogens to commercial pesticides have induced researchers to find an alternative and nature-safe method of crop protection. During recent decades, numerous bacteria and fungi were isolated and tested for their effectiveness as soil, seed, root and tuber inoculants in control of plant pathogens. However, the commercial use of the biocontrol preparations in practical agriculture is still limited. Single BCA typically has a relatively narrow spectrum of activity compared with synthetic pesticides, and it is strongly affected by various biotic and abiotic factors under natural field conditions. Thus, while effective in the laboratory or in controlled field experiments, BCAs rarely give consistent and satisfying results in practice. Despite the problems and limitations, the researchers spare no pain to find new active microorganisms and to develop the most effective methods of their application. However, studying past and present efforts in BCA's evaluation, it seems that a new outlook on biocontrol is needed. The natural environment is a very complicated and changeable system, therefore, an application of a single, even very active strain of the antagonist will never give as satisfactory result as a more condition-independent pesticide. Integration of several, complementary methods, e.g. application of BCA supported by favourable-for-microorganisms agrotechnic practices or organic amendments, could provide more reliable effects in plant protection against pathogens. Recently, the possible enhancement of the efficacy of BCAs by their combination was studied in many scientific laboratories. There are examples that some bacteria and fungi may interact with each other stimulating some beneficial aspects of their physiology. Moreover, bioprotection observed in naturally suppressive soils is usually attributed to the general activity of diverse indigenous microorganisms existing in these soils. Therefore, it is more likely that a community of several compatible microorganisms with multiple mechanisms of disease suppression and different requirements for growth conditions may broaden the spectrum of their activity and enhance the efficacy of biocontrol. The use of microbial mixtures would more closely mimic the situation in suppressive soils, and under natural changeable conditions one mechanism may compensate for the lack of activity of the other resulting in an additive or synergistic effect.

The review presents hitherto existing studies documenting the enhanced protection of plants treated with combined microorganisms, even against multiple pathogens. However, the reports describing a lack or negative effect of the microbial mixtures on plant development and health are also shown. The strategies in selection of the microorganisms for use in the mixtures, their possible formulation and methods of the application are considered. Also discussed are the problems resulting from the production and registration process of such multiple preparations and some potential areas for future research.

Introduction

The growing public concern of environmental pollution from pesticides and the awareness of harmful residues in food have induced researchers to find nature-safe means of plant pathogens control. The other factor determining the search for alternative methods is the withdrawal of methyl bromide as a fumigant [123]. Next, for postharvest crop protection only, a few chemicals are currently registered and their future use is questionable owing to declining effectiveness or problems with registration [87, 100, 172]. Moreover, agronomists experience a growing problem with the buildup of resistance of the pathogens to pesticides [47, 85] or, in some cases, effective preparations to control diseases are not available, as it was reported for bacterial blight of anthurium in Hawaii [59].

For several decades various bacteria such as genera *Rhizobium*, *Azospirillum*, *Bacillus*, *Pseudomonas*, or fungi such as *Trichoderma* spp. have been introduced to soil or seeds to improve growth of plants, to enhance N_2 fixation and also to suppress plant pathogens [32, 55, 74, 76, 116, 140, 160, 190, 192]. The biopesticides are generally pest specific, displays little or non-target toxicity and are considered less harmful to the environment than chemicals [85]. Also, the resistance of the pathogen to biocontrol is presumed not to develop or at least to develop relatively slowly. Handelsman and Stabb [71] suggested that most biocontrol agents suppress disease via more than one mechanism, and that resistance to multiple antagonistic traits should occur only at a very low frequency. Antagonistic microorganisms exert limited selection pressure since they operate in microsites, where only a fraction of the

* E-mail: mszczech@inwarz.skierniewice.pl

pathogen population is exposed during a short period of its life cycle [47]. The effects and modes of action of BCAs have been reviewed by many authors, among others by Benítez et al. [20], Cook and Baker [32], Fravel [53], Janisiewicz and Korsten [89], Pal et al. [140], Shoda [160], Weller [187] or Whipps [192]. Reading the reviews and numerous research papers we can realise how much effort has been devoted to finding microorganisms able to reduce the activity of the pathogens. The laboratories around the world have developed their own microorganisms. There are also lists and tables presenting the examples of registered active bacteria and fungi [55, 85, 93, 145, 181]. However, the commercial use of the BCAs in practical agriculture is still limited. Where is the problem?

The Reasons for Still Limited Commercial Application of Biopesticides

Inconsistent Performance

First of all, because of inconsistent performance, the microbial agents lose a competition with chemical pesticides that are more effective and stable under diverse environmental conditions. In many cases, the microorganisms effective on a laboratory scale fail to confirm their activity in the field [6, 102]. The nutritional status and other factors that affect growth and survival of the agents in nature are considerably different from those in a nutrient-rich culture media or simplified environments such as a growth chamber or greenhouse. After inoculation the BCA, a living organism, has to face a strongly heterogenous and unpredictable conditions of different soil types, plant species, variable temperatures, humidity or pH [14, 180]. Moreover, it has to compete with indigenous microflora of the site [1, 14, 49, 124, 184]. In such circumstances, the agents have great difficulties in finding a suitable niche to survive over a longer period and to activate the biocontrol mechanisms. Therefore, it is extremely hard to predict the final effect of inoculation and in many cases, the results of field or greenhouse applications are not satisfying and inconsistent [22, 71, 103, 187]. Guetsky et al. [66] have surveyed 64 greenhouse experiments conducted all over the world with *Trichoderma harzianum* T39 and have found that in aproximately 70% of them T39 suppressed *Botrytis cinerea* infections in tomato and cucumber as effectively as the chemical fungicide. However, in 20% of the experiments, control efficacy of BCA was significantly inferior to that of the fungicide, and in 10% of the experiments disease intensity in plots treated with T39 was not different from that in untreated control plots. In Australia take-all decline has been attributed to *Trichoderma* spp., which comprise a major proportion of the total microbial community in soils suppressive to *Gaeumannomyces graminis* var. *tritici* [45]. In field experiments in Southern Australia *Trichoderma koningii* strain significantly reduced take-all in three of six trials over a 4-year period and increased yield an average of 10% [103]. However, in other trials, it was found that performance of *T. koningii* varied among field sites and seasons, which is what delayed the commercialisation of this biocontrol agent [45]. In Washington state (US), a significant control of take-all and yield increase have been provided by fluorescent pseudomonads used as a seed treatment, but these bacteria were effective only about 60% of the time [187].

Narrow Spectrum of Activity

The next problem is that the biocontrol agents typically have a relatively narrow spectrum of activity compared with synthetic pesticides, and growers need to control several plant pathogens in the same crop. In the majority of screens for the biocontrol agents of plant diseases a large number of microbial isolates are tested against a single pathogen strain. Then, the promising agent is again tested in soil or a potting medium in controlled conditions for a given type of target pathogen in a specific stage of its life cycle. However, in a natural field environment there is a large variety of non-target pathogens that also may cause diseases of crop plants. If a biocontrol agent successfully suppresses a target pathogen but has no effect on non-target pathogens, that will subsequently limit plant development and yield, then the treatment will appear ineffective [34, 100, 124, 146, 187]. Mazzola and Cook [124] have found that *Rhizoctonia solani*, *Pythium* spp. and *G. graminis* var. *tritici* occur as mixtures in the same field and often on the same plant throughout wheat production areas of the Pacific Northwest. These pathogens had various effects on the ability of introduced fluorescent pseudomonads to persist and multiply in the wheat rhizosphere. While *G. graminis* var. *tritici* and *R. solani* supported larger amount of *P. fluorescens* 2-79RN$_{10}$ and Q72a-80R on wheat roots, *Pythium* sp. caused a rapid decline of these bacteria populations. Infection by *Pythium* resulted in loss of root hair, which can serve as a potential location for colonization of bacteria and loss of the hair due to the pathogen tended to reduce the root surface available for bacterial colonization [124]. *Pythium* had also modified the composition of root exudates in a manner that was more favourable for the fungus than for the growth of certain fluorescent pseudomonads [100, 124]. In this context, it would be worth considerating the study of the biocontrol activity of selected BCAs against a complex of different pathogens causing diseases of certain crops.

It is also worth emphasizing that most populations of the pathogens are not evolutionary static and harbour substantial genetic variation between strains of the same species [47, 100]. Significant genotypic variation in fungal pathogen populations exist among geographic regions, within a given field, or even between lesions on the same plant [57, 127]. Therefore, the strain or strains of target pathogen persisted in the site treated with the inoculant may differ in their susceptibility to BCA. Just as microbial antagonists utilise a diverse arsenal of mechanisms to dominate interactions with pathogens, pathogens also have diverse responses to counteract antagonism. This subject has been interestingly described by Duffy et al. [47] in their review. The pathogen responses include antibiotic resistance, repression of biosynthetic genes involved in biocontrol, active efflux of antibiotics, and detoxification [51, 126, 138]. It has been shown that for *Colletotrichum musae* even a broad host-range mycoparasites *Gliocladium* spp. and *Trichoderma* spp, which attacked two or more pathogen genera, discriminated significantly between *C. musae* strains [100]. Strain discrimination was correlated to differential susceptibility of *C. musae* to one or more minor mechanisms exhibited by antagonists. The variability in strain sensitivity was also observed with *Phytophthora* spp. [100]. Isolates of *G. graminis* var. *tritici* varied in sesitivity *in vitro* to the antibiothics phenazine-1-carboxylic acid (PCA) and 2,4-diacetylphloroglucinol (Phl) produced by fluorescent *Pseudomonas* spp, which was shown previously to have potential for biological control of this pathogen [126]. Approximately 12% of the total *G. graminis* var.

tritici population examined was classified as insensitive to PCA, and such isolates of this pathogen were present in collections from each of several wheat-producing areas of the world. PCA-producing strains of *Pseudomonas* exhibited a reduced or complete inability to suppress take-all caused by insensitive strains of *G. graminis* var. *tritici*. The same was observed for *P. fluorescens* Q2-87 produced Phl, which failed to control insensitive to this antibiotic strain of the pathogen [126]. Further, Gurusiddaiah et al. [68] and Jones and Pettit [92] showed variation in sensitivity of *Pythium* and *R. solani* species respectively to PCA or antibiotic gliotoxin. The studies by van Zyl et al. [185] reported a resistance of *Agrobacterium tumefaciens* strains to agrocins produced by *Agrobacterium radiobacter*.

There are some mechanisms of pathogen resistance to antimicrobial compounds. For example efflux pumps were identified to play a role in pathogen tolerance to antibiotics [47]. The pathogen can alter gene expression in the microbial agent reducing its adaptive capability in the environment and biocontrol activity. Fedi et al. [51] have found that five gene clusters of a biocontrol strain of *P. fluorescens* F113 were repressed in the presence of pathogenic *P. ultimum*. Therefore, the ecological fitness of the reporter mutants of F113 in the rhizosphere of seed-inoculated sugar beet was lower than of the wild type. It was also observed that the population size of *P. fluorescens* 2-79 in wheat rhizosphere was significantly reduced in the presence of three *Pythium* species [124]. Moreover, pathogens can have a direct negative impact on the mechanisms of biocontrol agents. For example *Fusarium* species may produce deoxynivalenol, which acts as a negative signal repressing the expression of *nag1*chitinase gene in *Trichoderma atroviride* - a producer of cell wall-degrading enzyme and competitor in crop residues [118]. Duffy and Défago [46] have shown that fusaric acid produced by *F. oxysporum* f. sp. *radicis-lycopersici* inhibited production by *P. fluorescens* CHA0 of the antibiotic 2,4-diacetylphloroglucinol, a key factor in the biocontrol activity of this strain.

The pathogens can also alter the environment to gain an ecological advantage over potential competitors or produce an array of toxins active against other microorganisms to improve ecological competiveness [47, 109, 166]. Structural and biochemical barriers are also possible such as melanin and hydropolysaccharides content protecting against enzymes produced by antagonists [18, 179].

Inadequate Colonization and Persistence of the Inoculum

The variable and not satisfactory effectiveness of biocontrol agents in commercial application is also due to inadequate colonization and persistence of the inoculum in soil, rhizosphere or phyllosphere [187]. In general, populations of introduced microorganisms decline more or less rapidly, especially following introduction into natural soil [184, 190]. It has been attributed to availability of nutrients and to numerous abiotic and biotic factors like water, temperature, indigenous microflora [31, 184]. The plant genotype can also play a significant role in shaping plant-associated microbial communities, and there are reports on differences among cultivars in the level of disease suppression obtained with a particular BCA [46, 65, 167]. Host plants may support disease suppression by enhancing growth of

BCA [167], but some plants release compounds inhibitory to the agent reducing its biocontrol activity as it was presented for *Daphne* plants and *Bacillus subtilis* Cot1 [146].

A linear relationship was observed between the population of *Pseudomonas putida* W4P63 and reduction of preemergence of potato seeds decay caused by *Ervinia carotovora* subsp. *atroseptica* [195] and between the population size of *P. fluorescens* 2-79 on wheat roots and the number of take-all lesions [24]. Szczech and Shoda [175] described the differences in tomato root colonisation by iturin A-producer *B. subtilis* RB14-C according to the method of bacterial application. When RB14C was introduced as a seed coating, the population of this bacterium was the largest near the crown of the plant and declined toward the root tip. Generally, the amount of the bacteria was lower on these roots than the more uniform population of RB14-C on the roots of plants grown in soil mixed with these bacteria. The higher degree of root colonisation was positively correlated with the percent of tomato seedlings that emerged in soil infested with *R. solani* [175]. Numerous abiotic and biotic factors contribute to the variable colonization of plant rhizosphere by biocontrol agents [146]. Their population size vary from root to root by several orders of magnitude, and some roots may be completely unprotected [124]. Very often there is no active or passive translocation of introduced inoculant from the site of inoculation to other sites [181]. Active translocation only occurs over short distances in soil or water application by irrigation to soil may induce active or passive vertical transport [111, 112]. Moreover, usually populations of introduced biocontrol agents decrease rapidly in time [184]. Therefore, if there is a long period between planting or the agent application and disease development, the biocontrol effect may not appear [135].

Limited Production of Antimicrobial Metabolites and Inactivation of Biocontrol Traits

In addition to the not effective colonisation, the BCA's are often not able to produce their antimicrobial metabolites or to activate the biocontrol traits in a new environment [146]. The changeable natural conditions cause fluctuations in production of antimicrobial compounds, which in many cases may be inadequate to control the pathogens [70, 139]. Raaijmakers et al. [151] described numerous biotic and abiotic factors that influence antibiotic production by BCAs. Moreover, even if the BCA is positioned in an infection court, the occurrence of a treshold concentration of a critical metabolite may be temporally separated from the site of infection or spread of the pathogen [146]. For example, in the spermosphere of cotton, expression in *P. fluorescens* Hv37a of the key biosynthetic gene for the antibiotic oomycin A occured 10 hr after planting bacteria-treated seed [82]; however, infection by *Pythium ultimum*, which is sensitive to the antibiotic occurred by 6 hr [69]. A similar situation was observed by Szczech and Shoda [175]. When tomato seeds were coated with *B. subtilis* RB14-C, the production of the antibiotic iturin A increased gradually around seeds, but during the strongest attack of *R. solani* its concentration was still to low to protect the seeds against infection. A significant influence on the antagonistic activity of the biocontrol agents also has an indigenous microflora. Production of antibiotics may be repressed by the bacterial or fungal extracellular metabolites [46, 159]. The competition for limited nutrients, which is an important mechanism of biological control, can be nullified by increasing the concentration of the relevant compound [28, 67].

The Origin of the Concepts of Microbial Mixtures

All the examples described above show that it is difficult to expect that a BCA will provide a satisfactory control every time and in every condition. Therefore, to achieve this goal, developing strain mixtures was proposed. The concept of multi-organism inoculants as an agricultural practice superior to single organism inoculation was not new [80]. There were examples of bacteria that might interact with other bacteria stimulating each other through physical or biochemical activities, that could enhance some beneficial aspects of their physiology [14]. The example is the association between *Bacillus* that degrades pectin and *Azospirillum* that can use products of this degradation as a carbon source [94]. Moreover, the *in vitro* studies have shown that *Azospirillum* could produce more phytohormones [90] or enhance nitrogen fixation [80] when it was grown in a mixed culture with other bacteria than in a pure culture. Lebsky et al. [104] described that microalga *Chlorella vulgaris* coinoculated with plant growth-promoting *Azospirillum brasilense* remained in a growth phase, which was advantageous for wastewater treatment. Mixed inoculation of bacteria and arbuscular-mycorrhizal fungi created positive interactions leading to significant increase in growth, in the phosphorous content of plants, enhanced mycorrhizal infection, and enhancement in the uptake of mineral nutrients [14]. When mangrowe seedlings were treated with a mixture of two bacterial species, the slow-growing N_2-fixing *Phyllobacterium* sp. and the fast-growing phosphate-solubilizing *Bacillus licheniformis*, nitrogen fixation and phosphate solubilization were enhanced, but bacterial multiplication did not increase [155].

It is also likely that naturally occurring biological control in suppressive soils results from mixtures of antagonists rather than from high populations of a single agent [45, 105, 108, 152, 189]. The suppressiveness of Chateaurenard soil in France and the soil from the Salinas Valley in California against *Fusarium* wilt was attribured to the activity of nonpathogenic *Fusarium* spp. and fluorescent *Pseudomonas* spp. [3, 157, 189]. Infestation of disease conductive soil with a disease-suppressing microorganism does not reach the level of suppression observed in the naturally suppressive soils, and the suppressive effect is often inconsistent [187]. Lemanceau et al. [106, 107] and Duijff et al. [48] have found that combination of fluorescent pseudomonads with nonpathogenic *Fusarium* suppressed *Fusarium* wilts more efficiently than the separate inoculation of the disease-suppressing organism. It was caused by reducing saprophytic growth of the pathogen throughout conjoined carbon and iron competition [106, 107]. Pierson and Weller [146] also suggested that natural take-all decline involves a community of microorganisms. For *Pythium* and *Phytophthora* root rots, the general suppression phenomenon, related to total microbial activity, was described as responsible for control of these pathogens in compost amended media [79]. Consequently, a combination of active microorganisms would increase the genetic diversity of the biocontrol system and may more closely mimic the natural situation in the environment.

Enhanced Efficacy of Microbial Combinations in Biocontrol

It was suggested that application of a mixture of biocontrol agents would be more effective in controlling diseases than single microorganisms, because the mixture offers more versatility in mechanisms of action, and the agents have different ecological requirements for survival and activity. Therefore, under changeable natural conditions different antagonistic traits may be expressed, and one mechanism may compensate for the other. Several agents could be able to suppress multiple pathogens. Moreover, their joint action may result in an additive or synergistic effect as in the case of *Trichoderma* produced cell wall degrading enzymes which increased the toxicity of antifungal metabolites secreted by *Pseudomonas* spp. antagonistic to *B. cinerea* [52, 193].

A previous report on combinations of microorganisms in biological control, that was found by the author, was presented by Sivasithamparam and Parker in 1978 [165]. They described that only mixture of bacteria, obtained from roots of wheat, reduced take-all caused by *G. graminis* var. *tritici*, while the single strains were not effective. Then Kwok et al. [101] have found that combinations of some bacterial antagonists with *Trichoderma hamatum* 382 were consistently more effective in controlling of *Rhizoctonia* damping-off in compost bark media than bacteria or fungal isolate alone. Combinations of fluorescent pseudomonad strains 2-79 and 13-79 were superior in control of take-all than separate treatments [187], and pseudomonads applied with avirulent species of *Fusarium* were better in control of *Fusarium* wilt of cucumber than either one alone [141].

The number of papers describing the superior efficacy of microbial mixtures to biocontrol agents used singly have increased since the 90s. Their majority show the effect of combinations of various bacteria, mostly genera *Pseudomonas* spp. Pierson and Weller [146] demonstrated the potential benefits of using several different strains of fluorescent pseudomonads to suppress take-all of wheat and to enhance growth and yield of these plants in fields infested with *G. graminis* var. *tritici*. Antibiotic producing *P. fluorescens* 224 and *P. cepacia* 233 (*Bulkholderia cepacia*) mixed with chitinase producing *Streptomyces* sp. 75 and *Bacillus cereus* 160 significantly reduced rice sheath blight (*R. solani*) compared to control and separate bacteria [173]. El-Tarabily et al. [50] also by combining of the chitinolytic bacteria *Serratia marcescens*, *Streptomyces viridodiasticus* and *Micromonospora carbonacea* obtained inhibition of the growth of *Sclerotinia minor,* which was caused by hyphal plasmolysis and cell wall lysis. Consequently, such treatment decreased basal drop of lettuce under controlled conditions. The other example of the efficacy of pseudomonads combinations was reduction of *Fusarium* wilt of radish by coinoculation of antagonistic, root-colonizing *Pseudomonas* spp. [36,37]. Sindhu et al. [164] have found that inoculation of *Pseudomonas* strains from the green gram rhizosphere, on chickpea seeds significantly improved the effectiveness of *Mesorhizobium* sp. *cicer* Ca181 in soil containing wilt-causing pathogens. There was no blackening of roots due to infection, and nodulation, plant dry weight and total plant nitrogen were enhanced compared to *Mesorhizobium* inoculated plants. The enhancement of N_2 fixation was also obtained by coinoculation of cowpea and soybean with ethylene-producing *Pseudomonas syringae* and N_2-fixing *Bardyrhizobium japonicum* [2]. Such stimulated plants caused an increased suicidal germination of the seeds of parasitic plant *Striga hermontica* (del.) Benth, which strongly limits cereal production in Africa. These legumes used in crop rotation reduced *S. hermontica* parasitism on subsequent maize crops [2].

The other group of antagonistic bacteria, which are recommended for biological control and often included in mixtures, belong to the genus *Bacillus*. These bacteria are described as biocontrol agents with broad spectrum of antifungal activity [21, 29, 160, 194]. The advantage of using *Bacillus* spp. in the mixture is its property to form spores resistant to unfavourable natural conditions, and therefore its tolerance to antimicrobial substances released by other microorganisms in soil and by coinoculant. The mixtures of plant growth-promoting rhizobacteria (PGPR) *B. pumilus* INR7 and *B. subtilis* GB03 coinoculated with *Curtobacterium flaccumfaciens* ME1 effectively suppressed various cucumber pathogens and enhanced consistency of control under greenhouse and field conditions [152, 153]. The bacteria were able to induce systemic resistance ISR [152]. Jetiyanon and Kloepper [91] then studied these strains together with other ISR bacteria for their ability to induce resistance against fungal (*Colletortichum gloeosporoides* – anthracnose of long cayenne pepper, *Rhizoctonia solani* – damping-off of green kuang futsoi), bacterial (*Ralstonia solanacearum* – bacterial wilt of tomato) and viral (cucumber mosaic virus) diseases. In these assays most tested mixtures showed greater disease suppression than individual PGPR strains, suggesting that combined systemic resistance-induced bacteria may provide protection of different plants against multiple pathogens. The above mentioned strain of *B. subtilis* GB03 was also used in other experiments together with various PGPR bacteria. Murphy et al. [134] combined GB03 with several *Bacillus* sp. strains and chitosan to control cucumber mosaic virus CMV. Tomato plants treated with such preparations appeared phenotypically and developmentally similar to nonbacterized plants that were 10 days older than plants treated with the mixtures. Moreover, CMV disease severity ratings and CMV accumulation in young leaves were significantly lower than in nonbacterized control plants. Domenech et al. [40] studied the effect of application of biological product LS213, which contains GB03, *B. amyloliquefaciens* IN937a and chitosan, with *B. licheniformis* CECT 5106, *P. fluorescens* CECT 5398 and *Chryseobacterium balustinum* CECT 5399 on growth promotion and biological control of soilborne pathogens of pepper and tomato (*Fusarium* and *R. solani*). When the individual bacteria and LS213 were put together, a synergistic growth promotion was obtained [40]. The mixtures also gave significantly higher percentages of healthy plants for both tomato and pepper than LS213 alone. The other antagonistic strain – *B. subtilis* RB14-C, which is a producer of lipopeptide antibiotics iturin A and surfactin [83, 138], was used with *Burkholderia cepacia* strain BY in control of *Rhizoctonia* damping-off of tomato plants [174]. *In vitro* test with mycelium inoculated on filter discs buried into soil added with the bacteria has shown, that single BY reduced the fungus growth but not completely. RB14-C had only a slight effect on pathogen growth, while the combined treatment completely inhibited *R. solani*. This effect was checked in pot experiments, where the best efficacy was obtained when BY was applied to the soil two days after RB14-C [174]. Also Lourenco et al. [115] have found that tomato plants colonized by rhizobacteria *B. cereus* and treated with *Cellulomonas flawgeria*, *Candida* sp. and *Cryptococus* sp. showed the lowest final severity of late blight caused by *Phytophthora infestans* compared to plants treated only with rhizobacteria.

The next examples of the superior efficacy of bacterial mixtures were presented by Fukui et al. [59, 60], who had isolated bacterial candidates for biological control form the guttation fluids of anthurium plants. Growth and survival of *Xanthomonas campestris* pv. *diffenbachiae* in guttations fluids were suppressed by several bacterial strains indigenous to leaves of various anthurium cultivars [60]. Inhibition of the pathogen was not observed in filter sterilised guttation fluids and was restored only by reintroducing of specific mixtures of bacteria. None of the individual strains inhibited *Xanthomonas* [59]. The mixture sprayed on the foliage of susceptible anthurium cultivars was also highly effective in suppressing wound invasion and leaf infection [59]. Next, De Boer et al. [38] have found that several soil bacteria that exhibited little or no visible antifungal activity on different agar media, when used in the mixture reduced growth of *Fusarium culmorum, R. solani* and *T. harzianum.* They suggested that non-antagonistic soil bacteria may be important contributors to soil suppressiveness and fungistasis in a community context.

Beside the reports describing the activity of bacterial mixtures there are also numerous studies on the effect of bacteria coinoculated with fungi, and especially *Pseudomonas* sp. with a non-pathogenic *Fusarium oxysporum.* Park et al. [141] demonstrated that such combination can give a satisfactory control of *F. oxysporum* f. sp. *cucumerinum,* and Alabouvette et al. [4] obtained a synergistic effect in controlling of *F. oxysporum* f. sp. *radicis-lycopersici.* The same, Duijff et al. [48] showed suppression of *Fusarium* wilt on flax with nonpathogenic *F. oxysporum* Fo47 and *P. putida* WCS358. According to the investigations, nonpathogenic *Fusarium* competed with the pathogen for carbon sources while the bacterial antagonist produced a siderophores competing for iron [48, 106, 107, 108]. Moreover, Leeman et al. [105] have found that roots of radish grown from seeds treated with *P. fluorescens* WCS374 suppressive against *Fusarium* wilt of radish (*F. oxysporum* f. sp. *raphani*), were more abundantly colonized by fungi than were roots of nonbacterized plants. Several isolates of these fungi suppressed *Fusarium* wilt of radish and supported activity of pseudomonads in the case when inoculum density of the antagonist was too low to reduce the disease in separate treatment.

Positive interactions between *Trichoderma* spp. strains and bacterial antagonists, such as *Pseudomonas syringae,* has been reported for combined applications in the control of plant pathogens [191]. Pierson and Weller [146] used several strains of fluorescent pseudomonads to suppress take-all of wheat caused by *G. graminis* var. *tritici.* Then the mixtures of these pseudomonads were examined for their control efficacy in combination with hypovirulent strain of *G. graminis* var. *graminis* [44] and *T. koningii* [45]. In both cases it was found that combined treatments consisting of antagonistic fungi introduced to the furrow and fluorescent pseudomonads applied to the seeds were significantly more suppressive to take-all than either treatment used alone. *T. virens* GL21 applied as a granular formulation, in combination with *Bulkholderia cepacia* BC-1 or *Bulkholderia ambifaria* BC-F applied as a seed treatment, significantly improved suppression of damping-off of cucumber caused by *R. solani* over individual applications [154]. In the same studies *B. ambifaria* BC-F with *T. virens* GL21 enhanced control of *Pythium* damping-off. Rudresh et al. [156] have shown that phosphate-solubilizing *Bacillus megaterium* and *Rhizobium* applied with *Trichoderma* spp. increased germination, nutrient uptake, nodulation, total biomass and yield of chickpea compared to either individual inoculations or an uninoculated control. The multiple combinations of antagonistic fungi, among other *Trichoderma* sp. and *Gliocladium* sp., with different active bacteria were also used to suppress diseases of cucumber, strawberry and potato [22, 154, 186]. However, among numerous treatments only combinations of *B. subtilis* with *T. virens* in control of stem canker of potato [22] and *T. harzianum* with *Gliocladium catenulantum* in control of *Phytophthora fragariae* on strawberry [186] were successful over control and single treatments.

There are also works on the positive effect of coinoculation of rhizobacteria or fungal agents with mycorrhizal fungi. A gram-positive bacterium *Paenibacillus* sp. strain B2, isolated from the mycorrhizosphere of *Sorghum bicolor* inoculated with *Glomus mossae*, has shown antagonistic activity against fungal pathogens and significantly stimulated mycorrhization [23]. The same, the mixture of mycorrhizal fungus *Glomus intraradices* with *B. subtilis* or *T. harzianum*, respectively, controlled *Rhizoctonia* stalk rot and root rot of celery [135] and fusarium crown and root rot of tomato plants [34] more effectively than these strains used individually. On the other hand, Probanza et al. [149], coinoculatig *Pinus pinea* with *Pisolithus tinctorius* and PGPR *Bacillus* strains, observed plant growth promotion. However, this effect was related on *Bacillus* activity and did not imply a synergistic effect with mycorrhizal infection.

Less reports were found on the improvement of biocontrol efficacy by combinations of different antagonistic fungi. However, Paulitz et al. [144] demonstrated an effective control for combination of *T. harzianum* protecting against infection by *P. ultimum* in the rhizosphere and *Pythium nuun* reducing inoculum density of the same pathogen in the soil mass. Then, in bioassays carried out by Krauss et al. [100] the mixtures of up to four fungal antagonists (*Gliocladium* and *Trichoderma* spp.) were increasingly more effective in controlling mixed infection evoked by four *Colletotrichum musae* strains. When these mixtures were used to treat banana clusters, which were infected naturally by a multitude of pathogens, biocontrol was best with inocula containing three or four antagonists.

The majority of researches on biocontrol have been performed for soilborne pathogens. However, there is a growing need to develop new strategies to reduce postharvest decays of fruits and vegetables. During storage, where the use of fungicides is restricted, it is necessary to find the agents with a broad spectrum of activity against major and minor pathogens. Effective biological control has been reported for postharvest diseases of apples, pears and citrus fruits [88, 89]. In recent decades several combinations of biocontrol agents were developed to reduce postharvest decay of apples [30, 86, 87, 89]. The mixtures of antagonists reduced variability and improved efficacy of biocontrol of postharvest diseases [66, 89]. It is advantageous that, the relative simplicity and stabilised conditions of storage systems offer an uncomplicated environment for survival and activity of introduced biocontrol agents. It is possible to add the agent in the site needing protection, at a required concentration, and regulate the environmental conditions to maintain protection. Furthermore, biotic interference is minimal so antagonists encounter only slight competition from indigenous microorganisms.

Why the Mixtures of Microorganisms Act Better than Single Agents?

More Consistent Efficacy of the Combined than Single BCAs

The big advantage of the mixtures, beside the additive or synergistic biocontrol effect, is enhanced consistency of their efficacy. Looking through the research papers, it is possible to find reports that there is a trend to obtain the positive results with combined biocontrol agents in repeated, consecutive treatments, while the activity of individual strains is variable, sometimes unpredictable [102, 186]. The combination of *P. fluorescens* 2-79 and 13-79 provided better suppression of take-all compared to 2-79 alone in about 50% of field trials in the Pacific Nortwest and the United Kingdom, and compared to 13-79 alone in six of six field trials in the Pacific Nortwest [25, 188]. Application of mixtures of biocontrol agents established inhibitory bacterial communities on anthurium leaves that were superior to individual strains in achieving effective and consistent biocontrol of bacterial blight over a range of anthurium cultivars [59]. Krauss et al. [100] successfully used the combination of three and four fungal antagonists in an attempt to overcome an inconsistent control of crown rot of banana. Szczech and Dyki [in publish] in all performed growth chamber experiments always observed suppression of *Rhizoctonia* damping-off of tomato plants by combinations of bacteria PT42, SZ141, PT60, 125 and 207. The same bacteria used separately sometimes were even more effective than the mixture but in other assays they did not exhibit any protective properties, their effect was variable. Similar results were obtained with the mixtures in greenhouse studies (Szczech unpublished). Guetsky et al. [66] demonstrated that application of *Bacillus mycoides* B16 and *Pichia guilermondi* Y2 as a mixture to strawberry leaflets not only effectively suppressed *Botrytis cinerea* but also significantly reduced variability of disease control under diverse conditions. Control efficacy achieved by the biocontrol agents used separately ranged between 39 and 98%, but their mixture suppressed *B. cinerea* effectively in 80 to 99.8% under all conditions.

Enhanced Effect of Microbial Combinations with Different Ecological Requirements

The improved effect of the mixture sometimes is related to the positive cooperation of two microorganisms with different ecological requirements. In a situation, when natural conditions are not favourable for one component of the microbial mixture, the other/s may find the environment tolerable to develop and act as biocontrol agents. The good example is described about superior efficacy of the mixture of *B. mycoides* B16 and *P. guilermondi* Y2 in control of *B. cinerea* [66]. It was found that below 15 °C the bacterium B16 multiplied more rapidly than the yeast. Therefore, it could compensate for the inability of the yeast to multiply under changing conditions. It was also important that both microorganisms reduced spore germination of the pathogen at different temperatures. *P. guilermondi* reduced spore germination at temperatures lower than 25 °C, whereas *B. mycoides* was more effective above 25 °C [66]. Moreover, under optimal conditions for the pathogen humidity regimes, efficacy of separated antagonists was decreased, while combined application of *P. guilermondi* and *B. mycoides* resulted in significant suppression in all regimes tested. Weller [187] reported that under alkaline conditions *Pseudomons fluorescent* plays a mayor role in take-all decline. However, in acidic soils or when cropping practices such as application of an ammonium form of nitrogen lower the rhizosphere pH, *Trichoderma* spp., which generally tolerate lower pH may protect plants better than bacteria [163]. *Trichoderma* is also active across a wider range of soil moisture [32] and provide protection in more arid regions or later in the growing season as moisture becomes less available [45]. On the other hand, endospores of *Bacillus* tolerant to extreme temperatures and pHs [6, 42], may remain viable for the period of unfavourable bacteria conditions and replace other antagonists. In the studies of Fukui et al. [60], it was shown that inhibitory effects of guttation fluids were considerably different among cultivars of anthurium. It suggested that indigenous bacterial communities may be specific to each cultivar and therefore, the effect of any single strain varies depending on the cultivar. It was proved that the protective effect of biocontrol agents often depends on the plant genus and cultivar [167]. Thus the most effective and reliable inoculum for achieving consistent biological control of *X. campestris* pv. *diffenbachiae* over a range of plants, was the mixture of strains isolated from various anthurium cultivars [59].

Several Biocontrol Mechanisms Contribute in the Activity of Microbial Mixtures

According to Pierson and Weller [146], the greater diversity of phenotypes within the mixture compared to single strains is likely to result in a greater variety of traits responsible for disease suppression, including a more diverse "arsenal" of secondary metabolites capable of suppression of both target and non-target pathogens. Further, the biosynthetic genes responsible for production of secondary metabolites involved in disease suppression may be regulated differently among strains and there is a greater probability that at least some of these genes will be expressed over a wider range of environmental conditions.

There are several microbial groups known to produce a broad spectrum of antimicrobial metabolites and to exhibit different traits of biocontrol. As the main types of biocontrol mechanisms antibiosis, competition, lysis, hyperparasitism and induction of systemic resistance have been reviewed by numerous authors [20, 31, 32, 53, 76, 81, 140, 160, 161, 183, 187, 190, 191, 192]. Mainly the biocontrol agents belong to bacteria genera *Pseudomonas* spp., and *Bacillus* spp. or fungi genera *Trichoderma* spp.

The biocontrol abilities of *Pseudomonas* spp. depend mostly on production of numerous diffusible and volatile antibiotics, induction of systemic resistance in plants and aggressive root colonization [11, 65, 70, 117, 151, 187, 190]. Typical antimicrobial secondary metabolites of pseudomonads are phenazines, 2,4-diacetylphloroglucinol, pyoluteorin, pyrrolnitrin, lipopeptides and HCN, which biosynthesis and biocontrol properties have been very well described by Haas and Keel [70] and Raaijmakers et al. [151]. Fluorescent *Pseudomonas* are also efficient competitors for iron [187, 190]. Siderophore-mediated competition for iron has been demonstrated to be responsible for suppression of several soilborne pathogens [9, 98, 106, 107, 110, 114, 128]. These bacteria have also the ability to promote plant growth, which may lead to greater host resistance to pathogenic invasion and environmental stresses [62, 190]. Some of the biocontrol strains are able to produce multiple bioactive metabolites and exhibit several modes of action [70, 190], that can compensate each other in the case of suppression of any of them.

The other "universal" group of biocontrol bacteria are *Bacillus* spp. Such as pseudomonads, they express numerous traits of antagonism including production of various antibiotics (iturin, bacillomycin, mycosubtilin, bacilysin, surfactin, mycobacillin) and siderophores, which are inhibitory to a broad range of plant pathogenic fungi [83, 121, 160, 196]. *Bacillus* spp. can also produce enzymes that lyse fungal cells [5, 146] and to induce systemic resistance [133, 183]. As a big advantage of these bacteria for use in biocontrol, their ability to spore formation should be emphasized. The endospores, tolerant to heat, dessication and starvation or other environmental stresses, may survive long, unfavourable for active cells periods and revive almost immediately when conditions improve [42]. Thanks to this ability *Bacillus* are well adapted to large-scale production methods, formulation and long storage [21].

Among biocontrol fungi, the most often used and the most versatile genus is *Trichoderma* spp., whose antagonistic properties are based on the activation of multiple mechanisms. These fungi exert protective properties either indirectly, by competing for nutrients and space, modifying the environmental conditions or promoting plant growth and plant defense mechanisms, or directly, by antibiosis and mycoparasitism [20, 72, 73, 74]. These mechanisms are complex, and what has been defined as biocontrol is the final result of different mechanisms acting synergistically to achieve disease control [81]. Benítez et al. [20] reported that 90% of application of biocontrol fungi has been carried out with different strains of *Trichoderma*. The success of *Trichoderma* as biocontrol agent is due to its high reproductive capacity, ability to survive under very unfavourable conditions, efficiency in utilisation of nutrients, capacity to modify the rhizosphere and strong aggressiveness against phytopathogenic fungi [20]. Moreover, this genus is spread in almost all habitats and occurs at high population densities [97].

Combination of such versatile BCAs, exhibited so many various mechanisms of disease suppression, may provide additive and consistent results in plant protection. In a changeable environment, when one or more mechanisms is not expressed, the others may compensate for the former absence, therefore the mixture may still be effective. Moreover, the activation of several mechanisms by combined agents may support each other resulting in more effective control. Good examples of such cooperation was the combination of *Pseudomonas* spp. with avirulent strains of *Fusarium* spp. or other antagonistic fungi [105, 106, 107, 141]. In combined control of *Fusarium* wilt of cucumber fluorescent pseudomonads competed with the pathogen for iron, while avirulent species of *Fusarium* induced resistance in host plant [141]. Similar example is suppression of *Fusarium* wilt of carnation by coinoculation *P. putida* WCS358 and nonpathogenic *F. oxysporum* Fo47 [106, 107]. In the presence of siderophores produced by WCS358 the pathogenic *Fusarium* was more sensitive to competition for carbon with nonpathogenic Fo47. Thus, the protective effect of the fungal-bacterial combination was higher than the effect of single Fo47. Similarly, Mazzola et al. [125] have found that *G. graminis* var. *graminis* was much less sensitive to phenazine-1-carboxylic acid produced by biocontrol fluorescent pseudomonads than pathogenic *G.*

graminis var. *tritici*. Thus, in addition to direct inhibition of *G. graminis* var. *tritici*, some *Pseudomonas* antibiotics may make the pathogen less competitive against *G. graminis* var. *graminis*.

The mixture of WCS358 with antibiotic producing *P. putida* RE8 gave additive control of *Fusarium* wilt of radish [37]. In the studies of Leeman et al. [105] suppression of *F. oxysporum* f. sp. *raphani* was a sum of siderophore-mediated competition for iron, induction of resistance due to lipopolysaccharides produced by *Pseudomonas* sp. strains WCS358, WCS374 and WCS417, and induced systemic resistance caused by coinoculated fungi *F. oxysporum* and *Verticillium lecani*. Sung and Chung [173] demonstrated that chitinase-producing strains of *Streptomyces* sp. 75 and *Bacillus cereus* 160 used in conjunction with antibiotic producing *P. fluorescens* 224 and *P. cepacia* 233 had a synergistic effect on the suppression of rice sheath blight caused by *R. solani*. The synergistic effect obtained also Woo et al. [193] and Fogliano et al. [52] with *Trichoderma* cell wall degrading enzymes and *Pseudomonas* spp. producing membrane-disrupting lipodepsipeptidases, syringotoxins and syringomycins. The results indicated that enzymatic degradation of the cell wall of *B. cinerea* permitted bacterial toxins, especially syringomycins, to reach its target, and alter cell membrane functions much more effectively than in the absence of fungal enzymes. Guetsky et al. [67] successfully used the mixture of yeast *P. guilermondi* and *B. mycoides*, exhibiting several mechanisms of biocontrol, to suppress germination of conidia of *B. cinerea*. *P. guilermondi* was able to compete with *B. cinerea* for glucose, sucrose, adenine, histidine and folic acid, and probably activated plant defense mechanisms. Due to its activity an extracellular matrix associated with germ tubes of *B. cinerea* seemed to be dissolved, thus germination and penetration ability of the pathogen conidia were suppressed. *B. mycoides* secreted volatile and nonvolatile inhibitory compounds and caused disortion and porosity of *B. cinerea* conidia, that were incapable for further development. When both antagonists were applied in the mixture, their activity reflected the sum of the biocontrol mechanisms of each agent separately.

It is also worth mentioning that different BCAs may attack the pathogen at various sites (e.g. in bulk soil, rhizosphere, protection of the whole plant against infection due to induction of resistance), or they can also suppress the development and kill different forms of the pathogen (e.g. destruction of mycelium, conidia, parasitism of sclerotia, inhibition of conidial germination). It may be illustrated by the combination of *T. harzianum* protecting against infection by *P. ultimum* in the rhizosphere and *P. nuun* reducing inoculum density of the same pathogen in the soil mass [142]. Subsequently, Duffy et al. [45] explained that the enhanced control of take-all on wheat obtained by combination of fluorescent pseudomonads with *T. koningii* resulted from antibiosis exhibited by *Pseudomonas* in the rhizosphere and a broad activity of *Trichoderma*. This fungus could promote plant growth, parasitise fungal hyphae and propagules and aggressively colonise crop residues. Such properties of *T. koningii* extended the sphere of protection provided by rhizosphere bacteria beyond the root zone by attacking inoculum in the soil and crop residues.

Combinations May Control Multiple Pathogens

The contribution of several modes of action may not only improve the biocontrol efficacy of the mixture against the target pathogen, but with broad spectrum of the activity, the combination may be capable of controlling multiple diseases. In controlled assays, when the agents are studied for their biocontrol abilities, they are tested against one selected target pathogen, which is at the same growth phase and uniformly distributed in soil or potting medium. In the field, such a situation does not happen. There are numerous fungal and bacterial pathogens or deleterious microorganisms, that attack the crop plant at the same time or consecutively during vegetation. They can be resistant to the metabolites or other biocontrol mechanisms exhibited by the agent. Therefore, when one or two pathogenic organisms are suppressed the other are still able to infect the protected plants causing severe losses. There is also a possibility that certain strains of the target pathogen is not sensitive to the agent. However, when the range of antagonistic mechanisms is used, the probability of more complete and effective protection increases. The optimal situation is when the mixture is composed of the agents with multiple mechanisms like *Pseudomonas* spp., *Bacillus* spp. and *Trichoderma* spp., or exhibiting strong activity like β-1,3-glucanase-producing bacterium *P. cepacia*, which decreased the incidence of diseases caused by *R. solani*, *Sclerotium rolfsii* and *P. ultimum* [56].

The best effects in suppressing of multiple pathogens were observed in control of postharvest diseases. Janisiewicz [86] used successfully the mixture of two antagonists active against blue and gray mold to control decay of apples. The combination of *Pseudomonas syringae* and yeast *Sporobolomyces roseus* also enhanced the control of these diseases on mature apple fruits [87]. Conway et al. [30] have obtained an effective protection of apple fruits against *Colletotrichum acutatum* and *Penicillium expansum* using a mixture of *Cryptococcus laurentii* and *Metschnikowia pulcherrima*.

The control of multiple pathogens after application of biocontrol mixtures was observed also in greenhouse and in field experiments. Mao et al. [119] used the mixture *of G. virens* Gl-3 and *B. cepacia* Bc-F to control a combination of several pathogens: *R. solani, P. ultimum, S. rolfsii, F. oxysporum* f. sp. *lycopersici* and *P. capsici* on tomato and pepper plants. Only the GL-3 + Bc-F treatment reduced damping-off of both plants to the level of plant stands similar to non-infested control. Similarly, when healthy seedlings were transplanted into pathogen-infested soiless mix/soil in greenhouse and field plots, and then drenched with GL-3, Bc-F or GL-3 + Bc-F suspensions, the combined application resulted in greater fresh weight, fruit yield and lower disease severity than those obtained with the agents used alone. The mixture of GL-3 + Bc-F also reduced a complex of diseases of corn caused by interaction of several pathogenic fungi [120].

The best way to control multiple pathogens seems to be induction of resistance in host plants. It was documented that PGPR strains suppressed various root and foliar diseases through the induction of systemic resistance, and once induced resistance was often maintained for the lifetime of the plant [183]. The studies of Rapauch and Kloepper [152, 153] have shown that, treatment of seeds with combination of bacteria *B. pumilus* INR7, *Curtobacterium flaccumfaciens* ME1 and *B. subtilis* GB03, each able to induce systemic resistance, caused growth promotion of cucumber plants and significant protection against angular leaf spot and anthracnose. In some field experiments, mixed infection was reduced to a level statistically equivalent to the synthetic elicitor Actigard applied as a spray [152]. Jetiyanon and Kloepper [91] selected mixtures of PGPR strains with the capacity to induce systemic resistance against diseases of several different plant hosts: bacterial wilt of tomato (*Ralstonia solanacearum*), anthracnose of long cayenne pepper (*Colletotrichum gloesporoides*), damping-off of green kuang futsoi (*R. solani*), and *cucumber mosaic virus* (CMV). Twenty one combinations of PGPR and seven individual strains were tested, and results indicated that four mixtures significantly reduced the severity of all four diseases compared to nonbacterized control.

Competitive Colonization of Rhizosphere and Infection Sites

The ability to colonize rhizosphere, phyllosphere or wounds is essential for microorganisms to function as BCAs [129, 135, 187, 190]. There is the positive relationship between colonization and pathogen suppression in many biocontrol systems [190], and inadequate population of the agent is often the cause of lack of reliable plant protection, as it was mentioned before in this paper. In general, populations of introduced microorganisms decline more or less rapidly, especially following introduction into natural soil [184, 190], and it has been attributed to many abiotic and biotic factors, especially to competition with indigenous microflora [7, 31, 46, 65, 167, 180, 184, 190]. Therefore, an efficient BCA should be an aggressive and competitive colonizer of different plant cultivars with the ability to survive a considerable time in a new environment over diverse conditions and geographical regions. Such requirements are rather a challenge for a single biocontrol agent, even if it belongs to a versatile group like *Trichoderma* spp. or aggressive colonizers like *Pseudomonas* spp. However, Harman [73] asserted that single strains of *Trichoderma* spp., especially the antagonistic strain T-22, are capable of controlling diverse diseases in different crops and applied in a different manner (seed/soil/foliar treatment, propagules suspension, granular formulation). On the other hand, there are reports of Duffy et al. [45], Guetsky et al. [66] and Smolińska [personal communication] about inconsistency of *Trichoderma* spp. applications in plant protection. In the case of microbial combinations, using several strains of BCAs can enhance the possibility that some of them will be able to spread and persist in a new environment. Here, multiple genes and traits may be involved in the process of colonization (e.g. ability to compete for or produce limiting resources, rapid growth rate, ability to survive physical or chemical stresses) [190]. They are affected by the plant species, soil type, environmental conditions, and the type of assays. In the mixture there is always a chance that at least one of the traits will be expressed after inoculation into a new environment. Pierson and Weller [146], demonstrating the potential benefits of using combinations of different strains of fluorescent pseudomonads to suppress take-all of wheat, suggested that greater diversity of introduced bacterial phenotypes resulted in a diverse and potentially more stable rhizosphere community, able to more throughly colonize roots and to survive the biological, chemical and physical changes that occurred during the growing season. Then Kim et al. [96] studied dynamics of population of *Bacillus* sp. strain L324-92_{12} (a rifampicin-resistant mutant of L324-92 suppressive to three root diseases of wheat: take-all, *Rhizoctonia* and *Pythium* root rot) and population of rhizosphere competents biocontrol agent *P. fluorescens* 2-79RN$_{10}$. In growth chamber studies, at the beginning the population of L324-92$_{12}$ was 1000-fold smaller than the amount of 2-79RN$_{10}$. However, in time population of *Bacillus* increased and then stabilized, while the population of *Pseudomonas* constantly decreased until the two were almost the same. L324-92$_{12}$ was recovered form the root section down to 3.5 cm below the seed, but 2-79RN$_{10}$ was transported to the dipper parts of roots. It was not investigated in the studies of Kim et al. [96], but in such disposition *Pseudomonas* could protect plants at the early growth stage, and then could be replaced by developing population of *Bacillus*, which generally stabilize its population in soil in spore form [180]. Thus, both bacteria might provide prolonged protection. Such effect was also suggested by Duffy and Weller [44] who applied *G. graminis* var. *graminis* in combination with fluorescent *Pseudomonas* to suppress take-all of wheat. According to their findings, bacterial seed treatments might provide better protection against take-all in the early stage of disease development, when seminal roots were attacked by *G. graminis* var. *tritici*, and biocontrol agent *G. graminis* var. *graminis* might provide better protection during later stages as the crown root system expanded. The other example is cooperation of rhizosphere-colonizing *Pseudomonas* and *T. koningii* colonizing soil and plant residues, that may extend the sphere of protection beyond the root zone [45].

It was also found that one component of the combination can support the development of the others. *P. putida* WCS358, limiting iron availability for pathogenic *Fusarium*, made it less competitive for carbon than nonpathogenic *F. oxysporum* Fo47 [106, 107]. Thus, development of the pathogen was reduced and colonization of the less sensitive biocontrol agent Fo47 was promoted. In some cases one agent may dominate and limit development of the other, but it is not always tantamount to reduced activity of the mixture. It was shown that population of the bacterium *P. syringae* increased in apple wounds and dominated the population of the yeast *Sporobolomyces roseus*, which was lower than that recorded after a single application [87]. However, both antagonists controlled blue mold (*Penicillium expansum*) on apple more efficiently when combined than the individual applications.

The advantage of the coinoculation is also that combined populations can significantly suppress disease even when their individual population density is too low to do so. Leeman et al. [105] have found that combinations of *F. oxysporum* and *Pseudomonas* strains WCS358, WCS374, WCS417 significantly suppressed fusarium wilt of radish (*F. oxysporum* f. sp. *raphani*) even if the microorganisms were applied in inoculum densities, which were ineffective in suppressing disease as separate inocula. A similar effect was presented by Schisler et al. [158] challenging *Gibberella pulicaris* in wounds on potatoes using paired inoculation with bacterial antagonists. Successful pairs of antagonists reduced disease by aproximately 70% versus control, a level of control comparable to that obtained with 100 times the inoculum dose of a single antagonist strain.

Mixtures of Microorganisms Do Not Always Give a Positive Effect

In the literature usually we can find the decriptions of the positive effects of combined microbial applications. In fact, many results of the studies, which were not satisfactory, have not been published. However, there are several examples of combinations of different bacteria and fungi providing no better or even worse plant growth or biocontrol than the isolates used singly [27, 36, 102]. Combination of *B. subtilis* and non-pathogenic *F. oxysporum* did not provide control of *Fusarium* wilt of chickpea (*F. oxysporum* f. sp. *ciceris*) either applied alone did [75], and biocontrol strain *Pseudomonas* spp. has been shown to have little or no effect on establishment and function of arbuscular mycorrhiza [143].

The complex interactions that can take place in the rhizosphere or spermosphere between BCAs and the indigenous microorganisms should be considered. For example, the groups of microorganisms that occupy the same ecological niche and have the same nutritional requirements may compete for nutrients [10, 87]. Raaijmakers et al. [150] demonstrated that siderophore-mediated competition for iron between two BCAs *P. putida* WCS358 and *P. fluorescents* WCS374 decreased colonization of radish roots by the latter strain. Hubbard et al. [84] reported that indigenous populations of fluorescent pseudomonads significantly reduced the biocontrol activity of *T. hamatum* applied to control *Pythium* seed rot of peas and that iron competition was the primary mechanism involved. Simon and Sivasithamparam [89, 162] also have shown that large populations of indigenous fluorescent pseudomonads were associated with decreased populations of *Trichoderma* spp. and decreased take-all suppression in Western Australia. A similar effect was observed by Bae and Knudsem [7], who have found that a higher level of soil microbial biomass increased interactions between introduced *T. harzianum* ThzID1-M3 and soil microorganisms, and microbial competition in soil favoured a shift from hyphal growth to sporulation in *T. harziaum* reducing its biocontrol activity. In contrast, Dandurand and Knudsen [33] reported that the combination of *P. fluorescens* 2-79 plus *T. harzianum* ThzID1 neither inhibited nor enhanced the biocontrol activity of the latter agent against root rot of pea caused by *Aphanomyces euteiches* f. sp. *pisi.*

In many cases the lack of an additive effect or even reduced growth of crop plants treated with the microbial mixtures may result from the incompatibility of strains used in the combination [8, 36]. Fukui et al. [58] suggested that competition for carbon was the primary factor affecting antagonism between pseudomonads on the sugar beet seeds, and the use of multiple bacterial strains was not superior to the use of the same individual strains for controlling of pericarp infection by *Pyhtium* spp. In this case several fluorescent *Pseudomonas* strains introduced at high inoculum density (apr. 10^7 cfu) to the same ecological niche (spermosphere of sugar beet) competed with or inhibited each other as they interacted with *Pythium* spp., resulting in no enhancement or even reduced efficacy in biocontrol. Raaijmakers et al. [150] demonstrated that one of the combined *Pseudomonas* strains could be outcompeted for iron by the other.

On the other hand, Pierson and Weller [146] have found that many of the strains that were components of mixtures effective in suppressing take-all of wheat, were either strongly inhibitory to or strongly inhibited by other members of the mixture in *in vitro* assays. The authors have stated in that case, that antagonism or incompatibility among strains may result in earlier and greater competition among introduced bacteria in the rhizosphere, and therefore, earlier and more consistent expressed traits involved in competition and disease control, espacially antibiotic production. Such relation between combined strains of *Pseudomonas* was also observed by de Boer et al. [36]. Incompatible strains RS56 and RS111 suppressed *Fusarium* wilt of radish significantly better as compared to the single strains.

Leeman et al. [105] suggest that the disease can not be reduced below a certain level. Therefore, sometimes microbial combination can not perform better than a single treatment, because the disease is already decreased to its maximum by one of the components of the mixture, and the combination simply can not further reduce the disease. As an example, a suppression of *Fusarium* wilt of radish can be obtained by coinoculated *P. fluorescens* WCS374 (or WCS358 or WCS471) and the antagonistic saprophytic *F. oxysporum*, but only if the bacterium and the fungus were applied at densities that did not suppress when applied on their own [105].

Strategies for Forming Mixtures of BCAs

The described above examples of negative efficacy of the mixtures suggest that the combination of biocontrol microbes should be realized according to defined strategies. The most important is selection of active microorganisms for combination. There are several systems proposed on how to combine the microorganisms. Usually mixtures of antagonists are composed or paired at random, but Janisiewicz [88] described the method for selection of antagonists to be combined in mixtures for control of postharvest diseases. The method is based on the nutritional profiles of the potential antagonists assessed on BIOLOG standard plates. Different utilisation of carbon and nitrogen sources allowed populations of several antagonists to colonise the same site without mutual competition for nutrients. Janisiewicz and Bors [87] reported that the broad nitrogen-utilizing *P. syringae* and carbon utilized yeast *Sporobolomyces roseus* allowed both antagonists to flourish in the same wound on apples, which caused greater depletion of nutrients essential for development *P. expansum* than by either antagonists alone. Antagonists selected for mixtures were also obtained from microbial succession at the wound site [88, 89]. Thus, the isolates were ecologically suited to the environment of protected fruits.

Besides the microorganisms utilizing various nutrient sources, the combination of microbes with different optimum temperature, pH or humidity may enhance consistency of the mixture effectiveness in changeable conditions as it was presented by Guetski et al. [66]. Especially taxonomically different microbes, indicating various conditional requirements and colonizing different sites, can improve protective properties of the mixture. According to Fukui et al. [58] seed protection by bacteria is inadequate when conditions are ideal for fungal infections. Therefore, in such circumstances the mixture that also contains antagonistic fungi may provide protection, while bacteria are not active. The fungi and bacteria may exhibit different colonization patterns, e.g. bacteria like *Pseudomonas* or *Bacillus* colonize the upper part of the roots and their concentration decline toward the root tip [35, 113, 190, 198]. Their populations also tend to drop in time [180, 190]. In contrast fungi, like

Trichoderma, may colonize the whole root surface for several weeks or months and remain functional for at least the life of an annual crop [73, 74]. It means that bacteria and fungi may protect different sites and at various growth stages of the plant. Thus, their combination may provide more complex protection during all vegetation.

Another strategy to select the best mixture was proposed by Fukui et al. [59] and Kraus et al. [100]. They suggested that it is more practical to start the evaluation of BCAs with a mixture of many antagonists and then eliminate the ineffective and incompatible ones. However, in most studies, the microorganisms are selected among the dominating microbial groups, e.g. in suppressive soils [3, 45, 108] or among already documented BCAs. In this case, each antagonist in the mixture should exhibit one or more mechanism of biocontrol such as antibiosis related to production of broad-spectrum antibiotics, competition for iron or other nutrients, parasitism and induction of systemic resistance, and a preferential attack on at least one pathogen. This may provide activity against multiple pathogens at different conditions, e.g. when antibiotic producing agents protect seedlings against damping-off diseases, the other may induce resistance or stimulate plant growth and prolong the protective effect against pathogens attacking plants at further stages of vegetation. However, at the same time, the antagonists should not suppress other coinoculants in the mixture. They should be compatible.

Alabouvette et al. [4] demonstrated that nonpathogenic *F. oxysporum* Fo47 was only little sensitive to pseudobactin-mediated iron competition caused by fluorescent *Pseudononas*, therefore, both agents could effectively outcompete pathogenic *Fusarium*. There was no indication of an antagonistic relationship among five biocontrol bacterial strains in anthurium guttation fluid and only the population of pathogenic bacterium *X. campestris* pv. *dieffenbachiae* diminished [59, 60]. A similar effect was observed by Krauss et al. [100] controlling *C. musae* with broad-host range fungal antagonists, which percentage recovery from inoculated banana discs increased, whereas the frequency of the pathogen reisolation decreased. In contrast, De Boer et al. [36] have found that the mixture of compatible *in vitro* two strains of the bacteria did not protect the plants better than separate strains, probably because both bacteria in the mixture suppressed the pathogen by the same mechanism.

Sometimes the mutual inhibition between biocontrol strains may be limited by spatial separation of the agents as in the case of *T. koningii* and the strains of fluorescent *Pseudomonas* spp. used to control take-all [45]. The majority of the bacteria used in this study produced metabolites that could inhibit *T. koningii*, but in growth chamber experiments and in the field the bacteria did not reduce the biocontrol activity of the fungus and *vice versa*. Szczech and Shoda [174] also observed, that when iturin A-produced *B. subtilis* RB14-C was applied to the soil as a mixture with the antagonistic *B. cepacia* BY, the suppression of damping-off of tomato plants was significantly lower, than when BY was added to the soil two days after RB14-C application. In these studies the positive effect of combining bacterial agents was clearly related to the order in which both agents were introduced. However, such temporal separation of the agents is rather complicated from the commercial point of view, because it requires additional application.

Another innovative approach for improving soilborne disease control could be the development of cocktails containing strains that communicate with each other to maximize antibiotic production and disease control [17].

Integration of BCAs with Agrochemicals and Agronomical Practices

As in the case of microbial mixtures, the combination of BCAs with several agronomical practices or chemicals has attracted attention in order to obtain an additive or synergistic effect against the target pathogen. There were efforts to combine microbial agents with fungicides or other chemical products (e.g. chitosan, calcium chloride), soil disinfestation, soil amendments, crop rotation and other cultural practices [90, 171]. Numerous studies were focused on the integration of biological and chemical tools to minimize environmental risk and improve plant protection. This kind of treatment can also reduce the possibility of the development of pathogen resistance. It was found, that application of reduced amounts of fungicides with BCAs has not only decreased the input of chemicals but also resulted in improved disease control [26, 54, 95, 99, 197]. The low dose of fungicide may effectively stress and weaken the pathogen to make it more susceptible to attack by the antagonists under variable climatic conditions and at a high level of disease preasure. However, the isolate of applied antagonist must be resistant to the used dose of pesticide. In the studies of Garibaldi et al. [61] and Minuto et al. [130], the resistance of non-pathogenic isolates of *Fusarium* to bezimidazole permitted the use of the antagonists with this chemical to obtain better protection of carnation and cyclamen against *Fusarium* wilt. Fungi genera *Trichoderma* spp. were found to be resistant to some fungicides, especially to metalaxyl [73, 76]. No chemical seed teratments inhibited root colonization by T-22 [73]. The other potential group of BCAs, with enhanced resistance to chemicals, is *Bulkholderia* spp. [113].

Integrated methods were effective in postharvest disease control, where biocontrol agents were mostly used with sodium bicarbonate [30, 89, 137, 170] or calcium chloride [43, 89, 176]. Both salts had an inhibitory effect on spore germination and subsequent developmet of a decay pathogens, thus created the space for antagonist development [43, 137]. A significant increase in biocontrol activity of used BCAs was also observed when the isolates were applied following hot water treatment [137, 170], where as the result of the pathogen disruption the antagonists have gained a competitive advantage over it.

The application of biocontrol agents with chitosan [19, 132, 173] seems to be very promising. Chitosan added to the suspension of chitinolityc bacteria or yeasts resulted in the increased population and activity of the bacteria and significantly enhanced the suppressive effect of the agents [89, 122, 148, 173]. Moreover, chitosan activates plant defense [19, 131, 136].

Some microorganisms such as *Trichoderma* spp. and *Gliocladium* spp. are less sensitive to fumigants than other organisms and they are capable ofcompetitive recolonization of the soil after disinfestation or solarization [171]. Combining fumigants or solarization with these agents may lead to an additive control effect and allow one to shorten the process of solarization or use it under broader climatic zones [171]. Also fumigation of soil with isothiocyanates, liberated from the *Brassica* plant residues incorporated into soil, caused a weakened structure of sclerotia of *Sclerotium cepivorum* [168] and made them more susceptible to attack of antagonistic fungi [169]. The other soil amendments, supported by the establishment and activity of specific BCAs populations in soil, are composts, which mostly provide the suppression of such pathogenic fungi as *R. solani, Pythium* sp.and *Phytophthora* spp. [39, 77, 78, 177].

The Mode of Formulation and Application of the Mixtures

Inoculation by BCAs is usually carried out by two main methods: the direct inoculation with bacterial or fungal culture, which is applied as seed coating, root dipping, soil drenching, foliar spraying, furrow application, or the other one consisting of a formulated, solid preparation of microorganisms based on different carriers supplemented with ingredients that protect and promote the active microbes [13, 85, 171]. In all experiments reviewed in this paper, the direct inoculation was used to introduce the combined microorganisms into the site of protection. Usually microbial cultures or propagule suspensions were mixed with soil or mixed together, mostly at equal inoculum densities, for soil drenching, seed coating or for dipping of the roots [37, 44, 48, 59, 146, 152]. Sometimes the bacterial cells or fungal propagules were mixed with peat or talc and carboxymethyl cellulose as an adhesive [173]. However, these are simple methods of inoculation, which do not provide any support or protection for an microbial agent introduced into a new environment. The proper method of formulation and application is a key factor to enhance efficacy of the BCAs and to gain a satisfactory control of crop plants. The formulation system and type of inoculant determine the potential success of the biopesticide in a new environment. According to Bashan [14] a major role of the inoculant is to provide a suitable condition to prevent the rapid decline of introduced BCA in the soil. A good inoculant: (i) should be abundantly produced in a cost-effective system; (ii) should be chemically and physically uniform, nearly sterile and easily manufactured by existing industry; (iii) inoculant should be nontoxic and biodegradable; (iv) a formulation process should secure a good survival and stability of processed microorganisms; (v) then inoculant should allow for ease of handling and application with standard agrotechnical machinery; (vi) the release of BCA into the environment should be controlled, and the inoculant has to help to establish a sufficient population of active agent under variable field conditions; (vii) the inoculant should also overcome the loss of viability of the formulated microorganisms during long shelf life, within the marketing distribution system over the range of temperatures of –5 to 30 ^{o}C, protecting the product against dessication, irradiation and humidity [14, 85, 171].

The important ingredient of the inoculant is carrier material. Bashan [14] grouped the carriers into four basic categories: (i) soils (peat, coal, clay); (ii) plant waste materials (composts, manure, soybean meal, plant oils, wheat bran, ground plant debris); (iii) inert materials (vermiculite, perlite, talc, ground rock, calcium sulfate, polyacrylamide gels, alginate beads; (iv) plain liophilized microbial cultures and oil-dried bacteria. The potential advantages and problems associated with the application of mainly used carriers was reviewed by van Elsas and Heijnen [181]. The most popular material is peat [14, 171, 181], which has been developed as an effective carrier for *Rhizobium* [181]. However, it has several drawbacks such as variability in the quality, that affects a final concentration of active propagules in the inoculant and may cause difficulties in dosage setting, poor organism survival under hot or dry conditions or easy contamination that can reduce the shelf life of the inoculant [14, 181]. The other, often used materials e.g. silt loam, montmorylonite clay or coal, are inexpensive, easily available and may even increase microbial survival [178, 181]. But they, as well as peat, seem to be not very useful as carriers for combined microorganisms.

The author has not found any reports about the effect of the long storage of such formulations on the efficacy of mixed inoculants, however, it is supposed that in those preparations, microorganisms may interact with each other inhibiting or significantly reducing the population of one or more of the components of the mixture. This is seen especially in peat, where microorganisms are still active and can multiply during storage [14]. An exception may be the mixture of endospore-forming bacteria such as *Bacillus* spp. with different modes of action, that at the dormant stage probably may not affect each other and are able to maintain viability for years in standard conditions of storage. Although, this theory should be proved experimentally. Kodiak®, containing the strain GB03 combined with fungicides, may serve as an example of a successful biocontrol product based on spore-forming *Bacillus* [21].

But what to do with microbial agents, that do not produce resistant propagules, such as bio-active strains *Pseudomonas* spp.? One of the possible ways is to formulate and store each strain separately, and then combine them at proper ratios just before application. However, growers expect an effective and easy to handle product. The necessity of preparation of uniform mixture, combined of properly rationed components, may discourage the farmers. Moreover, limited consistency and variability in quality of each preparation may result in changeable efficacy. The other, but close way may be the use of separately liophilized cells or propagules of the agents. But frozen cell pellets must be kept at low temperature until application, which requires additional equipment and remain a major obstacle for their large-scale use [171].

Among other methods of formulation, the most promissing for microbial mixtures seems to be the use of synthetic inoculant carriers immobilizing or entrapping active organisms. During the last decades, several polymer-based formulations have been evaluated [14]. The polymers encapsulate living cells protecting the microorganisms against many environmental stresses and release them to the environment gradually, when the polymers are degraded [14]. They can be dried and stored for prolonged periods, providing consistent quality and defined conditions for target biocontrol agents [13, 14, 15]. Bashan [13] entrapping plant growth-promoting bacterium *Azospirillum brasilense* in sodium alginate beads obtained high cell number per bead and a high level of bacteria survival during storage. A similar effect was obtained by van Elsas et al. [182] with cells of *P. fluorescens*. Encapsulated cells introduced into non sterile loamy sand survived better than cells added directly to the same soil. In both experiments, application of the alginate-bead preparations on wheat seeds resulted in effective establishments of bacterial inoculum on wheat roots [13, 182]. Addition of skim milk to the beads resulted in improved survival of the encapsulated cells [13, 16, 182]. Skim milk-amended beads were also more biodegradable in different types of soil [13]. It was also found that smaller beads, with diameters ranging from 100 to 200 μ, were more useful for application than bigger granules [16]. They can be better attached to the seeds and still contain the number of bacteria >10^6 cfu $bead^{-1}$ sufficient for successful inoculation [12, 16]. Moreover, O_2 diffusion into smaller beads is not limited as it was found for big granules, therefore, all cells through the microbeads are active [16]. Powder-like formulation requires only small-volume storage space, and its application do not require additional devices or agrotechnical procedures [13]. The beads remain undegraded the in field until it rains and germination of the seeds is accompanied by releasing of the microbes form the beads. What is also important, a slow release of the agents from the beads ensure a constant supply of the active microbes over a long period of vegetation.

From the point of view of biocontrol mixtures preparation, alginate microbeads seem to be the most useful formulation. First of all, they can be used for encapsulation both dormant - stress resistant structures, as well as for vegetative microbial cells. It is possible that cells of different BCAs, entrapped in separate alginate beads, may not affect each other, when mixed and stored together. Such a situation was not examined yet, but Gonzales and Bashan [64] observed positive relations between the fresh water microalga *Chlorella vulgaris* and plant growth promoting *A. brasilense* coimmobilized in the same alginate beads. The presence of bacterium significantly increased growth of the microalga. It would be interesting to check, if it is possible to immobilize two or more compatible BCAs colonizing independent space in the same beads. However, the more reliable way probably would be combining the defined rations of separately produced beads containing various agents or to store them in separate containers and then mix them just before application at recommended doses. Because, the beads provide good conditions for prolonged survival of entraped microbes at consistent density [13, 15], it may assure that the proportion between combined agents will not change during storage. Besides, quality control of such formulations is very simple.

A high concentration of antagonistic microbes in the inoculum is required for effective suppression of the pathogens. According to the numerous papers describing application of BCAs the cell concentration for bacterial inoculum ranges from 10^6 to 10^{10} cfu, and for antagonistic fungi from 10^3 to 10^7 cfu per ml of the suspension or g of the seeds/soil/potting mix. However, usually the concentrations of 10^8 cfu for bacteria and 10^5 cfu for fungi are used. Similar final amounts of the antagonists were applied in the form of the mixtures [44, 59, 102, 146, 173]. Some authors combined equivalent proportions of each active strain [44, 146], and then added a volume of inoculum like for individual strains, to obtain the same total propagule count for multiple and single treatments. Others added each strain in the combination at the level of a single application [36, 102]. In the first case, combining of several strains may reduce their individual concentration in the mixture compared with single treatments, although total cell density in the mixed inoculum will remain at the required level. In some cases, such “dilution” of the agents may reduce their reliable activity, and finally decrease expected efficacy of the mixture, especially when the agents belong to the same taxonomic group or exhibit similar mechanisms of suppression. Smolińska et al. [personal communication] mixed several strains of antagonistic *Trichoderma* spp. to control *Rhizoctonia* damping-off on cucumber, but the mixtures were not effective compared to individual isolates. Also Szczech [data not published] has found that the very active *Burkholderia* spp. strain CAT5 was less effective in the mixtures with other antagonistic bacteria, although in *in vitro* tests it was not affected by coinoculants. Increasing the concentrations of the coinoculants in the mixture may not always improve their effectiveness. Fukui et al. [58] monitored the growth and interactions of coinoculated strains of *Pseudomonas* spp. on sugar beet seeds. There was no interaction between and among strains when they were inoculated together at approximately 10^4 cfu per seed. Strains capable of producing antibiotics and siderophores did not inhibit sensitive strains in the spermosphere. Moreover, bacterial strains that were highly inhibitory to each other in vitro coexisted with no observable inhibitory effects when coinoculated on seeds. Scanning electron micrographs showed that in colonies that were spatially separated, competition was minimal. However, antagonism occured when at least one strain was inoculated at a higher density, because probably they had similar nutritional spectra and started to compete for carbon [58]. These data and observations suggest that proper determination of the cell/propagule concentration in the mixture as well as the proportions of each of the coinoculants are important factors to obtain a successful combination. The studies on the interactions between the coinoculated strains and these strains and indigenous microflora after introduction to the target sites, are very complicated but may help our knowledge of how to compose and use the mixtures properly.

Just as important as the selection of the set and dosage of the agents in the combination, the way of their application may also determine the strenght of biocontrol activity. Inoculation techniques should be practical and simple. Farmers are discouraged from inoculant use if they have to make additional treatments. Two main methods of inoculation of BCAs are currently used: seed inoculation and delivering the inoculant into the sowing furrow with seeds (14). In the experimental works, combined inoculants usually are mixed with soil or potting medium [37, 102, 173] or used as seed coating [44, 146, 156]. Seed coating, already developed for chemicals, is rather easy and does not require any special equipment to apply, but there may be a problem with sufficient-for-biocontrol distribution of the agents. Many antagonistic bacteria applied to the seeds colonize the upper zone of the roots, but transport to the dipper parts is limited and their population decrease with depth. Moreover, population of the agents is reduced during plant vegetation, as it was mentioned before. Therefore, this treatment may be effective in control of damping-off diseases but it may be inadequate to protect plants in later stages of growth, unless the agent induces systemic resistance. Such a situation was observed by Szczech [data not published], when the mixtures containing antagonistic bacteria and fungi significantly reduced damping-off caused by a "cocktail" consisting of several pathogenic fungi, but in later stages of plant growth these teratments did not protect plants against *Fusarium* wilts. Also Szczech and Shoda [175] treated tomato seeds with iturin A produced *B. subtilis* to control *Rhizoctonia* damping-off, but this application was significantly less effective than mixing of bacterial suspension with soil. Root colonization by *B. subtilis* added to soil was more abundant and uniform than after seed treatment, therefore, contact of the pathogen with bacterial antibiotic was more probable resulting in enhanced protection. However, seed inoculation could be effective if the combination of active rhizosphere colonizers is used such as *Pseudomonas* spp. or *Trichoderma* spp.

Sometimes, one component of the biocontrol combination is applied to the soil, while the other is applied to the seeds [45]. This way of inoculation could be effective, but it also seems to be not convenient for the commercial scale, because it needs additional treatments. The other method of application can be the use of combined microbial inoculants to grow more resistant transplants. Production of transplants may be carried out in controlled conditions with the use of limited volume of potting medium. In such circumstances, it is easy to mix the inoculum with medium, using small dosages, which provide a sufficient level of active cfu's. Thus, it may allow for abundant and uniform colonization of the root system of a young plant. For example, tomatoes were grown in a potting mix containing the granular formulation of *T. harzianum* T-22, which permitted roots to become colonized, then trasplanted to the field. *Fusarium* crown and root rot at harvest was reduced [34, 135]. The combination of T-22 and mycorrhizal fungus *Glomus intraradices* was even more effective [34]. Also pepper seedlings produced in the greenhouse with T-22 better survived transplanting into the infested field than the seedling that were not inoculated [73]. In the studies of Murphy et al. [134] tomato plants treated with PGPR bacteria appeared phenotypically and developmentally similar to nonbacterized control plants that were 10 days older. These plants were significantly less sensitive to infection by CMV virus. This suggests that application of not very high amounts of biocontrol organisms at the time of seeding of transplants can provide a season-long benefit of plant health. The other solution for the biocontrol agents application, although more troublesome for growers, can be dipping of plant roots in the microbial suspension before transplanting into soil. It may provide a uniform and abundant colonization of the rhizosphere resulting in better protection than in the case of seed inoculation.

An universal way of application for the biocontrol mixtures has not been developed yet, and the method of introduction should be rather adapted for a particular mixture. However, as for separate biocontrol agents, inoculation should be performed at the precise time needed by the plants and according to expected target pathogens; the use of mixtures probably may not be so strict, because of the versatility of the mechanisms and conditional requirements of the coinoculated agents.

Problems and Prospects

Till now no mixture of biocontrol microorganisms has been registered as a commercial product. First of all, it is difficult to compose a very effective combination of microorganisms and formulate them to get a consistent product. However, the alginate microbeads seem to be a promising method for production of mixed inoculant. It may limit the possible interactions between the components of the combination, and can provide good survival of entrapped microorganisms during prolonged storage. Besides formulation, also very important is the process of the production of different microbes providing for the mixture. The methods of mass production of various single agents were already developed [171]. However, in the case of mixtures, several kinds of the microorganisms have to be multiplied, and they require separate equipment and different parameters for fermentation. Moreover, certain precautions should be taken to avoid mutual contamination during mass production. Therefore, quality control methods for such preparations should be developed and standardized. The quality and possible contaminations should be determined during the mass production process as well as during storage, because farmers have to be assured that they always get a standard product. For the mixtures of organisms it is more complicated than for a single biocontrol agent, and all these requirements may enhance costs of the production. Thus, this may discourage potential producers and customers, which expect not only highly effective but also cheap and easy ways of handling the product. The combination of two or more antagonists also needs a multiple registration process, which also increases costs and creates difficulties in matching all the demands of legislation. According to Spadaro and Gullino [171] and Janisiewicz and Korsten [89], this option could be feasible with products already registered: a biofungicide based on different antagonistic strains may be labeled as compatible with each other and proposed for joint use. However, such an approach limits the range of potential agents, which are not as suppressive as a single treatment but may be effective components of the mixture.

The success of the disease control with the use of combined agents will also depend on the information of the sellers and customers about the product. Growers should be aware that microbial inoculants consist of living organisms, which need different management than chemicals. They should also know the possible impact of bioproducts on the environment and understand the mutual relations between introduced microorganisms, between the microorganisms and crop plant, and pathogens. Distributors should have a specialized knowledge about the inoculants to encourage farmers to test the product and to teach them how to store it, how to calculate doses, and how to apply the microorganisms. People understanding the environmental effects associated with releasing living organisms for biocontrol practices may build confidence in such products. It is very important because the chemical pesticide industry already faces a scheduled removal of many synthetic pesticides from the market and the best example is methyl bromide [123]. In several high value speciality markets such as flowers, organic fruits and vegetables chemicals are undesirable or their use is restricted [23]. Therefore, there is an urgent need for new pesticides that address environmental concerns and fulfill producers' expectations. As it was described before, single inoculants often cannot reach the level of efficacy of the pesticides. Perhaps mixed inoculants may fill this gap. Greenhouse crops, soil-less cultures or postharvest control may be a good target for such products since these types of cultivation and storage provide controlled conditions limiting interactions in an already complicated system. However, the biggest expectations are related to field application, where combined microorganisms should provide more consistent control of plant pathogens. Their effect may be additionally improved by genetic manipulation, which can result in increased production of toxic compounds, enzymes, improved competence, wider host range and enhanced tolerance of the strains to stresses [63]. Spadaro and Gullino [171] described numerous examples of superior activity of genetically modified microbes over their paternal strains. Some biocontrol agents may be additionally transformed by adding of genes encoding desired functions as in the case of *P. fluorescens* strain, in which is introduced the *chiA* gene encoding the chitinase of the *S. marcescens* [41]. On the other hand, there are difficulties in registering genetically modified microorganisms associated with potential risk related to potential allergenicity, toxicity to humans or nontarget organisms and transgene stability [171].

In conclusion, the combination of several microorganisms exhibiting different modes of action against various plant pathogens is a promissing alternative for chemical pesticides. However, their successful commercialization must be preceded by detailed studies of the behaviour of the coinoculated agents in the changeable environment. Especially the mutual relations between agents in the mixture and between them and the pathogens or indigenous organisms should be investigated. Knowledge of the producers about biological control measures which has been growing for decades and the consciousness about unfavourable changes in the natural environment may encourage for continuation of the studies and possibly help to shorten the time of product registration and introduction of the mixtures into commercial crop production. However, the future will show if the combined microorganisms are successful in commercial scale/use/production/market.

References

[1] Acea, M.J.; Moore, C.R.; Alexander, M (1988). Survival and growth of bacteria introduced into soil. *Soil Biol. Biochem.*, 20: 509 – 515.

[2] Ahonsi, M.O.; Berner, D.K.; Emechebe, A.M.; Lagoke, S.T.; Sanginga, N. (2003). Potential of ethylene-producing pseudomonads in combination with effective N_2-fixing bradyrhizobial strains as supplements to legume rotation for Striga hermonthica control. *Biol. Contr.*, 28: 1 – 10.

[3] Alabouvette, C. (1990). Biological control of *Fusarium* wilt pathogens in suppressive soils. In: D. Hornby (Ed.), Biological Control of Soilborne Plant Pathogens. *Wallingford*, UK: 27 – 43.

[4] Alabouvette, C.; Lemanceau, P.; Steinberg, C. (1996). Biological control of fusarium wilts: opportunities for developing a commercial product. In: R. Hall (Ed.), *Principles and Practice of Managing of Soilborne Plant Pathogens.* APS Press, St. Paul, United States: 192 – 212.

[5] Aono, R.; Hammura, M.; Yamamoto, M.; Asano, T. (1995). Isolation of extracellular 28- and 42-kilodalton β-1,3-glucanases and comparison of three β-1,3-glucanases produced by *Bacillus circulans* IAM1165. *Appl. Environ. Microbiol.*, 61: 122 – 129.

[6] Backman, P.A.; Wilson, M.; Murphy, J.F. (1997). Bacteria for biological control of plant diseases. In: N.A. Rechcigl & J.E. Rechcigl (Eds.), *Environmentally Safe Approaches to Crop Diseases Control.* CRC Lewis Publishers, Boca Raton, FL: 95 – 109.

[7] Bae, Y.S.; Knudsen, G.R. (2005). Soil microbial biomass influence on growth and biocontrol efficacy of *Trichoderma harzianum*. *Biol. Contr.*, 32: 236 – 242.

[8] Baker, R. (1990). An overview of current and future strategies and models for biological control. In: D. Hornby (Ed.), *Biological Control of Soil-Borne Plant Pathogens.* C.A.B. International, Wallingford: 375 – 389.

[9] Bakker, P.A.H.M.; Lamers, J.G.; Bakker, A.W.; Marugg, J.D.; Weisbeek, P.J.; Schippers, B. (1986). The role of siderophores in potato tuber yield increase by *Pseudomonas putida* in short rotation of potato. *Neth. J. Plant Pathol.*, 92: 249 – 256.

[10] Bakker, P.A.H.M.; Weisbeek, P.J.; Schippers, B. (1988). Siderophore production by plant growth-promoting *Pseudomonas* spp. *J. Plant Nutr.*, 11: 925 – 933.

[11] Bakker, P.A.H.M.; Pietrese, C.M.J.; van Loon, L.C. (2007). Induced systemic resistance by fluorescent *Pseudomonas* spp. *Phytopathol.*, 97: 239 – 243.

[12] Bashan, Y. (1986). Significance of timing and level of inoculation with rhizosphere bacteria on wheat plants. *Soil Biol. Biochem.*, 18: 297 – 301.

[13] Bashan, Y. (1986). Alginate beads as synthetic inoculant carriers for slow release of bacteria that affect plant growth. *Appl. Environ. Microbiol.*, 51: 1089 – 1098.

[14] Bashan, Y. (1998). Inoculants of plant growth-promoting bacteria for use in agriculture. *Biotechnol. Adv.*, 16: 729 – 770.

[15] Bashan, Y. & Gonzales, L.E. (1999). Long-term survival of the plant-growth-promoting bacteria *Azospirillum brasilense* and *Pseudomonas fluorescens* in dry alginate inoculant. *Appl. Microbiol. Biotechnol.*, 51: 262 – 266.

[16] Bashan, Y.; Hernandez, J.P.; Leyva L.A.; Bacilio, M. (2002). Alginate microbeads as inoculant carriers for plant growth-promoting bacteria. *Biol. Fertil. Soils,* 35: 359 – 368.

[17] Becker, D.M.; Kinkel, L.L.,; Schottel, J.L. (1997). Evidence for interspecies communication and its potential role in pathogen suppression in a naturally occuring disease suppressive soil. *Can. J. Microbiol.*, 43: 985 – 990.

[18] Bell, A.A.; Wheeler, M.H. (1986). Biosynthesis and functions of fungal melanins. *Annu. Rev. Phytopathol.*, 24: 411 – 451.

[19] Benhamou, N.; Kloepper, J.W.; Tuzun, S. (1998). Induction of resistance against Fusarium wilt of tomato by combination of chitosan with an endophytic becterial strain: ultrastructure and cytochemistry of the host response. *Planta,* 204: 153 – 168.

[20] Benítez, T.; Rincón, A.M.; Limón, M.C.; Codón, A.C. (2004). Biocontrol mechanisms of *Trichoderma* strains. *Inter. Microbiol.*, 7: 249 – 260.

[21] Brannen, P.M. & Kenney D.S. (1997). Kodiak® - a successful biological-control product for suppression of soil-borne plant pathogens of cotton. *J. Industr. Microbiol. Biotechn.*, 19: 169 – 171.

[22] Brewer, M.T. & Larkin, R.P. (2005). Efficacy of several biocontrol organisms against *Rhizoctonia solani* on potato. *Crop Protect.*, 24: 939 – 950.

[23] Budi, S.W.; van Tuinen, D.; Martinotti, G.; Gianinazzi, S. (1999). Isolation from the *Sorghum bicolor* mycorrhizosphere of a bacterium compatible with arbuscular mycorrhiza development and antagonistic towards soilborne fungal pathogens. *Appl. Environ. Microbiol.*, 65: 5148 – 5150.

[24] Bull, T.C.; Weller, D.M.; Thomashow, L.S. (1991). Relationship between root colonization and suppression of *Gaeumannomyces graminis* var. *tritici* by *Pseudomonas fluorescens* strain 2-79. *Phytopathol.*, 81: 954 – 959.

[25] Capper, A.L. & Higgins, K.P. 1993. Application of *Pseudomonas fluorescens* isolates to wheat as potential biological control agents against take-all. *Plant Pathol.*, 42: 560 – 570.

[26] Chand-Goyal, T. & Spotts, R.A. (1996). Postharvest biological control of blue mold of apple and brown rot of sweet cherry by natural saprophytic yeast alone or in combination with low doses of fungicides. *Biol. Contr.*, 6: 253 – 259.

[27] Chiarini, L.; Bevivino, A.; Tabacchioni, S.; Dalmastri, C. (1998). Inoculation of *Burkholderia cepacia, Pseudomonas fluorescens* and *Enterobacter* sp. on *Sorgum bicolor*: root colonization and plant growth promotion of dual strain inocula. *Soil Biol. Biochem.*, 1: 81 – 87.

[28] Chung, Y.R.; Hoitink, H.A.J.; Dick, W.A.; Herr, L.J. (1988). Effects of organic matter decomposition level and cellulose amendment on the inoculum potential of *Rhizoctonia solani* in hardwood bark media. *Phytopathol.,* 78: 836 – 840.

[29] Collins, D.P. & Jacobsen, B.J. (2003). Optimizing of *Bacillus subtilis* isolate for biological control of sugar beet cercospora leaf spot. *Biol. Contr.,* 26: 153 – 161.

[30] Conway, W.S.; Leverentz, B.; Janisiewicz, W.J.; Saftner, R.A.; Camp, M.J. (2005). Improving biocontrol using antagonist mixtures with heat and/or sodium bicarbonate to control postharvest decay of apple fruits. *Postharv. Biol. Technol.,* 36: 235 – 244.

[31] Compant, S.; Duffy, B.; Nowak, J.; Clément, C.; Ait Barka, E. (2005). Use of plant growth-promoting bacteria for biocontrol of plant diseases: principles, mechanisms of action, and future prospects. *Appl. Environ. Microbiol.,* 71: 4951 – 4959.

[32] Cook, J.R. & Baker K.F. (1983). The nature and practice of biological control of plant pathogens. *The Amer. Phytopathol. Soc.,* St. Paul, Minnesota, 539 pp.

[33] Dandurand, L.M.; & Knudsen, G.R. (1993). Influence of Pseudomonas fluorescens on hyphal growth and biocontrol activity of Trichoderma harzianum in the spermosphere and rhizosphere of pea. *Phytopathol.,* 83: 265 – 270.

[34] Datnoff, L.E.; Nemec, S.; Pernezny, K. (1995). Biological control of crown and root rot of tomato in Florida using *Trichoderma harzianum* and *Glomus intraradices. Biol. Contr.,* 5: 427 – 431.

[35] Davies, K.G. & Whitbread, R. (1989). Factors affecting the colonization of a root system by fluorescent pseudomonads: the effects of water, temperature and soil microflora. *Plant Soil,* 116: 247 – 256.

[36] De Boer, M.; van der Sluis, I.; van Loon, L.C.; Bakker, P.A.H.M. (1999). Combining fluorescent *Pseudomonas* spp. strains to enhance suppression of fusarium wilt of radish. *Europ. J. Plant Pathol.,* 105: 201 – 210.

[37] De Boer, M.; Bom, P.; Kindt, F.; Keurentjes, J.J.B.; van der Sluis, I.; van Loon, L.C.; Bakker, P.A.H.M. (2003). Control of fusarium wilt of radish by combining *Pseudomonas putida* strains that have different disease-suppressive mechanisms. *Phytopathol.,* 93: 626 – 632.

[38] De Boer, W.; Wagenaar, A.M.; Klein Gunnewiek, P.J.; van Veen, J.A. (2007). In vitro suppression of fungi caused by combinations of apparently non-antagonistic soil bacteria. *FEMS Microbiol. Ecol.,* 59: 177 – 185.

[39] De Ceuster, T.J.J. & Hoitink, H.A.J. (1999). Using compost to control plant diseases. *BioCycle,* June 1999: 61 – 64.

[40] Domenech, J.; Reddy, M.S.; Kloepper, J.W.; Ramos, B.; Gutierrez-Mañero, J. (2006). Combined application of the biological product LS213 with *Bacillus, Pseudomonas* or *Chryseobacterium* for growth promotion and biological control of soil-borne pathogens in pepper and tomato. *BioControl,* 51: 245 – 258.

[41] Downing, K.J. & Thomson, J.A. (2000). Introduction of the *Serratia marcescens* chiA gene into an endophytic *Pseudomonas fluorescens* for the biocontrol of phytopathogenic fungi. *Can. J. Microbiol.,* 46: 363 – 369.

[42] Driks, A. (2002). Overview: development in bacteria: spore formation in *Bacillus subtilis. Cell Mol. Life Sci.,* 59: 389 – 391.

[43] Droby, S.; Wisniewski, M.E.; Cohen, L.; Weiss, B.; Touitou, D.; Eilam, Y.; Chalutz, E. (1997). Influence of $CaCl_2$ on *Penicillium digitatum*, grapefruit peel tissue, and biocontrol activity of *Pichia guilliermondii*. *Phytopathol.*, 87: 310 – 315.

[44] Duffy, B.K. & Weller D.M. (1995). Use of *Gaeumannomyces graminis* var. *graminis* alone and in combination with fluorescent *Pseudomonas* spp. to suppress take-all of wheat. *Plant Dis.*, 79: 907 – 911.

[45] Duffy, B.K.; Simon, A.; Weller, D.M. (1996). Combination of *Trichoderma koningii* with fluorescent pseudomonads for control of take-all on wheat. *Phytopathol.*, 86: 188 – 194.

[46] Duffy, B.K. & Défago G. (1997). Zinc improves biocontrol of fusarium crown and root rot of tomato by *Pseudomonas fluorescens* and represses the production of pathogen metabolites inhibitory to bacterial antibiotic biosynthesis. *Phytopathol.*, 87: 1250 – 1257.

[47] Duffy, B.; Schouten, A.; Raaijmakers, J.M. (2003). Pathogen self-defense: mechanisms to counteract microbial antagonism. *Ann. Rev. Phytopathol.*, 41: 501 - 538.

[48] Duijff, B.J.; Recorbet, G.; Bakker, P.A.H.M.; Loper, J.E.; Lemanceau, P. (1999). Microbial antagonism at the root level is involved in the suppression of *Fusarium* wilt by the combination of nonpathogenic *Fusarium oxysporum* Fo47 and *Pseudomonas putida* WCS358. *Phytopathol.*, 89: 1073 – 1079.

[49] Dupler, M. & Baker, R (1984). Survival of *Pseudomonas putida*, a biological control agent, in soil. *Phytopathol.*, 74: 195 – 200.

[50] El-Tarabily, K.A.; Soliman, M.H.; Nassar, A.H.; Al-Hassani, H.A.; Sivasithamparam, K.; McKenna, F.; Hardy, G.E.St.J. (2000). Biological control of *Sclerotinia minor* using a chitinolytic bacterium and actinomycetes. *Plant Pathol.*, 49: 573 – 583.

[51] Fedi, S.; Tola, E.; Moënne-Loccoz, Y.; Dowling, D.N.; Smith, L.M.; O'Gara, F. (1997). Evidence for signaling between the phytopathogenic fungus *Pythium ultimum* and *Pseudomonas fluorescens* F113: *P. ultimum* responses the expression of genes in *P. fluorescens* F113, resulting in altered ecological fitness. *Appl. Environ, Microbiol.*, 63: 4262 – 4266.

[52] Fogliano, V.; Ballino, A.; Gallo, M.; Woo, S.; Scala, F.; Lorito, M. (2002). *Pseudomonas* lipodepsipeptides and fungalcell wall-degrading enzymes act synergistially in biological control. *Mol. Plant-Microbe Interact.*, 15: 323 – 333.

[53] Fravel, D.R. (1988). Role of antibiosis in the biocontrol of plant diseases. *Annu. Rev. Phytopathol.*, 26: 75 – 91.

[54] Fravel, D.R.; Deahl, K.L.; Stommel, J.R. (2005). Compatibility of the biocontrol fungus *Fusarium oxysporum* strain CS-20 with selected fungicides. *Biol. Contr.*, 34: 165 – 169.

[55] Fravel, D.R. (2005). Commercialization and implementation of biocontrol. *Annu. Rev. Phytopathol,*. 43: 337 – 359.

[56] Fridlender, M.; Inbar, J.; Chet, I. (1993). Biological control of soilborne plant pathogens by a β-1,3 glucanase-producing *Pseudomonas cepacia*. *Soil Biol. Biochem.*, 25: 1211 – 1221.

[57] Fry, W.E.; Goodwin, S.R.; Matuszak, J.M.; Spielman, L.J.; Milgroom, M.G. (1992). Population genetics and intercontinental migrations of *Phytophthora infestans*. *Annu. Rev. Phytopathol.,* 30: 107 – 129.

[58] Fukui, R;, Schroth, M.N.; Hendson, M.; Hancock, J.G. (1994). Interaction between strains of *Pseudomonas* in sugar beet spermospheres and their relationship to pericarp colonization by *Pythium ultimum* in soil. *Phytopathol.,* 84: 1322 – 1330.

[59] Fukui, R.; Fukui, H.; Alvarez, A.M. (1999). Comparisions of single versus multiple bacterial species on biological control of anthurium blight. *Phytopathol.,* 89: 366 – 373.

[60] Fukui, R.; Fukui, H.; Alwarez, A.M. (1999). Suppression of bacterial blight by a bacterial community isolated from the guttation fluids of anthuriums. *Appl. Environ. Microbiol.,* 65: 1020 – 1028.

[61] Garibaldi, A.; Gullino, M.L.; Aloi, C. (1990). Biological control of *Fusarium* wilt of carnation. Proceedings of the Brighton Crop Protection Conference: Pest and diseases. 19 – 22 November 1990. Brighton, *Great Britan,* 3A-1: 89 – 95.

[62] Glick, B.R. (1994). The enhancement of plant growth by free-living bacteria. *Can. J. Microbiol.,* 41: 109 – 117.

[63] Glick, B.R. & Bashan Y. (1997). Genetic manipulation of plant growth-promoting bacteria to enhance biocontrol of phytopathogens. *Biotechnol. Adv.,* 15: 353-378.

[64] Gonzales, L.E. & Bashan Y. (2000). Increased growth of the microalga *Chlorella vulgaris* when coimmobilized and cocultured in alginate beads with the plant-growth-promoting bacterium *Azospirillum brasilense*. *Appl. Environ. Microbiol.,* 66: 1527 – 1531.

[65] Gottlieb, M. (2002). Factors affecting the ability to colonisation of root system by bacteria from the genus *Pseudomonas*. *Post. Microbiol.,* 41: 277 – 297 [in Polish, summary in English].

[66] Guestky, R.; Shtienberg, D.; Elad, Y.; Dinoo,r A. (2001). Combining biocontrol agents to reduce the variability of biological control. *Phytopathol.,* 91: 621 – 627.

[67] Guetsky, R.; Shtienberg, D.; Elad, Y.; Fischer, E.; Dinoor, A. (2002). Improving biological control by combining biocontrol agents each with several mechanisms of disease suppression. *Phytopathol.,* 92: 976 – 985.

[68] Gurusiddaiah, S.; Weller, D.M.; Sakar, A.; Cook, R.J. (1986). Characterization of an antibiotic produced by a strain of *Pseudomonas fluorescens* inhibitory to *Gaeumannomyces graminis* var. *tritici* and *Pythium* spp. Antimicrob. *Agents Chemother.,* 29: 488 – 495.

[69] Gutterson, N.; Howie, W.; Suslow, T. (1990). Enhancing effects of biocontrol agents by use of biotechnology. In: R.R. Baker & P.E. Dunn (Eds.), *New Directions in Biological Control: Alternatives for Suppressing Agricultural Pests and Diseases.* Alan R. Liss, New York: 749 – 765.

[70] Haas, D. & Keel, C. (2003). Regulation of antibiotic production in root-colonizing *Pseudomonas* spp. and relevance for biological control of plant disease. *Annu. Rev. Phytopathol.,* 41: 117 – 153.

[71] Handelsman, J. & Stabb, E.V. (1996). Biocontrol of soilborne plant pathogens. *Plant Cell,* 8: 1855 – 1869.

[72] Harman, G.E. & Kubicek, C.P. (1998). *Trichoderma* and *Gliocladium*, Vol. 2: *Enzymes, biological control and commercial applications.* Taylor & Francis Ltd, London: 129 – 204.

[73] Harman, G.E. (2000). Myths and dogmas of biocontrol. Changes in perceptions derived form research on *Trichoderma harzianum* T-22. *Plant Dis.*, 84: 377 – 393.

[74] Harman, G.E.; Howell, C.R.; Viterbo, A.; Chet, I.; Lorito, M. (2004). *Trichoderma* species – opportunistic, avirulent plant symbionts. *Nature Rev./Microbiol.*, 2: 43 – 56 (http://www.nature.com/reviews/micro).

[75] Hervás, A.; Landa, B.; Datnoff, L.E.; Jiménez-Díaz, R.M. (1998). Efects of commercial and indigenous microorganisms on *Fusarium* wilt development in chickpea. *Biol. Contr.*, 13: 166 – 176.

[76] Hjeljord, L. & Tronsmo A. (1998). *Trichoderma* and *Gliocladium* in biological control: an overview. In: G.E. Harman & C.P. Kubicek (Eds.), *Trichoderma* and *Gliocladium*. Vol. 2. *Enzymes, Biological Control and Commercial Applications.* Taylor & Francis Ltd., London: 131 – 151.

[77] Hoitink, H.A.J.; Inbar, Y.; Boehm, M.J. (1991). Status of compost-amended potting mixes naturally suppressive to soilborne diseases of floricultural crops. *Plant Dis.*, 75: 869 – 873.

[78] Hoitink, H.A.J.; Stone, A.G.; Grebus, M.E. (1995). Suppression of plant diseases by composts. In: M. De Bertoldi; P. Sequi; B. Lemmes; T. Papi (Eds.), *The Science of Composting.* Blackie Academic & Professional, London: 373 – 381.

[79] Hoitink, H.A.J. & Boehm, M.J. (1999). Biocontrol within the context of soil microbial communities: a substrate-dependent phenomenon. *Annu. Rev. Phytopathol.*, 37: 427 – 446.

[80] Holguin, G. & Bashan, Y. (1996). Nitrogen-fixation by *Azospirillum brasilense* Cd is promoted when co-cultured with a mangrove rhizosphere bacterium (*Staphylocccus* sp.) *Soil Biol. Biochem.*, 28: 1651 – 1660.

[81] Howell, C.R. (2003). Mechanisms employed by *Trichoderma* species in the biological control of plant diseases: the history and evolution of current concepts. *Plant Dis.*, 87: 4 – 10.

[82] Howie, W.J. & Suslow, T.V. (1991). Role of antibiotic biosynthesis in the inhibition of *Pythium ultimum* in the cotton spermosphere and rhizosphere by *Pseudomonas fluorescens*. *Mol. Plant-Microbe Interact.*, 4: 393 – 399.

[83] Huang C.C., Ano T., Shoda M., 1993. Nucleotide sequence and characteristics of the gene, *lpa*-14 responsible for biosynthesis of the lipopeptide antibiotics iturin A and surfactin from *Bacillus subtilis* RB14. *Journal of Fermentation and Bioengineering* 76: 445 – 450.

[84] Hubbard, J.P.; Harman, G.E.; Hadar, Y. (1983). Effect of soilborne *Pseudomonas* spp. on the biological control agent, *Trichoderma hamatum*, on pea seeds. *Phytopathol.*, 73: 655 – 659.

[85] Hynes, R.K. & Boyetchenko, S.M. (2006). Research initiatives in the art and science of biopesticide formulations. *Soil Biol. Biochem.*, 38: 845 – 849.

[86] Janisiewicz, W.J. (1988). Biocontrol of postharvest diseases of apples with antagonist mixtures. *Phytopathol.*, 78: 194 –198.

[87] Janisiewicz, W.J. & Bors, B. (1995). Development of a microbial community of bacterial and yeast antagonists to control wound-invading postharvest pathogens of fruits. *Appl. Environ. Microbiol.*, 61: 3261 – 3267.

[88] Janisiewicz, J. (1996). Ecological diversity, niche overlap, and coexistence of antagonists used in developing mixtures for biocontrol of postharvest diseases of apples. *Phytopathol.*, 86: 473 – 479.

[89] Janisiewicz, W.J. & Korsten, L. (2002). Biological control of postharvest diseases of fruits. *Annu. Rev. Phytopathol.*, 40: 411 – 441.

[90] Janzen, R.A.; Rood, S.B.; Dormaar, J.F.; McGill, W.B. (1992). *Azospirillum brasilense* produces gibberellin in pure culture on chemically-defined medium and in co-culture on straw. *Soil Biol. Biochem.*, 24: 1061 – 1064.

[91] Jetiyanon, K. & Kloepper, J.W. (2002). Mixtures of plant growth-promoting rhizobacteria for induction of systemic resistance against multiple plant diseases. *Biol. Contr.*, 24: 285 – 291.

[92] Jones, R.W. & Pettit, R.E. (1987). Variation in sensitivity among anastomosis groups of *Rhizoctonia solani* to the antibiotic gliotoxin. *Plant Dis.*, 71: 34 – 36.

[93] Kabaluk, T. & Gazdik, K. (2004). Directory of microbial pesticides for agriculturals crops in OECD countries 2004, www.agr.gc.ca/env/pdf/cat_e.pdf, internet April 2005.

[94] Khammas, K.M. & Kaiser, P. (1992). Pectin decomposition and associated nitrogen fixation by mixed cultures of *Azospirillum* and *Bacillus* species. *Can. J. Microbiol.*, 38: 794 – 797.

[95] Kiewnick, S.; Jacobsen, B.J.; Braun-Kiewnick, A.; Eckhoff, J.L.A.; Bergman, J.W. (2001). Integrated control of *Rhizoctonia* crown and root rot of sugar beet with fungicides and antagonistic bacteria. *Plant Dis.*, 85: 718 – 722.

[96] Kim, D.S.; Weller, D.M.; Cook, R.J. (1997). Population dynamics of *Bacillus* sp. L324-92R_{12} and *Pseudomonas fluorescens* 2-79RN_{10} in the rhizosphere of wheat. *Phytopathol.*, 87: 559 – 564.

[97] Klein, D. & Eveleigh, D.E. (1998). Ecology of *Trichoderma.* In: G.E. Harman & C.P. Kubicek (Eds.), *Trichoderma* and *Gliocladium,* Vol. 2: *Enzymes, biological control and commercial applications.* Taylor & Francis Ltd, London: 57 – 74.

[98] Kloepper, J.W.; Leong, J.; Teintze, M.; Schroth, M.N. (1980). *Pseudomonas* siderophores: a mechanism explaining disease suppressive soils. *Curr. Microbiol.*, 4: 317 – 320.

[99] Kondoh, M.; Hirai, M.; Shoda, M. (2001). Integrated biological and chemical control of damping-off caused by *Rhizoctonia solani* using *Bacillus subtilis* RB14-C and flutolanil. *J. Biosci. Bioeng.*, 91: 173 – 177.

[100] Krauss, U.; Soberanis, W.; Matthews, P. (1999). The use of antagonist mixtures in biocontrol. In: U. Krauss & H. Prakash (Eds.), *Workshop Manual – Research Methodology for the Biological Control of Plant Diseases with Special Reference to Fungal diseases of Cocoa.* (CATIE, Turrialba, Costa Rica) 28 June – 4 July: http://www.cabi-commodities.org/Acc/ACCrc/PDFFiles/W-BPD/Ch10.pdf

[101] Kwok, O.C.H.; Fahy, P.C.; Hoitink, H.A.J.; Kuter, G.A. (1987). Interactions between bacteria and *Trichoderma hamatum* in suppression of *Rhizoctonia* damping-off in bark compost media. *Phytopathol.*, 77: 1206 – 1212.

[102] Larkin, P.R. & Fravel, D.R. (1998). Efficacy of various fungal and bacterial biocontrol organisms for control of *Fusarium* wilt of tomato. *Plant Dis.,* 82: 1022 – 1028.

[103] Larkin, R.P.; Roberts, D.P.; Grazia-Garza, J.A. (1998). Biological control of fungal diseases. In: D. Hutson & Miyamoto J. (Eds.), *Fungicidal Activity-Chemical and Biological Approaches to Plant Protection.* Wiley, New York, NY: 141 – 191.

[104] Lebsky, V.K.; Gonzales-Bashan, L.E.; Bashan, Y. (2001). Ultrastructure of interaction in alginate beads between the microalga *Chlorella vulgaris* with its natural associative bacterium *Phyllobacterium myrsinacearum* and with plant growth-promoting bacterium *Azospirillum brasilense*. *Can. J. Microbiol.,* 47: 1 – 8.

[105] Leeman, M.; Den Ouden, F.M.; Van Pelt, J.A.; Cornelissen, C.; Matamala-Garros, A.; Bakker, P.A.H.M.; Schippers, B. (1996). Suppression of fusarium wilt of radish by co-inoculation of fluorescent *Pseudomonas* spp. and root-colonizing fungi. *Europ. J. Plant Pathol.,* 102: 21 – 31.

[106] Lemanceau, P.; Bakker, P.A.H.M.; de Kogel, W.J.; Alabouvette, C.; Schippers, B. (1992). Effect of pseudobactin 358 production by *Pseudomonas putida* WCS358 on suppression of fusarium wilt of carnations by nonpathogenic *Fusarium oxysporum* Fo47. *Appl. Environ. Microbiol.,* 58: 2978 – 2982.

[107] Lemanceau, P.; Bakker, P.A.H.M.; de Kogel, W.J.; Alabouvette, C.; Schippers, B. (1993). Antagonistic effect of nonpathogenic *Fusarium oxysporum* Fo47 and pseudobactin 358 upon pathogenic *Fusarium oxysporum* f. sp. *dianthi*. *Appl. Environ. Microbiol.,* 59: 74 – 82.

[108] Lemanceau, P. & Alabouvette, C. (1993). Suppression of fusarium wilts by fluorescent pseudomonads: mechanisms and applications. *Biocontr. Sci. Techn.,* 3: 219 – 234.

[109] Lewis, J.A. & Lumsden, R.D. (1995). Do pathogenic fungi have the potential to inhibit biocontrol fungi? *J. Phytopathol.,* 143: 585 – 588.

[110] Leong, J. (1986). Siderophores: their biochemistry and possible role in biocontrol of plant pathogens. *Annu. Rev. Phytopathol.,* 24: 187 – 208.

[111] Liddel, C.M.; Parke, J.M. (1989). Enhanced colonization of pea taproots by a fluorescent pseudomonad biocontrol agent by water infiltration into soil. *Phytopathol.,* 79: 1327 – 1332.

[112] Liu, Z.L. & Sinclair, J.B. (1992). Population dynamics of *Bacillus megaterium* strain B153-2-2 in the rhizosphere of soybean. *Phytopathol.,* 82: 1297 – 1301.

[113] Liu, Z.L. & Sinclair, J.B. (1993). Colonization of soybean roots by *Bacillus megaterium* B153-2-2. *Soil Biol. Biochem.,* 25: 849 – 855.

[114] Loper, J.E. (1988). Role of fluorescent siderophore peoduction in biological control of *Pythium ultimum* by a *Pseudomonas fluorescens* strain. *Phytopathol.,* 78: 166 – 172.

[115] Lourenco Junior, V.; Maffia, L.A.; da Silva Romeiro, R.; Mizubuti, E.S.G. (2006). Biocontrol of tomato late blight with the combination of epiphytic antagonists and rhizobacteria. *Biol. Contr.,* 38: 331 – 340.

[116] Lucy, M.; Reed, E.; Glick, B.R. (2004). Applications of free living plant growth-promoting rhizobacteria. *Antonie van Leeuwenhoek,* 86: 1 – 25.

[117] Lugtenberg, B.J.J. & Dekkers, L.C. (1999). What makes *Pseudomonas* bacteria rhizosphere competent? *Environ. Microbiol.,* 1: 9 – 13.

[118] Lutz, M.; Feichtinger, G.; Défago, G.; Duffy, B. (2003). Mycotoxigenic *Fusarium* and deoxynivalenol production repress chitinase gene expression in the biocontrol agent *Trichoderma atroviridae*. *Appl. Environ. Microbiol.,* 69: 3077 – 3084.

[119] Mao, W.; Lewis, J.A.; Lumsden, R.D.; Hebbar, P.K. (1998). Biocontrol of selected soilborne diseases of tomato and pepper plants. *Crop Protect.,* 17: 535 – 542.

[120] Mao, W.; Lumsden, R.D.; Lewis, J.A.; Hebbar, P.K. (1998). Seed treatment using preinfiltration and biocontrol agents to reduce damping-off of corn caused by species of *Pythium* and *Fusarium*. *Plant Dis.,* 82: 450 – 455.

[121] Maget-Dana, R. & Peypoux, F. (1994). Iturins, a special class of pore-forming lipopeptides: biological and physiological properties. *Toxicol.,* 87: 151 – 174.

[122] Manjula, K. & Podile, A.R. (2001). Chitin-supplemented formulations improve biocontrol and plant growth promoting efficiency of *Bacillus subtilis* AF 1. *Can. J. Microbiol.,* 47: 618 – 625.

[123] Martin, F.N. (2003). Development of alternative strategies for management of soilborne pathogens currently controlled with methyl bromide. *Annu. Rev. Phytopathol.,* 41: 325 – 350.

[124] Mazzola, M. & Cook R.J. (1991). Effects of fungal root pathogens on the population dynamics of biocontrol strains of fluorescent pseudomonads in the wheat rhizosphere. *Appl. Environ, Microbiol.,* 57: 2171 – 2178.

[125] Mazzola, M.; Fujimoto, D.K.; Cook, R.J. (1994). Different sensitivity of *Gaeumannomyces graminis* populations to antibiotics produced by biocontrol fluorescent pseudomonads. (Abstr.) *Phytopathol.,* 84: 1091.

[126] Mazzola, M.; Fujimoto, D.K.; Tomashow, L.S.; Cook, R.J. (1995). Variation in sensitivity of *Gaeumannomyces graminis* to antibiotics produced by fluorescent *Pseudomonas* spp. and effect on biological control of take-all of wheat. *Appl. Environ. Microbiol.,* 61: 2554 – 2559.

[127] McDonald, B.A.; Miles, A.J.; Nelson, N.R.; Pettway, R.E. (1994). Genetic variability in nuclear DNA in field populations of *Stagonospora nodorum*. Phytopathol., 84: 250 – 255.

[128] Meyer, J.M.; Halle, F.; Hohnadel, D.; Lemanceau, P.; Ratefiarivelo, H. (1987). Siderophores of *Pseudomonas*: biological properties. In: D. Van Der Helm, J. Neilands, & G. Winkelmann (Eds.), *Iron Transport in Microbes, Plants and Animals.* VCH, Weinheim, Germany: 189 – 205.

[129] Milus, E.A. & Rothrock, C.S. (1993). Rhizosphere colonization of wheat by selected soil bacteria over diverse environments. *Can. J. Microbiol.,* 39: 335 – 341.

[130] Minuto, A.; Migheli, Q.; Garibaldi, A. (1995). Evaluation of antagonistic strains of *Fusarium* spp. in the biocontrol and integrated control of *Fusarium* wilt of cyclamen. *Crop Prot.,* 14: 221 – 226.

[131] Molloy, C.; Cheah, L.H.; Koolaard, J.P. (2004). Induced resistance against *Sclerotinia sclerotiorum* in carrots treated with enzymatically hydrolysed chitosan. *Postharv. Biol. Technol.,* 33: 61 – 65.

[132] Monte, E.; Gómez, M.; Guerra, I.; Liobell, A.; Bautista, J. (2003). Increased antifungal activity of *Trichoderma atroviridae* induced by sonolyzed crawfish chitin. In: L.B. Orlikowski & C. Skrzypczak (Eds.), Chitosan and Other Natural Compounds in the

Control of Plant Diseases. *Proceedings of XII Conference of the Section for Biological Control of Plant Diseases,* Polish Phytopathological Society. Skierniewice, Poland, April 3 – 4, 2003: pp. 20.

[133] Murphy, J.F.; Zehnder, G.W.; Schuster, D.J.; Sikora, F.J.; Polston, J.E.; Kloepper, J.W. (2000). Plant growth-promoting rhizobacterial mediated protection in tomato against *Tomato Mottle Virus*. *Plant Dis.*, 84: 779 – 784.

[134] Murphy, J.F.; Reddy, M.S.; Ryu, C.M.; Kloepper, J.W.; Li, R. (2003). Rhizobacteria-mediated growth promotion of tomato leads to protection against *Cucumber mosaic virus*. *Phytopathol.,* 93: 1301 – 1307.

[135] Nemec, S.; Datnoff, L.E.; Strandberg, J. (1996). Efficacy of biocontrol agents in planting mixes to colonize plant roots and control root diseases of vegetables and citrus. *Crop Protect.,* 15: 735 - 742.

[136] Nemtsev, S.V.; Varlamov, V.P.; Ozeretskovskaya, O.L.; Vasyukova N.I.; Skryabin, K.G. (2003). Stimulation of plant growth and induction of potato resistance to diseases by low molecular weight chitosan. In: L.B. Orlikowski & C. Skrzypczak (Eds.), Chitosan and Other Natural Compounds in the Control of Plant Diseases. *Proceedings of XII Conference of the Section for Biological Control of Plant Diseases,* Polish Phytopathological Society. Skierniewice, Poland, April 3 – 4, 2003: pp. 19.

[137] Obagwu, J. & Korsten, L. (2003). Integrated control of citrus green and blue molds using *Bacillus subtilis* in combination with sodium bicarbonate or hot water. *Postharv. Biol. Technol.,* 28: 187 – 194.

[138] Ohno, A.; Ano, T.; Shoda, M. (1992). Production of a lipopeptide antibiotic surfactin with recombinant *Bacillus subtilis*. *Biotech. Lett.,* 14: 1165 – 1168.

[139] Ownley, B.H.; Weller, D.M.; Alldredge, J.R. (1991). Relation of soil chemical and physical factors with suppression of take-all by *Pseudomonas fluorescens* 2-79. In: C. Keel; B. Koller; G. Défago (Eds.), *Plant Growth-Promoting Rhizobacteria-Progress and Prospects.* IOBC/WPRS Bull., 14: 299 – 301.

[140] Pal, K.K. & McSpadden Gardener, B. (2006). Biological control of plant pathogens. The Plant Health Instructor DOI: 10.1094/PHI-A-2006-1117-02: 1 – 25.

[141] Park, C.S.; Paulitz, T.C.; Bake,r R. (1988). Biocontrol of *Fusarium* wilt of cucumber resulting from interactions between *Pseudomonas putida* and nonpathogenic isolates of *Fusarium oxysporum*. *Phytopathol.,* 78: 190 – 194.

[142] Paulitz, T.C. (1990). Biochemical and ecological aspects of competition in biological control. In: R.R. Baker & P.E. Dunn (Eds.), *New Directions in Biological Control: Alternatives for Suppressing Agricultural Pests and Diseases.* New York, Alan R. Liss: 713 – 724.

[143] Paulitz, T.C. & Linderman, R.G. (1989). Interactions between fluorescet pseudomonads and VA mycorrhizal fungi. *New Phytol.,* 113: 37 – 45.

[144] Paulitz, T.C.; Ahmad, J.S.; Baker, R. (1990). Integration of *Pythium nuun* and *Trichoderma harzianum* isolate T-95 for the biological control of *Pythium* damping-off of cucumber. *Plant Soil,* 121: 243 – 250.

[145] Paulitz, T.C. & Bélanger, R.R. (2001). Biological control in greenhouse systems. *Annu. Rev. Phytopathol.,* 39: 103 – 133.

[146] Pierson, E.A. & Weller, D.M. (1994). Use of mixtures of fluorescent pseudomonads to suppress take-all and improve the growth of wheat. *Phytopathol.,* 84: 940 – 947.

[147] Podile, A.R. & Prakash A.P. (1996). Lysis and biological control of *Aspergillus niger* by *Bacillus subtilis* AF1. *Can. J. Microbiol,.* 42: 533 – 538.

[148] Prévost, K.; Couture, G.; Shipley, B.; Brzezinsky, R.; Beaulieu, C. (2006). Effect of chitosan and a biocontrol streptomycete on field and potato tuber bacterial communities. *BioControl,* 51: 533 – 546.

[149] Probanza, A.; Mateos, J.L.; Lucas Garcia, J.A.; Ramos, B.; de Felipe, M.R. (2001). Effects of inoculation with PGPR *Bacillus* and *Pisolithus tinctorius* on *Pinus pinea* L. growth, bacterial rhizosphere colonization, and mycorrhizal infection. *Microb. Ecol.,* 41: 140 – 148.

[150] Raaijmakers, J.M.; van der Sluis, I.; Koster, M.; Bakker, P.A.H.M.; Weisbeek, P.J.; Schippers, B. (1995). Utilization of heterologous siderophores and rhizosphere comperence of fluorescent *Pseudomonas* spp. *Can. J. Microbiol.,* 41: 126 – 135.

[151] Raaijmakers, J.M.; Vlami, M.; de Souza, J.T. (2002). Antibiotic production by bacterial biocontrol agents. Antonie van Leeuwenhoek Int. *J. Gen. Mol. Microbiol.,* 81: 537 – 547.

[152] Rapauh, G.S. & Kloepper, J.W. (1998). Mixtures of plant growth-promoting rhizobacteria enhance biological control of multiple cucumber pathogens. *Phytopathol.,* 88: 1158 – 1164.

[153] Rapauch, G.S. & Kloepper, J.W. (2000). Biocontrol of cucumber diseases in the field by plant growth-promoting rhizobacteria with and without methyl bromide fumigation. *Plant Dis.,* 84: 1073 – 1075

[154] Roberts, D.P.; Lohrke, S.M.; Meyer, S.L.F.; Buyer, J.F.; Bowers, J.H.; Baker, C.J.; Li, W.; de Souza, J.T.; Lewis, J.A.; Chung, S. (2005). Biocontrol agents applied individually and in combination for suppression of soilborne diseases of cucumber. *Crop Protect.,* 24: 141 – 155.

[155] Rojas, A.; Holguin, G.; Glick, B.R.; Bashan, Y. (2001). Synergism between *Phyllobacterium* sp. (N_2-fixer) and *Bacillus licheniformis* (P-solubilizer), both from semiarid mangrove rhizosphere. *FEMS Microbiol. Ecol.,* 35: 181 – 187.

[156] Rudresh, D.L.; Shivaprakash, M.K.; Prasad, R.D. (2005). Effect of combined application of *Rhizobium*, phosphate solubilizing bacterium and *Trichoderma* spp. on growth, nutrient uptake and yield of chickpea (*Cicer aritenium* L.). *Appl. Soil. Ecol.,* 28: 139 – 146.

[157] Scher, F.M. & Baker, R. (1982). Effect of *Pseudomonas putida* and synthetic iron chelator on induction of soil suppressiveness to *Fusarium* wilt pathogens. *Phytopathol.,* 72: 1567 – 1573.

[158] Schisler, D.A.; Slininger, P.J.; Bothast, R.J. (1997). Effect of antagonist cell concentration and two-strain mixtures on biological control of fusarium dry rot of potatoes. *Phytopathol.,* 87: 177 – 183.

[159] Schnider-Keel, U.; Seematter, A.; Maurhofer, M.; Blumer, C.; Duffy, B.; Gigot-Bonnefoy, C.; Reimmann, C.; Notz, R.; Défago, G.; Haas, D.; Keel, C. (2000). Autoinduction of 2,4-diacetylphloroglucinol biosynthesis in the biocontrol agent

Pseudomonas fluorescens CHA0 and repression by the bacterial metabolites salicylate and pyoluteorin. *J.Bacteriol.*, 182: 1215 – 1225.

[160] Shoda, M. (2000). Bacterial control of plant diseases. J. Biosc. Bioeng., 89: 515 – 521.

[161] Simon, A. (1989). Biological control of take-all of wheat by *Trichoderma koningii* under controlled environmental conditions. *Soil Biol. Biochem.*, 21: 323 – 326.

[162] Simon, A.; Sivasithamparam, K. (1988). Interactions among *Gaeumannomyces graminis* var. *tritici, Trichoderma koningii,* and soil bacteria. *Can. J. Microbiol.*, 34: 871 – 876.

[163] Simon, A.; Sivasithamparam, K.; MacNish, G.C. (1988). Effect of application to soil of nitrogenous fertilizers and lime on biological suppression. *Trans. Br. Mycol. Soc.*, 91: 287 – 294.

[164] Sindhu, S.S.; Suneja, S.; Goel, A.K.; Parmar, N.; Dadarwal, K.R. (2002). Plant growth promoting effects of *Pseudomonas* sp. on coinoculation with *Mesorhizobium* sp. *cicer* strain under sterile and "wilt sick" soil conditions. *Appl. Soil Ecol.*, 19: 57 – 64.

[165] Sivasithamparam, K. & Parker, C.A. (1978). Effects of certain isolates of bacteria and actinomycetes on *Gaeumannomyces graminis* var. *tritici* and take-all of wheat. *Austr. J. Bot.*, 26: 773 – 782.

[166] Slininger, P.J. & Shea-Wilbur, M.A. (1995). Liquid-culture pH, tempareture, and carbon (not nitrogen) source regulate phenazine productivity of the take-all biocontrol agents *Pseudomonas fluorescens* 2-79. *Appl. Microbiol. Biotechnol.*, 37: 388 – 392.

[167] Smith, K.P. & Goodman R.M. (1999). Host variation for interactions with beneficial plant-associated microbes. *Annu. Rev. Phytopathol.*, 37: 473 – 491.

[168] Smolińska, U. & Dyki, B. (2002). Viability and micromorphology of *Sclerotium cepivorum* sclerotia in field soil after addition of *Brassica juncea* and *Brassica napus* plant residues. *Phytopathol. Pol.*, 23: 39 – 51.

[169] Smolińska, U.; Dyki, B.; Kwaśna, H. (2002). Activity of fungi towards sclerotia of *Sclerotium cepivorum* as influenced by cruciferous plant residues. *Phytopathol. Pol.*, 24: 5 – 16.

[170] Spadaro, D.; Garibaldi, A.; Gullino, M.L. (2004). Control of *Penicillium expansum* and *Botrytis cinerea* on apple combining a biocontrol agent with hot water dipping and acibenzolar-S-methyl, baking soda, or ethanol application. *Postharv. Biol. Technol.*, 33: 141 – 151.

[171] Spadaro, D. & Gullino, M.L. (2005). Improving efficacy of biocontrol agents against soilborne pathogens. *Crop Protect.*, 24: 601 – 613.

[172] Spotts, R.A. & Cervantes, L.A. (1986). Populations, pathogenicity, and benomyl resistance of *Botrytis* spp., *Penicillium* spp., and *Mucor piriformis* in packing houses. *Plant Dis.*, 70: 106 – 108.

[173] Sung, K.C. & Chung, Y.R. (1997). Enhanced suppression of rice sheath blight using combination of bacteria which produce chtinases or antibiotics. In: A. Ogoshi; K. Kobayashi; Y. Homma; F. Kodama; N. Kando; S. Akino (Eds.), *Plant Growth-Promoting Rhizobacteria Present Status and Future Prospects.* Nakanishi Printing, Sapporo, Japan: 370 – 372.

[174] Szczech, M. & Shoda, M. (2004). Biocontrol of *Rhizoctonia* damping-off of tomato by *Bacillus subtilis* combined with *Burkholderia cepacia*. *J. Phytopathol.*, 152: 549 – 556.

[175] Szczech, M. & Shoda, M. (2006). The effect of mode of application of *Bacillus subtilis* RB14-C on its efficacy as a biocontrol agent against *Rhizoctonia solani*. *J. Phytopathol.*, 154: 370 – 377.

[176] Tian, S.P.; Fan, Q.; Xu, Y.; Jiang, A.L. (2002). Effects of calcium on biocontrol activity of yeast antagonists against the postharvest fungal pathogen *Rhizopus stolonifer*. *Plant Pathol.*, 51: 352 – 358.

[177] Tuitert, G.; Szczech, M.; Bollen, G.J. (1998. Suppression of *Rhizoctonia solani* in potting mixtures amended with compost mede from organic household waste. *Phytopathol.*, 88: 764 - 773.

[178] Van Dyke, M.I. & Prosser, J.I. (2000). Enhanced survival od *Pseudomonas fluorescens* in soil following establishment of inoculum in a sterile soil carrier. *Soil Biol. Biochem.*, 32: 1377 – 1382.

[179] Van Eck, W.H. (1978). Chemistry of cell walls of *Fusarium solani* and the resistance of spores to microbial lysis. *Soil Biol. Biochem.*, 10: 155 – 157.

[180] Van Elsas, J.D.; Dijkstra, A.F.; Govaert, J.M.; van Veen, J.A. (1986). Survival of *Pseudomonas fluorescens* and *Bacillus subtilis* introduced into two soils of different texture in field microplots. *FEMS Microbiol. Ecol.*, 38: 151 – 160.

[181] Van Elsas, J.D. & Heijnen, C.E. (1990). Methods for the introduction of bacteria into soil: a review. *Biol. Fertil. Soils,* 10: 127 – 133.

[182] Van Elsas, J.D.; Trevors, J.T.; Jain, D.; Wolters, A.C.; Heijnen, C.E.; van Overbeek, L.S. (1992). Survival of, and root colonization by, alginate-encapsuled *Pseudomonas fluorescens* cells following introduction into soil. *Biol. Fertil. Soils,* 14: 14 – 22.

[183] Van Loon, L.C.; Bakker, P.A.H.M.; Pietrese, C.M.J. (1998). Systemic resistance induced by rhizosphere bacteria. *Annu. Rev. Phytopathol.*, 36: 453 – 483.

[184] Van Veen, J.A.; van Overbeek, L.S.; van Elsas, D.J. (1997). Fate and activity of microorganisms introduced into soil. *Microbiol. Mol. Biol. Rev.*, 61: 121 – 135.

[185] Van Zyl, F.G.H.; Strijdom, B.W.; Staphorst, J.L. (1986). Susceptibility of *Agrobacterium tumefaciens* strains to two agrocin-producing *Agrobacterium* strains. *Appl. Environ. Microbiol.*, 52: 234 – 238.

[186] Vestberg, M.; Kukkonen, S.; Saari, K.; Parikka, P.; Huttunen, J.; Tainio, L.; Devos, N.; Weekers, F.; Kevers, C.; Thonart, P.; Lemoine, M.-C.; Cordier, C.; Alabouvette, C.; Gianinazzi, S. (2004). Microbial inoculation for improving the growth and health of micropropagated strawberry. *Appl. Soil Ecol.*, 27: 243 – 258.

[187] Weller, D.M. (1988). Biological control of soilborne plant pathogens in the rhizosphere with bacteria. *Annu. Rev. Phytopathol.*,26: 379 – 407.

[188] Weller, D.M. & Cook, R.J. (1983). Suppression of take-all of wheat by seed treatments with fluorescent pseudomonads. *Phytopathol.*, 73: 463 – 469.

[189] Weller, D.; Raaijmakers, J.M.; McSpadden Gardener, B.B.; Tomashow, L.S. (2002). Microbial populations responsible for specific soil suppressiveness to plant pathogens. Annu. Rev. *Phytopathol.*, 40: 309 – 48.

[190] Weller, D.M. (2007). *Pseudomonas* biocontrol agents of soilborne pathogens: looking back over 30 years. *Phytopathol.*, 97: 250 – 256.

[191] Whipps, J.M. (1997). Developments in the biological control for soilborne plant pathogens. *Adv. Bot. Res.*, 26: 1 – 134.

[192] Whipps, J.M. (2001). Microbial interactions and biocontrol in the rhizosphere. *J. Exper. Bot.,* 52: 487 – 511.

[193] Woo, S.; Fogliano, V.; Scala, F.; Lorito, M. (2002). Synergism between fungal enzymes and bacterial antibiotics may enhance biocontrol. *Antonie van Leeuwenhoek,* 81: 353 – 356.

[194] Wulff, E.G.; Mguni, C.M.; Mortensen, C.N.; Keswani, C.L.; Hockenhull, J. (2002). Biological control of black rot (*Xanthomonas campestris* pv. *campestris*) of brassicas with an antagonistic strain of *Bacillus subtilis* in Zimbabwe. Eur. *J.Plant Pathol.,* 108: 317 – 325.

[195] Xu, G.W.; Gross, D.C. (1986). Selection of fluorescent pseudomonads antagonistic to *Ervinia carotovora* and suppressive of potato seed piece decay. *Phytopathol.,* 76: 414 – 422.

[196] Yu, G.Y.; Sinclair, J.B.; Hartman, G.L.; Bertagnolli, B.L. (2002). Production of iturin A by *Bacillus amyloliquefaciens* suppressing *Rhizoctonia solani. Soil Biol. Biochem.,* 34: 955 – 963.

[197] Zablotowicz, R.M.; Press, C.M.; Lyng, N.; Brown, G.L.; Kloepper, J.W. (1992). Compatibility of plant growth promoting rhizobacterial strains with agrochemicals applied to seeds. *Can. J. Microbiol.,* 38: 45 – 50.

[198] Zheng, X.Y. & Sinclair, J.B. (2000). The effects of traits of *Bacillus megaterium* on seed and root colonization and their corelation with the suppression of Rhizoctonia root rot of soybean. *BioControl,* 45: 223 – 243.

In: Progress in Environmental Microbiology
Editor: Myung-Bo Kim, pp. 111-149
ISBN: 978-60021-940-5

Chapter III

Heavy Metals and Microorganisms in the Environment: Taking Advantage of Reciprocal Interactions for the Development of a Wastewater Treatment Process

***Diana L. Vullo*[*], *Helena M. Ceretti*,**
Silvana A. M. Ramírez* and *Anita Zalts
Área Química, Instituto de Ciencias, Universidad Nacional de General Sarmiento,
J.M. Gutiérrez 1150, (B1613GSX) Los Polvorines, Buenos Aires, Argentina

Abstract

Anthropic activities have been responsible for the introduction of increasing amounts of heavy metals in the environment. Metal production, leather and tanning processes, gas and electricity production, sewage and waste disposal and related activities, contribute to the presence of copper, cadmium, zinc, lead, chromium and nickel in soil and surface and ground waters if waste products are not properly treated before discharged. Exposition to heavy metals causes irreversible damage to living organisms; their presence above certain limits is a potential risk to the environment and human health. In order to evaluate this risk, total metal concentration is a poor indicator because reactivity, bioavailability and toxicity depend on the distribution of the different metal species in that particular environment. A physicochemical understanding of metal speciation is required.

Microorganisms from different habitats have developed several strategies in order to cope with metal toxicity. Thorough studies on microbes-metal interactions can help to

[*] Tel.: 54-11-4469-7542, Fax: 54-11-4469-7506, Email: dvullo@ungs.edu.ar

understand detoxifying mechanisms that can be applied to wastewater treatment. An important advantage of these innovative metal removal technologies, particularly if they are to be employed in developing countries, is the cost-effectiveness of using autochthonous bacteria, since they may be isolated from local polluted environments.

Buenos Aires Metropolitan Area presents one of the most polluted watersheds in Argentina: the Reconquista River. It receives high amounts of both faecal and industrial wastes without previous treatment, leading to high loads of pathogen microorganisms and metals in sediments, surface and pore waters of Reconquista basin. Autochthonous microorganisms, able to grow in the presence of copper, zinc, cadmium and chromium, were isolated from water and sediment samples taken from this basin, and used in metal biosorption studies under different experimental conditions to improve metal retention. Cadmium has been chosen as model metal because its toxicity limits bacterial growth. Cadmium complexing capacity (CC) of culture media and electroplating effluents was evaluated in terms of total ligand concentration (L_t) and conditional stability constants (K_f'), assuming 1:1 Cd-ligand complexes are formed. In these systems total ligand concentration is in the μM range, far from the typical results obtained for seawater (nM), where most speciation studies were performed. As only moderate strength ligands were detected ($4<\log K_f'<8$), it is expected that the isolated bacteria would compete favorably with the ligands for metal retention.

The aim of this article is to describe how understanding of both, the physicochemistry of aqueous systems and microbial strategies to survive in polluted environments, contribute to a proper design of wastewater treatment process.

1. Introduction

1.1. Sources of Metal Pollution

The geosphere is the original source of all metals, except those that enter the atmosphere from space in the form of meteorites and cosmic dust; it is also their ultimate sink. Life evolved in the presence of metals and most living forms depend on them. The salient feature of metals is that unlike organic compounds they do not degrade in the environment and barely move from one environmental matrix to another. A basic human instinct has always been to transform and tame nature in pursuit of a better lifestyle. Global population growth, coupled with increasing consumption, has led to dramatic changes in material flows due to anthropogenic activities. The study of frozen samples from the Greenland and Antarctic ice caps has given a wealth of fascinating information on the history of the composition of the atmosphere of our planet. Investigation of the changes in the occurrence of metals in dated snow and ice from central Greenland has allowed to document, for example, early large scale pollution of the atmosphere of the northern hemisphere for Pb and Cu two millennia ago and the recent history of metal pollution from the Industrial Revolution to present (van de Velde et al., 1999, and references therein). Nowadays, the increased availability of products, often leading to better standards of living, produces the related material flows which are marked by such problems as high-energy consumption, natural resource depletion and increased disposal. Massive mining operations are designed to handle virgin sources, and capture economies of scale. When comparing virgin and recycled resources, a complete analysis must take into account the economics of both processes. It may be the case that costs of mining and refining the anticipated lower grade future virgin materials may eventually surpass the costs of mining discard streams and repositories. If material scarcity becomes more pressing in the future, discard repositories such as landfills may increasingly be considered "mines" (Johnson et al., 2007).

According to the World Urbanization Prospects (2005), prepared by the United Nations, the world's population is expected to grow from 6.1 billion in 2000 to 8.3 billion in 2030. The statistics indicate that nearly all of this expected growth will take place in urban areas, with almost no growth in the rural population. The urban environment will soon become the most dominant human habitat for the first time in history. Human activities include municipal, industrial, commercial and agricultural operations, releasing a variety of toxic or potentially toxic pollutants to the environment. Within urban environments, human interventions are especially intense, emissions are vastly accelerated, inevitably rendering the urban environment particularly susceptible to environmental degradation and contamination. The situation is more dramatic in less developed regions: in 2005, the urban population in these areas was 2.3 billion people, about 7 times larger than in 1950. Over the next 25 years, this urban population is projected to continue to increase fast, reaching 3.9 billion people by 2030. In advanced industrialized countries there has been a substantial reduction in the rate of discharge and hence in ambient concentrations of trace metals. By contrast, the emissions of metals in developing countries continue to rise sharply with the growth of industrial activity and transportation infrastructures (Nriagu, 1997). So, as part of the world is dealing with incentives, information, and research designed to prevent instead of merely control pollution, in other places, the basic foundation of environmental regulatory controls has not been firmly established. A dense population and a high level of productivity, primarily driven by non-agricultural activities as industrial processes, gas and electricity production and distribution, building and road construction (and corrosion of the employed materials), traffic-related activities (fossil fuel combustion, wear and tear of vehicular parts, leakage of metal containing motor oils), sewage and waste disposal and related activities (incineration and landfill), contribute to the presence of copper, cadmium, zinc, lead, chromium and nickel in the urban environment. In the case of industrial and domestic wastewater, soil, surface and ground waters are highly affected if waste products are not properly treated before discharged. Agriculture and industrial development, metal production, and other activities that take place outside the urban environment are highly related to urban areas, their needs and most of their environmental problems. Industry has traditionally focused on production rather than waste management.

In the case of metals, human interventions take these compounds from their stable and non bioavailable geological matrix into situations of biological accessibility (Moore, 2004). These activities transform metals into persistent environmental contaminants, since they can neither be destroyed nor degraded. Elevated emissions and their deposition over time can lead to anomalous enrichment, causing metal contamination of the surface environment. The prolonged presence of the contaminants in the urban environment, particularly in urban soils, and their close proximity to the human population can significantly amplify the exposure of the urban population to metals via inhalation, ingestion, and dermal contact (Wong et al.,

2006). Trace metal contamination of the urban environment include the subsequent migration of the pollutants to receiving water bodies via urban runoff, resulting in the trace metal enrichment of sediments. Nowadays metals are among the most important pollutants in source and treated water and are becoming a severe public health problem. The quality of aquatic ecosystems is affected and increase in the body loadings of aquatic organisms through bioaccumulation and biomagnification is expected, potentially causing trace metal contamination of the food chain.

The studies of health and environmental problems associated to metals in the 1970s and the 80s were driven by the process of abatement of lead in the environment. Humans have clear and identifiable sources of exposure to lead from fuels, food, and leaded water pipes. The interventions that started at that time have dramatically lowered the human lead exposure. Attention has now shifted to other metals, in particular cadmium (Moore, 2004). Cadmium is mainly extracted from zinc ores, where it is found as a minor constituent. Over one-half of the refined cadmium is used in batteries. Anthropogenic pathways by which Cd enters the environment are through industrial waste from processes such as electroplating, manufacturing of plastics, mining, paint pigments, alloy preparation and batteries that contain this metal. Household appliances, automobiles and trucks, agricultural implements, industrial tools and fasteners of all kind (e.g., nuts, bolts, screws, nails) are commonly Cd coated. Cd is also used for luminescent dials, in photography, rubber curing and as fungicides. Cadmium is an impurity in several high-volume commodities: coal, oil and phosphate fertilizer. Collectively, these commodities account for about one-half as much cadmium as in the product stream. Cadmium in phosphate fertilizer is a particular concern because this metal in food plants is the main exposure route for humans (Thomas et al., 1997). Tobacco concentrates Cd, leading to human exposure to this metal through smoking (Kirkham, 2006, and references therein).

Chromium (III) occurs naturally and is an essential nutrient. Chromium (VI), a much more dangerous species, rarely occurs naturally, but is usually released from anthropogenic sources due to the extensive use of Cr in many industries such as electroplating, steel production, wood preservation and leather tanning (Demir et al., 2007).

1.2. Metal Bioavailability, Speciation and Complexing Capacity in Aquatic Environments

The United Kingdom's Contaminated Land Regulations under Part IIA of the Environmental Protection Act of 1990 defines *contaminated land* as "land that appears to the local authority to be in such a condition, by reason of substances in, on, or under the land, that significant harm is being caused, or there is significant possibility that harm is being caused". Thus, the presence of substances of concern is not sufficient to describe a contaminated site; harmful interaction with a receptor must be a possibility. Because toxic effects require that an organism takes up the contaminant, the extent to which substances are bound to soil particles or are available to cause harm needs to be considered.

Not all of the different chemical species in the environment are biologically available. Many definitions are given in the literature to describe what *bioavailability* is, depending greatly from the author's approach (from the environmental, pharmacological, biological or toxicological point of view, for example) and the targets to whom harm may be caused. On the other hand, definitions of bioavailability are related to each other: the amount present as free metal ion (Brand et al., 1986; Bergman et al., 1983; Anderson et al., 1978; Campbell, 1995), the amount that is taken up (Turner, 1986; Campbell, 1995) or the amount that is available for a biological action (Rand, 1995). According to Morel (1983), trace metal interactions with an organism can be understood by considering it as an assemble of potential ligands, and the interaction between these ligands and a particular metal species is reflected in a biological effect. Semple at al (2004) proposed to use the literal definition of available (capable of being employed, at one's disposal, at hand) to define what a bioavailable compound is (that which is freely available to cross an organism's cellular membrane from the medium the organism inhabits at a given time). The same is done for the term accessible (capable of being approached or reached, approachable, attainable) to define a bioaccessible compound as that which is available to cross an organism's cellular membrane from the environment, if the organism has access to the chemical. Bioaccessibility encompass what is actually bioavailable now, plus what is potentially bioavailable. This creates the need for a better understanding of the existing chemical species.

From a physiological point of view, metals can be classified in three main groups: a) essential and basically non-toxic, b) essential, but harmful at higher concentrations and c) toxic. This classification of metals can be linked to their Lewis acidity, a good tool to predict both metallic ion preferred ligands and their general trend in the properties of metal complexes. The covalent index X_m^2r, where X_m is the metal ion electronegativity and r is its radius, describes the ability of the metal to accept electrons from a donor ligand. Elements as calcium and potassium, which serve as macronutrients for microorganisms, plants and animals, present a low covalent index (class A in Nieboer and Richardson classification) and are non-toxic. Most biological micronutrients including manganese, copper and zinc, are found in the borderline group: they form stable complexes with a variety of electron donor atoms including oxygen, nitrogen, and sulfur. These metals are important to sustain life, but in minute amounts. At higher concentrations they are toxic. Type B metals include several that are known to be toxic to organisms because they have a strong affinity for sulfur donor atoms, form more stable complexes with oxygen-donating compounds and more stable methylated derivatives than do borderline or A type metals (vanLoon et al., 2000). It must be recognized that the classification refers in each case to a specific ion, so that in cases where the metal may exist in more than one oxidation state, each ionic form must be treated differently (Duffus, 2002).

In natural waters, the bioavailability of trace metals, including their toxicity, is related to their ability to cross biological barriers. Trace metals are present in water as different chemical species, depending on their concentration and on the physicochemical characteristics of the aquatic environment. To be bioavailable, mass transport is the first requisite: metal ions and their complexes must travel from the external media (bulk of the solution, sediments, minerals) to the biological surface. Metal complexes are frequently dynamic, able to dissociate and reassociate in the time that it takes to diffuse. If the metals form complexes that are strong enough (with active sites in the surface of sediments or with dissolved species), they may not be bioavailable and will not cause harm, at least temporally.

To have an effect, the metal must react with a sensitive site at the cells surface. Since for several trace metals in natural waters, complexes may account for a great proportion of total metal concentration, the identification of metal complexes and their ability to contribute fully, partially or not at all to biological internalization is crucial (Worms et al., 2006). For a given compound in a specific environment, distinction between total, bioaccessible and bioavailable concentration is critical. For quite a long time, the effect of a metal on the biota was usually related to the total metal concentration. More recent environmental studies demonstrated that toxicity is not directly correlated to total metal content. For some trace metals like copper, lead, nickel and zinc, the speciation is dominated by the presence of organic matter (Donat et al., 1994; Capodaglio et al., 1990; Kozelka et al., 1998; Martino et al., 2003; Morel, 1983; Davey et al., 1973).

Speciation and bioavailability are keywords in the relation between the total metal content in a system and the effect on the biota. *Speciation* can be defined as the distribution of a metal over all possible chemical forms, the species, in a particular sample or matrix (Templeton et al., 2000). This distribution is the result of a number of processes such as adsorption, complexation and precipitation, all of them depending on the nature of the metal and the environmental conditions (pH, salt content, organic matter, redox potential, etc.).

How can we determine the concentration of a certain free metal or metal complex? One possibility is to make calculations, provided the total metal concentration, ligand concentration and the stability constant (K_f) of the complex are known. However, calculation of equilibrium composition is difficult because of the existence of a variety of natural and/or anthropogenic ligands in aqueous systems; also the discrimination of free metal ion from the metal associated with a certain ligand is difficult.

Under these circumstances, the term *complexing capacity* is introduced (Davey et al., 1973; Shuman et al., 1973; Kunkel et al., 1973) and defined as "the total concentration of ligands that form complexes with the metal ion in the sample". If we analyze the definition of complexing capacity through the knowledge on bioavailability and toxicity that has been developed, the expression "form complexes" is not specific enough: an idea of the strength of these complexes is needed.

A variety of methodologies has been proposed to address speciation studies, including electrochemical and non electrochemical techniques (Buffle, 1988; Mota et al., 1996). These techniques point to an instrumental evaluation of metal speciation, rather than to a biological point of view. Some of them directly measure metal ion activity; others involve a preliminary separation step to select specific forms of the element.

In the first group of methodologies, *ion selective electrodes* (ISE) have been explored in the measurement of copper. First results showed a detection limit higher than the environmental Cu(II) concentration found in most natural waters, limiting its applicability. A serious disadvantage in the case of seawater samples is the interference by chloride ion when present at concentrations greater than 0.1M (Belli et al., 1993). All these limitations have been overcome (Zirino et al., 1998).

Different strategies have been developed for *fractioning*, where metal ions are separated from the matrix solution by means of a competitive equilibrium established at a solid phase. The complexing/sorbing properties of the solid phase are crucial and determine which form or forms will be preconcentrated. Van den Berg (1982[a]) proposed the use of MnO_2 as solid phase to adsorb copper. In this case copper ions from weak inorganic complexes are adsorbed to the solid, while the organic complexes are considered to remain in the dissolved fraction. Copper concentration in any fraction may be determined by any standard method for the measurement of total copper (AAS Atomic Absorption Spectrometry, ICP-MS Inductively Coupled Plasma Mass Spectrometry, IC Ion Exchange Chromatography, etc.), provided it is sensitive enough.

Ligand/resin titrations are variations of this procedure. Solid MnO_2 is replaced by a competing analytical reagent, an ion exchanger or complexing resin. These materials can be synthesized to meet different sorbing properties allowing to "tune" which complexes are stable enough to remain in the solution and which will decompose. The feasibility of the determination was demonstrated in the case of model solutions of known composition (Pesavento et al., 1995; 2000[a]) and also in natural water samples (Pesavento et al., 2000[b] Itabashi et al., 2000; Lohan et al., 2005; Manouchehri et al., 2006). *Solid phase extraction* (SPE) is another variation which involves the adsorption of organic ligand - metal complexes onto C_{18} SepPak cartridges (Hanson et al., 1993; Mills et al., 1981). This method proved to be effective when calibrated against the addition of EDTA as a competing ligand (Sunda et al., 1987), but may underestimate the extent of organically complexed copper in oceanic surface waters (Donat et al., 1986).

Diffusive gradients in thin films (DGT) combines the use of resins with a layer of a hydrogel. The principle of this procedure is that metal ions would bind to the resin after diffusion through a well defined layer of gel. Metal complexes that are weak enough to dissociate in the diffusion zone in the hydrogel reach the resin and will be measured. *In situ* measurements are possible, opening the possibility of obtaining information of natural or polluted systems (Scally et al., 2003).

Due to their high sensitivity and ability to distinguish between the different forms of a dissolved metal, *electroanalytical methods* are widely used to address metal ion speciation studies. These have been applied to different aquatic environments: model solutions (Town et al., 2002; van Leeuwen et al., 2002; Rodriguez Presa et al., 1998; Raspor et al., 1973, 1975), seawater (Donat et al., 1992, 1994; Rue et al., 1995; van den Berg, 1982[b]; van den Berg et al., 1992; Nuester et al., 2005; Tuschall et al., 1981; van Veen et al., 2001, 2002[a,b]; Bruland et al., 1992; Jurado-Gonzalez et al., 2003; Donat et al., 1994; Campos et al., 1994), freshwater (Xue et al., 1999, 1998; Gardner et al., 2000; Meylan et al., 2004), pore water (Skrabal et al., 2000; Adams et al., 1999), wastewater (Van Veen et al., 2002; Alonso Alvarez et al., 1998) and also to a number of biological applications (Guéguen et al., 2003; Gonzalez-Gil et al.; 2003, Ceretti et al., 2006).

Basically, two electroanalytical methodologies are mentioned in the literature for determination of total metal content: *anodic* (ASV) and *cathodic stripping voltammetry* (CSV). Based on CSV, a strategy using *competitive ligand equilibration/cathodic stripping voltammetry* (CLE/CSV) is used in speciation studies. ASV and CLE/CSV are among the best methods for determining the complexation characteristics and speciation of dissolved trace metals in natural waters. A common feature to both strategies is that low detection limits can be achieved because of a first preconcentration step, which makes them also suitable for total metal determination in clean environments. Furthermore, since preconcentration occurs *in situ*, sample contamination is kept to a minimum.

In ASV, preconcentration of metal ions is achieved by reduction at a suitable potential for a fixed time. The reduced metal accumulates at the working electrode, usually a mercury drop or a mercury film electrode. In the quantification step, potential is scanned in positive direction to promote oxidation of the reduced metal form. This can be done in different fashions, depending on the required sensivity: *linear scan* (LSASV), *derivative pulse* (DPASV) and *square wave* (SWASV) are widely applied in the literature (Mylon et al., 2003; Town et al., 2002; van Leeuwen et al., 2002; Donat et al., 1994; Cobelo-García et al., 2003, 2004; van den Berg et al., 1992). The resulting current is recorded as a function of potential. Deposition potential and time are two crucial parameters that need optimization in order to concentrate the relevant species.

When dealing with speciation studies, ASV combines selection and quantification (similar to an ISE). In speciation studies an ASV monitored titration is performed assuming that this technique allows the detection of metal complexes that are labile toward reduction on the electrode (M′). A titration curve is obtained measuring peak current after each metal addition. The advantage of this approach is that the whole determination is achieved in one step without a previous fractionation stage. Careful attention should be paid to the deposition potential and time since both should allow discrimination of the labile from the complexed species. The drawback is that if any dissociation of a complex takes place in the diffusion layer (a kinetically labile complex), this leads to an overestimation of the inorganic (labile, M′) fraction. DPASV has been used extensively to study the complexation and speciation of copper (Donat et al., 1994, 1986; Coale et al., 1988, 1990; Bruland et al., 2000; Cobelo-García et al., 2004).

In CSV, preconcentration is achieved by adsorption of an "electrochemically detectable" complex, formed by adding a suitable ligand. There is usually no charge transfer during the preconcentration step, and the adsorbed layer is then reduced as potential is scanned in the negative direction. It is quite common to combine CSV with a differential pulse scan, leading to DPCSV method. Once again, optimization of adsorption potential and time is required.

When CSV is used in speciation studies, a well characterized ligand is added to the sample. The sample is titrated with the metal M and a competing equilibrium is established between the natural ligands and the added ligand (AL) for the metal M. An experimental data point is taken when equilibration is reached after each metal addition (this can take several hours). The concentration of the complex formed between M and the added competing ligand AL is determined by CSV. The resulting concentration of the metal complex M-AL depends on the concentration of AL, and the strength of the interaction between M and AL and also between M and other natural ligands present in the sample.

CLE/DPCSV has been used successfully for the speciation of copper (Donat et al., 1994; Donat et al., 1992; Gardner et al., 2004; Meylan et al., 2004; van Veen et al., 2001), nickel (Donat et al., 1994; Xue et al., 2001), iron (Rue et al., 1995; Wu et al., 1995) with a variety of added ligands like salicylaldoxime (Campos et al., 1994; van den Berg, 1984; van den Berg et al., 1992), 8-hydroxyquinoline (Donat et al., 1994, 1992; van den Berg et al., 1990), tropolone (Donat et al., 1992), dimethylglyoxime (Donat et al., 1994) and benzoylacetone (Moffet, 1995), among others (Xue et al., 2002).

Recently, a new method was proposed to address speciation studies by varying the concentration of a competing ligand throughout the titration (Nuester et al., 2005). It can be described as a *reverse titration* where the concentration of the free metal ion is gradually lowered from its natural level after every addition of the competing ligand. This method is carried out in the electrochemical cell and uses only one sample aliquot. The main advantage is that it may be especially useful to detect low ligand concentrations, because according to the authors′ conclusions, the sensivity is greatest at the lowest free metal ion concentration.

1.3. Metal-Microbe Interactions Useful for Wastewater Treatments

Living organisms inhabit the Earth environment, interacting with its component parts. There is a continuous exchange of matter and energy among the geosphere, hydrosphere, atmosphere and biosphere. The requirement of living systems to both acquire and reject metals has led to the selection of a whole repertoire of mechanisms of interaction which ensure the adaptation of microorganisms to a changing and frequently hostile environment Trace metal bioavailability depends on physical, biological and chemical factors that are highly complex and interdependent. While physicochemical factors may limit trace metal internalization, organisms can adapt resulting in modifications to the rate limiting flux and thus to the chemical availability of metal complexes. Essential trace metals are often highly regulated in order to avoid both situations of trace metal deficiency and overload. Strategies are varied, depending on the organism, the metal and the physicochemical environment. None of the processes are entirely independent since each process can lead to modifications in the different fluxes that control trace metal bioavailability in organisms (Worms et al., 2006). Different organisms exhibit diverse responses to toxic ions, which confer upon them a certain range of metal tolerance (Figure 1). Eukaryotes are more sensitive to metal toxicity than bacteria and their typical mechanism for regulating intracellular metal ion concentrations is the expression of metal chelating proteins (metallothioneins). Bacteria show different responses to deal with metals. Some microorganisms are adapted to grow at high metal concentration producing sulfide, for example, as a secondary outcome of their metabolism that immobilizes metals by precipitation (Valls et al., 2002).

The term *biosorption* is frequently used to describe the uptake or binding of metals (often reversibly) to cellular components (Figure 2). In many cases biosorption is considered a passive process of metal uptake (Volesky, 2001); this definition will be used in the present work. Although, for other authors (Johnson, 2006) metal incorporation can be mediated either by active metabolic processes or via a physicochemical mechanism such as ion-exchange, adsorption or entrapment. The use of living material and biopolymers has been recently incorporated into the concept of biosorption. Biosorbents are natural ion-exchange materials that contain weakly acidic and basic groups, the chelation process being unspecific. Biomass capabilities to immobilize metals depend on the type of biomass, the chemical composition of the solution and the physicochemical process environment. After loading, the metals can be stripped from the matrix and the system may be regenerated for further sorption-desorption cycles. Comparing with synthetic ion-exchange resins, biosorption methods seem to be more effective in removing dissolved metals at low concentrations (below 2-10 mg/L). The higher specificity of biosorbents avoids the problem of resin overload, particularly when high concentrations of alkaline-earth metals are present in wastewater. Finally, biological systems offer the potential of genetic modification to further increase the specificity towards certain metal ions.

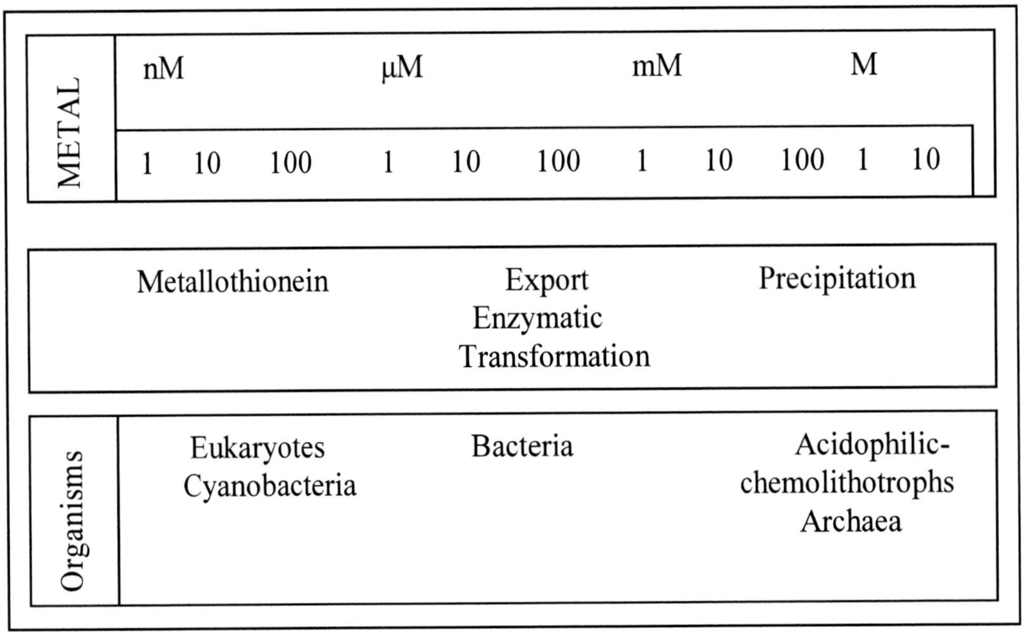

Figure 1. Bacterial capacities and mechanisms of tolerance to toxic metals such as Cd(II) or Ni(II); metal concentration range is only indicative since metal resistance varies within taxonomic groups and the metal considered (adapted from Valls et al., 2002).

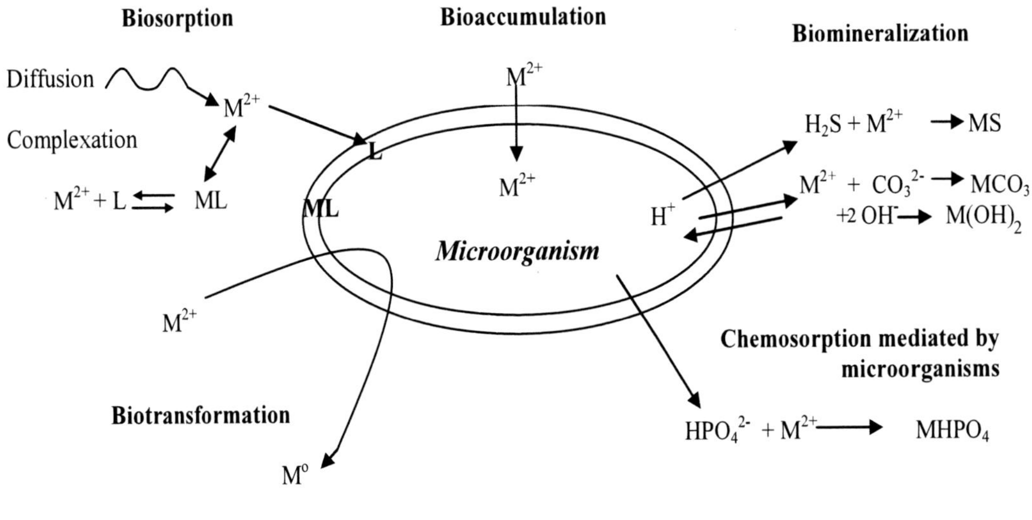

Figure 2. Conceptual model of some important metal-aquatic microorganism interactions and related physicochemical processes (L: ligand, M: metal).

Accumulation of metals inside microbial cells usually involves an energy dependent transport system. In the first place, metals bind to the extracellular cell surface and then they are actively transported inside the cell membrane. There is a variety of functional groups in the biomolecules in the cell walls (carboxylate, phosphate, for example) that are likely to be affected by solution pH; thus, the microorganisms have net negative surface charge for most of the cases.

Microorganisms may promote the mobilization or immobilization of metals producing their dissolution or precipitation. This can be achieved through pH changes or excreting organic acids; some of these acids are capable of complexing metals. Microbial mediated precipitation of metals can result from formation of extracellular insoluble hydroxides, sulfides, carbonates, phosphates, etc. Microorganisms are capable of biotransforming metals with multiple oxidation states by catalyzing oxidation and/or reduction reactions. In some cases the redox reaction is accompanied by a precipitation: this is typical for Cr(VI) / Cr(III) system, in which the reduction of Cr(VI) transforms a soluble species to a more insoluble form. Other biotransformations include methylation or dealkylation reactions (Johnson, 2006). Due to the complexity of biomaterials, it is possible that some of these mechanisms are acting simultaneously to different extent according to the specific conditions of the system: the nature of the active sites and the overall operating conditions.

1.4. Remediation Strategies: Physicochemical Procedures vs. Biological Treatments

Water scarcity is one of the greatest concerns of this century. There is an increasing trend to require a more efficient use of water resources, both in urban and rural environments. Removal of pollutants from water is becoming more important with the increase of industrial activities. The stricter environmental regulation on the acceptable levels of metals make it necessary to develop different technologies for their removal, depending on whether the water is intended to be discharged as treated effluent in the aquatic environment, or reused, mainly for irrigation purposes. While many countries have guidelines for the use of reused water, these guidelines tend to focus on the environmental risk from microbial pathogens, viruses and nutrients. Maximum total trace metal content is usually given (Toze, 2006).

The use of metals has become so pervasive that few industrial operations currently do not discharge metal-containing wastes. Due to the use of large amounts of metallic compounds, electroplating industry is one of the most hazardous among the chemical intensive industries. The treatment of metal contaminated wastewater prior to its discharge is paramount, so it is very important to develop cost-efficient technologies that allow metal recovery and produce an environmentally acceptable, high quality effluent.

There are many different methods for treating wastewaters, both to decrease the amount of wastewater produced and to improve the quality of the treated effluent. Current methods include precipitation, adsorption, coagulation/flotation, sedimentation, filtration, membrane processes, electrochemical techniques, ion-exchange, biological processes and chemical reactions. In many cases, these methods are combined to get better results or to adapt to special requirements.

Overall, pH adjustment to the basic conditions (pH = 11) is the major parameter that significantly improves metal removal by chemical precipitation. Due to its availability in most countries, lime or calcium hydroxide is the most commonly employed precipitant agent. No expensive equipment is required and the operation is safe. Although the process is simple, it has several drawbacks: it requires large amounts of chemicals to reduce metals to an acceptable level for discharge, it is slow and large amounts of sludge are produced. This sludge requires pH adjustment (pH ≤ 9) according to environmental legislation and must be treated for disposal, leading to long-term environmental impacts (Kurniawan et al., 2006). Lime precipitation can be employed to effectively treat inorganic effluents with a metal load of 1000 mg/L.

A combination of coagulation-flocculation processes, based usually on pH adjustment and addition of ferric/alum salts as the coagulant, can be employed to remove metals. This approach has limitations such as high operational costs due to chemical consumption and production of large amounts of sludge.

Membrane filtration has received considerable attention for inorganic effluent treatment. Depending on the pore size and molecular weight of the separating compounds, various types of membrane filtration such as ultrafiltration, nanofiltration or reverse osmosis can be employed for metal removal. All these methods result in high operational costs (e.g., membrane fouling and biodegradation, need of pressurizing systems). For reversal osmosis, instead of pH, pressure is the major parameter that affects the extent of metal removal; the higher the pressure, the higher the metal removal, but also higher energy consumption.

Biosorption technology is an innovative wastewater treatment to remove metals from aqueous solutions by using economical biomasses such as seaweeds, agricultural and industrial wastes, and specially propagated biomasses of bacteria, algae, yeast and fungi. The most effective configuration for biosorption reactors is the flow-through bed column. The performance depends strongly on the biomaterial sorption capabilities towards metals, so the accomplishment of product optimization and continuous process design are the main challenges for biosorption application in large-scale plants (Volesky, 2001). This is strictly related to a deeper understanding of the specific metal-site interactions, that can vary greatly, and to the fact that different kind of interactions can take place simultaneously. Microorganisms have been successfully exploited to deal with metal pollution and emerging technologies in this area rely on enhancing metal biosorption or precipitation related to microbial metabolism.

Cost and benefit analysis for treatment techniques is the most common method used in decision making and in determination of criteria for social benefit, such as environmental protection and foreseeing of possible environmental effects through the application of these processes. Cost and benefit analysis is an estimation of benefit from a project during its economic life time and expected cost followed by comparison between different methods. Demir et al. (2007) compared chemical and biological Cr(VI) removal methods with regard to their cost and percentage in metal removal. From the comparison of methods based on ion exchange resins (Amberlite IRA 904 and mixed resin) and biological techniques (vegetative cell and endospore crystal and endotoxine methods), they concluded that although ion-exchange provides high benefit in Cr(VI) removal the costs are high when resin regeneration step is included. When biological techniques are used, process and investment costs are lower; although the recovery is also lower, these methods are preferable because of their environmental benefits.

Microbial metal transformations are useful tools for the development of bioremediation technologies and the environmental biogeochemistry comprehension. Microbial immobilization of metals shows a clear potential for bioremediation either *in situ* or *ex situ* processes (Gadd, 2000). Biosorption and bioaccumulation, metal-binding biomolecules, reductive precipitation, sulphide and phosphate precipitations are alternative ways to improve removal processes applied to metal-polluted effluent treatments. A deeper understanding of the specific microorganism-metal interactions: adsorption, ionic exchange, surface complexation, surface precipitation, etc. can also be achieved by using an equilibrium modeling approach. The models can be characterized by different degrees of complexity and/or accuracy. Of course, the predicting capacity of these models depends strongly on the initial choice of the set of reactions among active sites and metals. Mechanistic modeling performed in a batch system by considering the adsorbent characterization and equilibrium biosorption can be extended to dynamic simulations of continuous biosorption processes (Pagnanelli et al., 2005).

The aim of this chapter is to remark the relevance of an interdisciplinary and meticulous study on the physicochemistry of aqueous systems and microbial strategies to survive in heavy metal polluted environments. This knowledge could be used to develop a proper design of wastewater treatment processes.

2. Physicochemical Studies

2.1. Speciation Analysis in Action: ASV Monitored Titration

2.1.1. Titration Procedure

ASV monitored titrations were performed throughout this study. After every addition of a certain volume of Cd(II) standard solution to the sample being analyzed, the solutions were stirred for 15 minutes. Once equilibrium was established, labile cadmium (Cd′) peak current was measured by SWASV.

Measurements were performed with an Autolab PGStat10 (EcoChemie) and a Metrohm 663 VA polarographic stand (hanging mercury drop electrode mode). Instrument settings for cadmium detection were: preconcentration potential (E_{prec}) -0.75V vs. Ag/AgCl (3M KCl), equilibration time 5 s; scan range -0.8 V to -0.25 V, step potential 5.1 mV, pulse amplitude 19.95 mV. A suitable preconcentration time (t_{prec}) was selected for each system being studied.

Cd(II) standard solutions were prepared from a certified 1,000 ppm solution (Merck Certipur) by dilution. KCl 0.01M (Merck) was used as supporting electrolyte and nitrogen was bubbled to deoxygenate solutions.

2.1.2. Preliminary Aspects - Complexing Capacity in Model Systems

When metal speciation studies are carried out in natural or model systems, it is mandatory to keep solution pH in a narrow range. This is particularly relevant during ASV titration experiments because metal - ligand interactions are pH dependent. The choice of a suitable pH buffer must take into account not only the desired pH range, but also its ability to complex the metal. Many of the currently used buffers are also complexing agents and may disturb the solution equilibrium state. Good (Good et al., 1966) introduced zwitterionic N-substituted aminosulfonic acids to avoid some of the chemical limitations of the usual buffers. Nowadays, Good´s buffers are widely applied in biological and chemical research due to their low complexing capacity. Nonetheless, previous evaluation of buffer complexing capacity is needed (Soares et al., 1999; Yu et al., 1997; Vasconcelos et al., 1998), taking into account that there is evidence that some Good´s buffers are capable of weakly binding a variety of metals (Mash et al., 2003).

In this work, we studied the behavior of two Good´s buffers, MES (2-[N-morpholino]ethansulfonic acid, pKa = 6.1) and HEPES (N-[2-hydroxyethyl]piperazine-N´-[2-ethansulfonic] acid, pKa = 7.5) and borate buffer towards Cd(II) in order to evaluate if Cd(II) binding by the buffer takes place. When there is no Cd(II) binding, the shape of the titration curve is linear ($Ip = m\ Cd_t + b$, where m is the slope and b is the y-intercept; Cd_t is Cd(II) total concentration added corresponding to each datapoint). MES and HEPES (0.01 M, pH = 5.5 and 7.5, respectively) were used previous dissolution in KCl 0.01 M. Titration curves are shown in Figure 3. The linear shape of the titration curves revealed no Cd(II) binding when compared to the titration curve in KCl (pH = 5.0, $y = 0.1467\ Cd_t - 0.0007$, $R^2 = 0.9996$).

Slow complexation kinetic effects in Cd-MES or Cd-HEPES solutions were disregarded performing batch experiments with MES and HEPES after 48 hs equilibration.

Although cadmium is known to form weak complexes with chloride and hydroxyl ($\log K_f = 1.52$ and $\log K_1 = 4.00$, respectively, Martell et al., 2001), both present in the supporting electrolyte, results showed in Figure 3 indicate that Cd(II) complexes behave at least as electrochemically labile under our experimental conditions. Also, these results indicate that both buffers, MES and HEPES, are suitable for cadmium speciation studies. The slope of the buffer titration curve is the reference value for all further titrations.

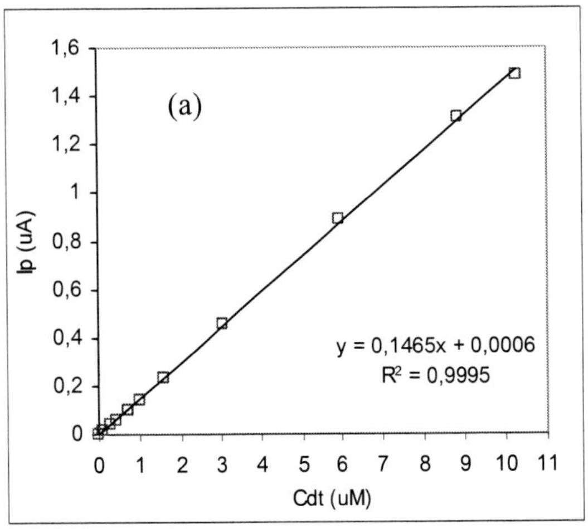

Figure 3. Titration curves of 0.01M HEPES (a) and 0.01M MES in KCl 0.01M (b) with Cd(II). E_{prec}= -0.75V, t_{prec}=120s.

As we mentioned previously, it is important to select proper E_{prec} and t_{prec} in order to detect the relevant species. Pseudopolarograms are diagnostic tools that allow selection of E_{prec} to ensure labile cadmium detection and measurement. A pseudopolarogram is a plot of the current signal obtained by ASV when E_{prec} is varied (for t_{prec} constant). When no reduction takes place, a small current signal is obtained; as the E_{prec} applied is moved in the negative direction, eventually a reduction process can be observed as a sudden increase in current. The shape of the pseudopolarogram is sigmoideal, and the E_{prec} is selected from the potential values corresponding to the plateau zone.

In a first attempt to simulate a real sample, solutions of two polyamminocarboxylic acids, EDTA (ethylenediaminetetraacetic acid, Merck) in 0.1M Borax (Merck) and NTA (nitrilotriacetic acid, Merck) in 0.01M HEPES were titrated. The use of model ligands provides a reference that allows characterization of the solution in terms of ligand concentration and stability constants (Martell et al., 2001).

Figure 4 shows the pseudopolarogram obtained for a solution containing Cd(II) without and with EDTA. When no EDTA is present, for $E_{prec} < -0.65V$ uncomplexed Cd(II) deposition takes place. On the other hand, for Cd(II)-EDTA (2:1) mixtures the uncomplexed Cd(II) plateau is obtained due to Cd(II) excess, but at approximately -1.20V, a second plateau corresponding to the deposition of Cd(II) from the EDTA-Cd complex is observed. Taking into account these pseudopolarograms, $E_{prec} = -0.75V$ was selected for Cd-EDTA complexing studies.

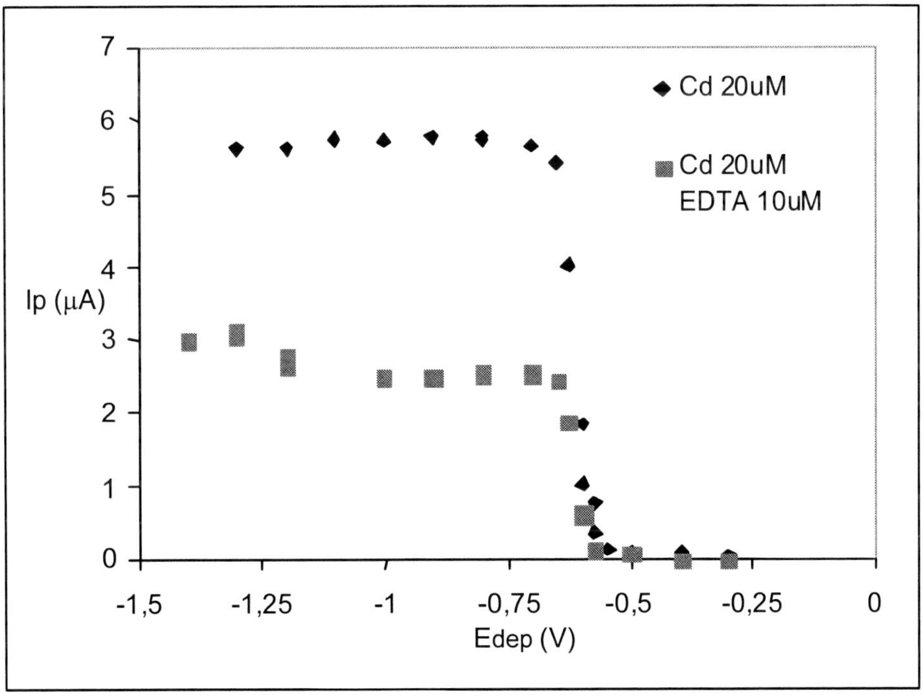

Figure 4. EDTA as a model ligand. Pseudopolarogram obtained in 0.1M borate buffer (pH = 9.3), t_{prec} = 120s.

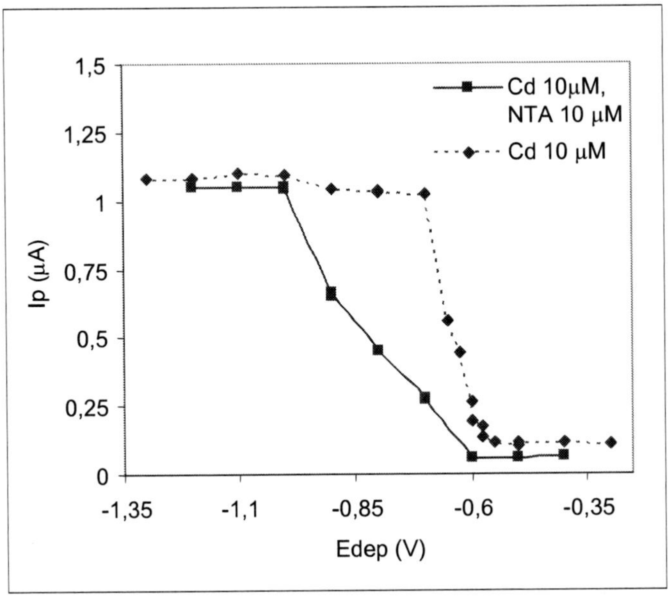

Figure 5. NTA as a model ligand. Pseudopolarogram obtained in 0.1M borate buffer (pH = 9.3), t_{prec} = 120 s.

Different t_{prec} were explored. The t_{prec} should be the result of a compromise between two opposite contributions: on one hand, longer t_{prec} could be applied to produce significant current signals; on the other hand, if t_{prec} is too long contribution of metal released by the decomposition of the complex in the diffusion layer, a kinetic effect, might occur. As a consequence, t_{prec} was evaluated for each particular ligand; typical values ranged between 15 and 120 seconds.

NTA was also used as a model ligand. It is interesting to observe the pseudopolarogram of a 1:1 Cd-NTA system (Figure 5, Cd 10μM - NTA 10μM curve). A plateau is observed starting at approximately -1.0V, which compared to Figure 4 (without EDTA) can not be assigned to the uncomplexed cadmium but to the reduction of NTA-Cd complex. The current response in the region between -0.6V and -1.0V is also different from what is observed in Figure 4. As NTA and Cd(II) are in 1:1 ratio and $\log K_f' = 9.2$, at pH = 9.3 almost zero current would be expected in the potential range were uncomplexed cadmium would occur. However, there is a measurable current signal due to Cd(II) reduction in that potential interval. This result is an evidence of kinetic decomposition of the complex during the preconcentration time which is referred in the literature as kinetic lability. This is an effect that needs to be recognized because it leads to an overestimation of the labile cadmium fraction and to an underestimation of the stability constant.

A typical plot that helps detection of kinetic decomposition of the complex in the diffusion layer is shown in Figure 6. In the case of EDTA (Figure 6(a)), no change in peak potential is observed; this behavior is consistent with an inert complex (van Leeuwen et al., 1989). So Cd(II) bound to EDTA is not released from the complex during t_{prec}. NTA has a different behavior, as can be observed in Figure 4b, where the peak potential shifts in the positive direction. NTA behavior is referred to as labile (van Leeuwen et al., 1989).

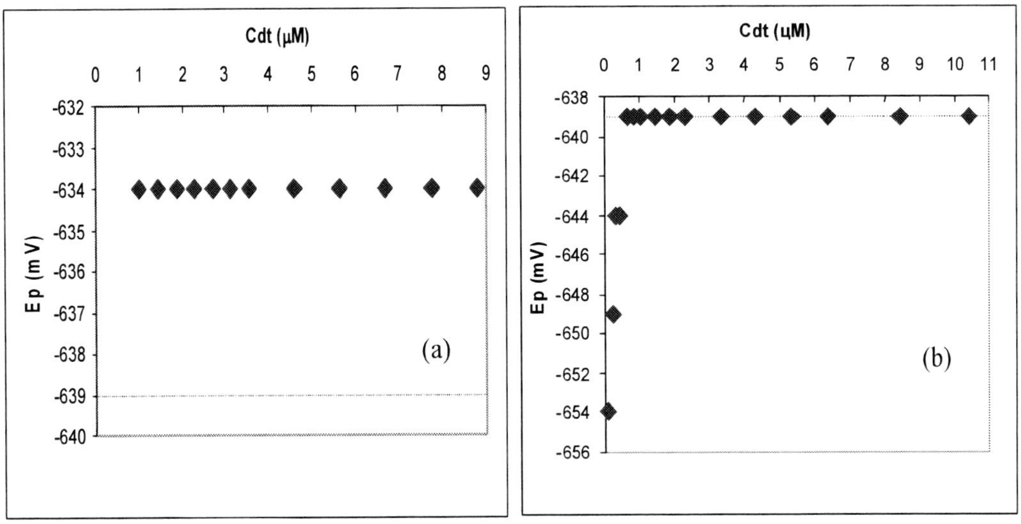

Figure 6. Ep vs Cd_t plot in borate 0.1M buffer (pH = 9.3), (E_{prec} = -0.75 V) (a) EDTA 1µM (dep. time 120 s); (b) NTA 1 µM (dep. time 45 s); (---) borax 0.1M.

2.1.3. Data Treatment

The aim of determining CC of an aqueous system, either natural or model, is to obtain information from the shape of the titration curve about its complexing capability in terms of two complexation parameters: the conditional stability constant K_f', and total ligand concentration L_t. K_f' is related to the thermodinamic stability constant (Martell et al., 2001; Scarponi et al., 1996) and L_t is the analytical concentration of ligand L.

The product of both parameters, $K_f'L_t$, is referred in the literature as the analytical window. There are practical lower and upper limits for $K_f'L_t$ ($K_f'L_t = 10^{-2}$ and $K_f'L_t = 10^2$, respectively), highlighting the influence of the ligand concentration on the interval of stability constant values that can be determined by the titration procedure (Scarponi et al., 1996).

As an example, the titration curve of 1.3 µM EDTA solution (borax 0.1M, pH = 9.3, t_{prec}= 120 s) with Cd(II) is shown in Figure 7(a). Clearly, two regions can be observed. The first portion of the curve, where almost no labile cadmium is detected, corresponds to EDTA in excess respect to cadmium. For higher Cd(II) concentrations, the curve has a linear behavior due to excess cadmium. Interception of this linear portion with the x-axis allows estimation of total ligand concentration but no information is obtained on the conditional stability constant through this procedure.

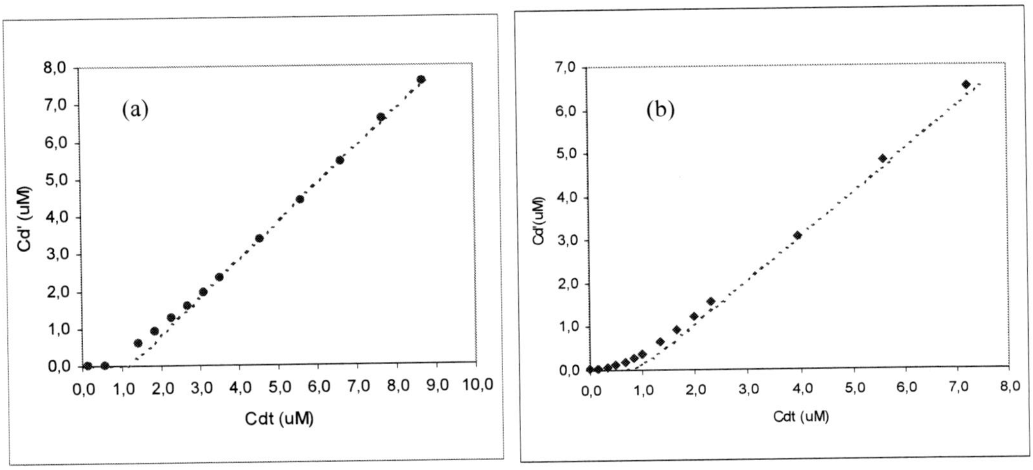

Figure 7. Titration curve for (a) 1μM EDTA (0.1M borate buffer, pH = 9.3); (b) 1μM NTA (HEPES/KCl pH = 7.5). E_{prec} = -0.75V t_{prec} = 120s (EDTA), 15s (NTA). (---) calculated titration curve using eq. 4.

Once the titration has been performed and the excess region has been reached, the slope of the linear portion (S) of the curve is determined. Labile metal concentration, M', is calculated for all data points in the titration curve by means of the expression:

$$M' = \frac{I_p}{S} \tag{1}$$

where Ip is the current signal obtained after each cadmium addition. The complexed metal concentration, ML, can be obtained from the mass balance for the metal. In the case of a 1:1 complex:

$$M_t = M' + ML \tag{2}$$

Considering also mass balance for the ligand and the expression for K_f'

$$Lt = L' + ML \tag{2'}$$

$$K_f' = \frac{[ML]}{[M'][L]} \tag{3}$$

a mathematical expression describing the titration curve can be derived

$$Mt_{sim} = \frac{K`M`^2 + M`(1 + K`Lt)}{K`M`+1} \tag{4}$$

Equation 4 allows calculation of M_t, provided M′ and K_f' are known. Every M′ value is calculated from a titration point using K_f' value obtained from the literature. It was used to produce the fitting curves (continuous lines) shown in Figure 7(a) and (b). A good fit is observed for EDTA ($\log K_f' = 15.5$); in NTA case, experimental data points correspond to higher currents than predicted by the equation 4. This behavior is in agreement with the kinetic lability of the complex during preconcentration: if the complex decomposes, the labile fraction detected in the experiment increases, resulting in higher currents.

L_t and K_f' can be obtained by linearization of the titration data. Ruzic linearization (Ruzic, 1982) is a simplistic approach which enables determination of L_t and K_f' from a unique plot. The model assumes that only 1:1 complexes are formed, which is the case of EDTA-Cd(II) and NTA-Cd(II). However, secondary phenomena, like adsorption, stoichiometries different from 1:1, kinetic effects, etc., may incorporate deviations in the experimental results. In our case, the evolution of peak potential during EDTA titration is in agreement with the response observed for an inert complex, as already mentioned (van Leeuwen et al., 1989). For NTA, evolution of peak potential is similar to the response of a labile complex (van Leeuwen et al., 1989; Raspor et al., 1975), proving the need of minimizing preconcentration time to avoid kinetic decomposition during preconcentration step.

Considering metal and ligand mass balances (eqs. 2 and 2′) and the expression for K_f', Ruzic′s linearization is obtained (Ruzic, 1982):

$$\frac{[M']}{[ML]} = \frac{[M']}{[L_t]} + \frac{1}{Kf'[L_t]} \tag{5}$$

plotting [M′]/[ML] as a function of [M′], L_t can be obtained from the slope and K_f' from the y-intercept.

Ruzic linearization applied to the titration of 1.3μM EDTA solution (borate buffer 0.1 M, pH = 9.3, 120 s preconcentration time) gives total ligand concentration, $L_t = (1.21 \pm 0.06)$μM. This is in agreement with $L_t = (1.2 \pm 0.1)$μM, obtained by extrapolation. In this case, since ligand concentration range is μM, K_f' remains undefined, although greater than 10^8. To achieve a better K_f' estimation it should be necessary to work with extremely dilute solutions (nM range, Scarponi et al., 1996), which is beyond our scope.

In the case of the titration data of 1μM NTA solution (HEPES 0.01M, KCl 0.01M, pH = 7.5, 15 s preconcentration time), L_t obtained from Ruzic linearization, $L_t = (0.76 \pm 0.01)$μM, is in good agreement with ligand concentration obtained by extrapolation of the upper linear portion of the titration curve to the abscissa, $L_t = (0.78 \pm 0.15)$μM. The conditional stability constant for NTA (pH = 7.5) obtained from Ruzic linearization ($\log K_f' = 7.1$) was in fair agreement with calculated values ($\log K_f' = 7.7$) from thermodynamic data (Martell et al., 2001). The occurrence of a kinetic effect could help explaining the differences observed in L_t and also the slightly smaller value of $\log K_f'$.

2.2. Application: Industrial Effluent Complexing Capacity

When thinking of a metal retention strategy using microorganisms, one condition must be fulfilled for a successful application: metals need to be bioavailable, in a chemical form that can interact with the microbiota.

An interesting system from an environmental point of view and also to be microbiologically treated is the second static rinse that is performed after an alkaline noncyanide electroplating bath ("second rinse bath", Murphy, 1998). Bath rinses are basically water that has been used to rinse metallic pieces after they have been treated in an electrolytic bath. In this process, the metallic pieces are rinsed twice, and thus there are a first and a second bath rinse. The purpose of rinsing the pieces is to wash away components of the electrolytic bath, mainly metals and organic compounds that work as surface modifiers and that are added to produce metal deposits with the desired characteristics. Comparing the first and the second rinses, the latter has lower concentrations of both metals and organic compounds; for this reason, first rinse waters are transferred back to the electrolytic bath, while second rinse bath waters are changed 3-4 times a day.

"Second rinse bath" waters are of environmental concern because it is necessary to deal with relative big volumes (approximately ten thousand liters per day) of a diluted solution containing *borderline* or *type B metals* (presented in 1.2.). Total Zn(II) (0.23μM), Cd(II) (< 0.001μM), Pb(II) (< 0.001μM) and Cu(II) (0.09μM) were measured in an independent aliquot by SWASV after UV-photooxidation and pH acidification (pH = 2). Second rinse bath water must meet environmental regulations before disposal.

The methodology discussed in 2.1.2. was applied to evaluate Cd(II) CC in an aqueous solution taken from a "second rinse bath" of a zinc electroplating industrial process. Results of the monitored titration are shown in Figure 8. In this case dilution was avoided, pH of the solution was 11, and no buffer was needed during titration. An estimation of total ligand concentration (Figure 8(a)) was obtained by extrapolation, $L_t = (1.2 \pm 0.1)\mu M$.

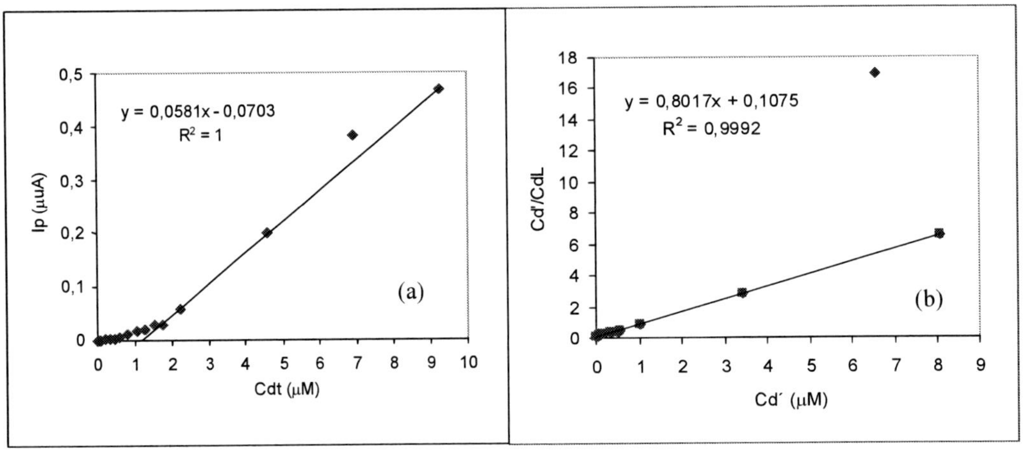

Figure 8. Electroplating second rinse bath: (a) titration curve with Cd(II) (E_{dep} = –0.75V, dep. time 120 s) (b) Ruzic linearization.

Also Ruzic linearization was applied (Figure 8(b)). The system behaved as a single ligand system with 1:1 stoichiometry. Total ligand concentration = (1.25±0.02)μM and log K_f'= (6.9 ± 0.2) were determined for this solution. Extrapolated and Ruzic values for total ligand concentration are in good agreement.

The value obtained for K_f' indicates that ligands present in this solution are moderate in strength. As L_t is approximately 1.2μM, any Cd(II) concentration greater than 1.2μM can be considered bioavailable. Moreover, if microorganisms were able to use these organic ligands as nutrients the treatment would be more effective.

3. Microbiological Studies

3.1. Screening of Heavy Metal Resistance/Tolerance in Aquatic Ecosystems

Considering microorganisms that are able to survive in polluted environments, pollutant resistance and tolerance concepts arise. The term "resistance" defines the ability to survive and grow through a heavy metal inducing mechanism encoded either in plasmids or chromosomes. On the other hand, the term "tolerance" indicates the presence of constitutive cell structures or biological excretion reactions that decrease the heavy metal local bioavailability; for example this can be achieved by metal complexation or precipitation. So, testing growth in culture media supplemented with Cd, Cu, Pb and Zn gives only the information about microbial surveillance but not about the strategy involved. Reconquista river and its affluent Las Catonas stream were chosen as study case of aquatic ecosystems and sample sources because of its high pollution levels of several contaminants from both industrial and faecal discharges. Physicochemical and bacteriological characterization of Las Catonas basin has been performed (Vullo et al., 2005). Bacterial resistance/tolerance to heavy metals were tested in samples taken from surface and pore waters from two sites: LC4 (34° 36' 36.7" S, 58° 46' 22.8" W) and LC5 (34° 36' 49.4" S, 58° 45' 36.8" W), which presented the highest industrial activity and population density. Culturable strains were screened and a microbial enumeration was carried out as follows: 0.1 mL from each sample (or its 10 fold dilutions) were spread on Plate Count Agar (PCA) (casein peptone 5 g/L, yeast extract 2.5 g/L, D(+)-glucose 1 g/L, agar 14 g/L) and on metal supplemented PCA. Six different supplemented plates were prepared with: 1) 0.5mM Cu(II), 2) 0.5mM Cd(II), 3) 0.5mM Zn(II), 4) 0.5mM Pb(II), 5) 0.5mM Cu(II), Zn(II) and Pb(II) and 6) 0.5mM Cu(II), Cd(II), Pb(II) and Zn(II). Incubation was performed at 30°C, from 24 to 96 hours. As a selection step, colonies grown in presence of the four heavy metals were also tested on plates prepared with PCA supplemented with a mixture of Cu(II), Cd(II), Zn(II) and Pb(II), containing 1mM, 2mM, 4.8mM and 9.6mM as final concentration in each metal.

In agreement with a higher concentration of heavy metals in pore than in surface water, a higher percentage of heavy metal resistant (tolerant) culturable population in pore water was obtained (Vullo et al., 2005). An example of the obtained results is resumed in Table 1. Screening results show that pore water samples contained a majority of culturable multiresistant microbial population due to the higher environmental selection pressure resulting from a higher heavy metal concentration in that phase. Every culturable strain exhibited resistance to Cu, Pb and Zn under the culture conditions. Cd presented the highest toxic effects as shown by the lowest number of resistant (tolerant) colonies obtained. Cd resistant (tolerant) isolates were also resistant (tolerant) to Cu, Zn and Pb. We may therefore conclude that Cu, Cd, Pb and Zn multiresistance (multitolerance) is clearly related to Cd.

Table 1. Examples of bacterial counts exposed to heavy metals from polluted samples (extracted from Vullo et al., 2005)

Metal (0.5mM each in PCA)	% resistant/tolerant Colony-Forming Units (CFU)	
	Surface water 1.24×10^5 CFU/mL	Pore water 8.4×10^5 CFU/mL
Cu	69	100
Zn	89	100
Pb	80	100
Cd	4.7	14
Cu, Zn and Pb	23	100
Cu, Zn, Pb and Cd	2.9	14

Surface water: 38μM Zn, 0.018μM Cd, 0.068μM Pb, 0.33μM Cu.
Pore water: 17.1μM Zn, 0.125μM Cd, 0.125μM Pb, 9.6μM Cu.

3.2. Isolation and Characterization of Bacterial Strains from Polluted Habitats

The microbial strategies to survive in presence of heavy metals are expressed in natural polluted environments, where the selection pressure is present at any moment. Reconquista river watershed is one of these environments with high pollution levels. In order to isolate heavy metal resistant/tolerant strains with potential application in wastewater treatment, water and sediment samples were taken from different sites of Reconquista river watershed. For that purpose, 0.1 mL of each water sample were spread on plates prepared with PCA spiked with 0.5mM Cu(II), 0.5mM Cd(II) and 0.5mM Zn(II). When sediments samples were treated, 0.1 mL of a 1:10 suspension in NaCl 150mM were spread in the same medium. Plates were incubated at 32°C for at least one week. Isolated colonies were purified three times in the same medium and tested for Minimal Inhibitory Concentration (MIC). MIC is the lowest metal concentration that does not allow growth for a specific strain. It can be estimated just by checking bacterial development by turbidity measurement in presence of increasing heavy metal concentrations in a culture broth. In this case, MIC was determined by duplicate in PYG broth (casein peptone 2.5 g/L, yeast extract 1.25 g/L, glucose 0.5 g/L) supplemented with 0 to 10mM Cu(II) or Zn(II) or Cd(II). Cultures were incubated at 32°C for at least one week, checking bacterial growth by measuring absorbance at 600 nm every 24 hours. MIC was estimated as the first dilution which completely inhibits the bacterial growth, in PYG medium. Although eleven strains were able to grow in presence of Cu, Cd and Zn in the semisolid medium, five of them were chosen due to their highest MIC for each heavy metal assayed and the highest biomass yield in shortest incubation times. Table 2 resumes these results.

Clearly, for four of five of the isolates, Cd(II) is again the most toxic metal assayed. Two factors can be considered to account for Cd toxicity: a) this metal is not an essential trace element for organisms and b) as it is shown in the next section, the low complexing capacity of the employed culture medium leaves cadmium almost completely bioavailable.

Working with indigenous bacteria requires an identification of the isolates, especially when a bioremediation process depends on them. Biochemical characterization can be performed by API multitest system (Bio Mérieux), but it did not have a good identification level with the isolates. To get the best approach, molecular characterization was performed by 500 bp 16S r-RNA gene sequencing (MIDI Labs, USA) or 1500 bp 16S r-RNA (MacroGen, Korea). Table 2 shows the isolation results with the final both molecular and biochemical characterization of the obtained strains.

3.3. Cd(II) CC Evaluation in a Microbiological Culture Medium

When metal retention is addressed by microorganisms it is possible to need the addition of nutrients to maintain cells in a biologically active state. Typical organic components of the culture medium (peptides, growth factors and carbohydrates) are potential Cd(II) ligands and changes in culture medium composition occur along with cell development. When PYG medium is inoculated (time = 0 hours), its composition corresponds to a fresh mixture, containing peptides, growth factors and glucose. After 24 hours (exponential growth phase) bacterial metabolism has modified the initial composition: initial components are still available to the bacteria, plus metabolic products. At t = 48 hours (stationary phase) microorganisms stopped growing as a consequence of nutrient consumption and metabolization. At this stage a great amount of metabolic products, basically ammonia and amines, are present. For this reason it is important to have information about changes in the complexing capacity of the system and to characterize metal-ligands interactions in terms of L_t and K_f'.

Table 2. Isolation and selection of Cu, Cd and Zn resistant/tolerant strains

Strain	MIC in PYG broth (mM)		
	Cu	Cd	Zn
Pseudomonas veronii 2E	2	0.5	10
Ralstonia taiwanensis M2	2	5	10
Delftia acidovorans AR	2	1	5
Klebsiella ornithinolytica 1P	2	0.25	0.5
Klebsiella oxytoca P2	1	0.25	0.5

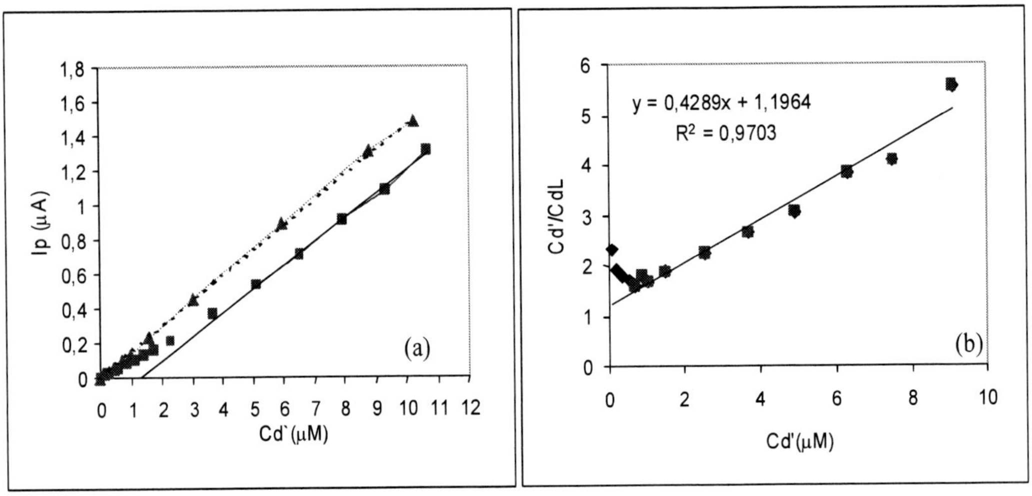

Figure 9. Titration curves and data treatment. (a) (—) Fraction t_0 (1:50 dilution in buffer MES 0.01M, pH = 5.5); (---) buffer MES (0.01M MES, 0.01M KCl, pH = 5.5). (b) Ruzic linarization of fraction t_0.

The methodology previously described was applied to evaluate Cd(II) CC in PYG medium. *Pseudomonas veronii* 2E was inoculated in PYG broth, in absence of Cd(II). According to biomass evolution and pH changes associated to bacterial growth, three representative samples were taken in order to evaluate changes in CC during the batch culture. Each sample corresponds to a particular physiological state: initial (t_0: incubation time 0 hours, pH = 7.0), exponential phase (t_{exp}: incubation time 24 hours, pH = 7.5) and stationary phase (t_{st}: incubation time 48 hours, pH = 8.0). In order to avoid adsorption on the electrode surface which interferes with the signal measurement, a 1:50 dilution of each fraction in buffer MES (pH = 5.5) and HEPES (pH = 7.5) was used (Ceretti et al., 2006). Experimental data were obtained and Ruzic linearization was applied.

Table 3. Total ligand concentration and conditional stability constant values for fractions t_0, t_{exp} and t_{st}

Fraction	pH = 5.5	pH = 7.5
t_0 (0 hours)	L_{ext} = (1.7 ± 0.6)μM	L_{ext} = (1.2 ± 0.4)μM
pH = 7.0	L_t R = (2.3 ± 0.1)μM	L_t R = (1.1 ± 0.1)μM
	log K_f' = (5.5 ± 0.1)	log K_f' = (5.9 ± 0.2)
t_{exp} (24 hours)	L_{ext} = (0.33 ± 0.04)μM	L_{ext} = (0.9 ± 0.3)μM
pH = 7.5	L_t R = (0.35 ± 0.02)μM	L_t R = (1.5 ± 0.3)μM
	log K_f' = (6.0 ± 0.4)	log K_f' = (5.9 ± 0.3)
t_{st} (48 hours)	L_{ext} = (0.11 ± 0.04)μM	L_{ext} = (1.7 ± 0.6)μM
pH = 8.0	L_t R = (0.27 ± 0.02)μM	L_t R = (1.3 ± 0.2)μM
	log K_f' = (6.3 ± 0.3)	log K_f' = (5.8 ± 0.2)

Figure 9 shows the titration curve of t_0 fraction, performed in buffer MES. Total ligand concentration was estimated by extrapolation of the titration data, $L_{ext} = (1.7 \pm 0.1)\mu M$, (Figure 9(a)). The plot of Cd′/CdL vs Cd′ corresponding to Ruzic′s linearization is also obtained (Figure 9(b)). Using equation (5), $L_t = (2.3 \pm 0.1)\mu M$ and $\log K_f' = (5.5 \pm 0.1)$ were calculated.

The same procedure was applied to the other samples in order to evaluate complexing capacity at t_{exp} and t_{st} at pH = 5.5 (buffer MES) and 7.5 (buffer HEPES). Changes in complexing capacity parameters as function of *Pseudomonas veronii* 2E development are summarized in Table 3. Results obtained for the different fractions are consistent with a single ligand system with 1:1 stoichiometry. Evolution of peak potential during all titrations is similar to the response of a labile complex (Raspor et al., 1975).

When total ligand concentration is evaluated along microbial development and pH is regulated at 5.5, it is clear that L_t decreases. As a consequence of microbial metabolism (t_{exp}, t_{st}) a mixture rich in amino compounds is produced. If pH is fixed at 5.5, amine groups are mainly protonated, they are not be able to bind Cd(II) and thus L_t decreases. On the other hand, when pH is fixed at 7.5, total ligand concentration remains nearly constant during bacterial development. The values of $\log K_f'$ obtained for the different culture phases at both pH values were within one order ($5.5 < \log K_f' < 6.5$), corresponding to moderate ligands, so there is no evidence of a dramatic change in the complexing capacity during microbial evolution. Although we cannot identify the cadmium complexing ligands in the different fractions of the culture medium, reported values for typical Cd-biological ligands are in the same complexing stability constants range (Martell et al., 2001) as our experimental results. These ligands do not necessarily represent different chemical compounds but rather less defined ligand families.

It is interesting to note that in a 50-fold dilution of this culture medium, maximum L_t is about 1μM. In the case of solely considering dilution factor, total ligand concentration available for Cd(II) complexation can be estimated ca. 50μM. When fully saturation of ligands occurs, the major fraction of Cd(II) would remain free and bioavailable to the microorganisms, and would not interfere in the biological treatment.

It is important to stress that this study was performed on a culture medium where microorganisms developed with no Cd(II) addition. Although alterations in primary metabolism of the microorganism due to Cd(II) spiking were not detected in previous studies, some cation mediated induction or repression effects on the synthesis of other cellular products with potential complexing capacity are expected. Also a different pH evolution is observed when Cd(II) is present, starting at 6.0 (t_0), 7.2 (t_{exp}) and finally 7.8 in the stationary phase. These differences in pH values are possibly related to a different metabolic course and to a different behavior of complexing capacity during the experiment.

When *Pseudomonas veronii* 2E is considered for wastewater treatment of, for example, an industrial effluent containing Cd(II) in the range 250μM > Cd(II) > 9μM, the use of PYG medium is a good option for nutrient supply. For Cd(II) concentrations greater than 250μM inhibition of microorganism development has been observed. On the other hand, Cd(II) concentrations lower than 9μM (equivalent to 0.1ppm Cd(II)) fulfill argentinean environmental regulations (Decree 999/92, Environmental Secretary, Argentina) and are considered safe to be discharged without further treatment.

3.4. Heavy Metal Retention in Bacterial Growth and Non-Growth Conditions

Heavy metal retention was evaluated during different physiological states in bacterial cultures in PYG broth and in non growth conditions by contacting a bacterial suspension with each metal without nutrients.

For metal retention in batch culture studies, 100 mL of PYG culture medium plus Cu(II), Cd(II) and Zn(II) were inoculated with 10 mL of a previous culture. Each initial metal concentration was 0.5 or 0.25mM according to the bacterial MIC values previously obtained. Samples were taken at different times during 1 week incubation (200 rpm and 32°C). Growth was tested measuring absorbance at 600 nm (A_{600nm}). All samples were centrifuged and supernatants analyzed for total Cu, Cd and Zn to determine total retention (%). Cell free medium was tested simultaneously as control.

Batch cultures of each strain demonstrated that the retention depends on the bacterial growth phase. Figure 10 shows the bacterial growth in batch cultures in comparison with Cu(II), Cd(II) and Zn(II) retention at initial state, exponential and stationary phases. pH evolution was also monitored at each point. *Pseudomonas veronii* 2E retained 35% Cu(II), 47% Cd(II) and 41% Zn(II) at stationary phase of growth, which represent the maximal retention yields in presence of the three metal for all the selected bacteria for this experiment. In all cases, pH increased from 5-6 to 7.5-8, in accordance to the culture medium components the accumulation of metabolites and different metal interactions. As seen in Section 3.3., for this pH range no important changes in Cd(II) bioavailability are expected since experiments in complexing capacity showed that only ca. 10% Cd(II) would be complexed by ligands present in the culture medium (Ceretti et al., 2006).

Cu(II), Cd(II) and Zn(II) retention in non growth conditions (biosorption), was tested for each bacterial isolate and heavy metal separately. For that purpose, similar bacterial suspensions were prepared using the same biosorbent mass. For bacterial suspension preparation 50 mL early stationary phase cultures (A_{600nm}= 1.1-1.2) in PYG broth were centrifuged (3,000 x g, 15 min.). Cells were washed and resuspended in water (18 MΩcm, Millipore) in order to obtain suspensions containing approximately 3 g dry weight /L living biomass. The biosorption assay was carried out as follows: the biosorption mixture [5 mL cellular suspension, 0.5mM of each metal, 10mM buffer and water (18 MΩcm, Millipore) to a final volume of 10 mL] was incubated at 32°C and 200 rpm during 24 hours. Suspensions were centrifuged (3,000 x g, 20 min.) and filtered through 0.45 μm-pore diameter cellulose membrane. Cu(II), Cd(II) and Zn(II) concentrations in supernatants were determined and their decrease was estimated by comparison to a cell free control mixture. A parameter to take care of was the pH of the mixture to avoid changes during the biosorption process which can modify final interpretations, so in order to keep pH constant, biosorption experiments were carried out at pH = 5.5, 6.2 and 7.5, regulated by buffer solutions of MES, PIPES (piperazine-N, N′-bis[2-ethanesulfonic acid], pKa = 6.8) and HEPES. Results are summarized on Table 4.

Table 4. Heavy metal biosorption by bacterial isolates

Strain	% Heavy metal biosorbed							
	pH = 5.5 (MES)			pH = 6.2 (PIPES)			pH = 7.5 (HEPES)[a]	
	Cu	Cd	Zn	Cu	Cd	Zn	Cd	Zn
***Pseudomonas veronii* 2E**	**43**	**5**	**7**	**40**	**0**	**0**	**50**	**53**
Ralstonia taiwanensis M2	25	6	3	42	0	5	33	38
Delftia acidovorans AR	27	0	0	51	3	4	64	78
Klebsiella ornithinolytica 1P	29	0	0	26	0	0	49	61
Klebsiella oxytoca P2	22	0	0	19	0	48	42	45

[a] Cu(II) biosorption was not evaluated at this pH due to chemical precipitation of copper compounds.

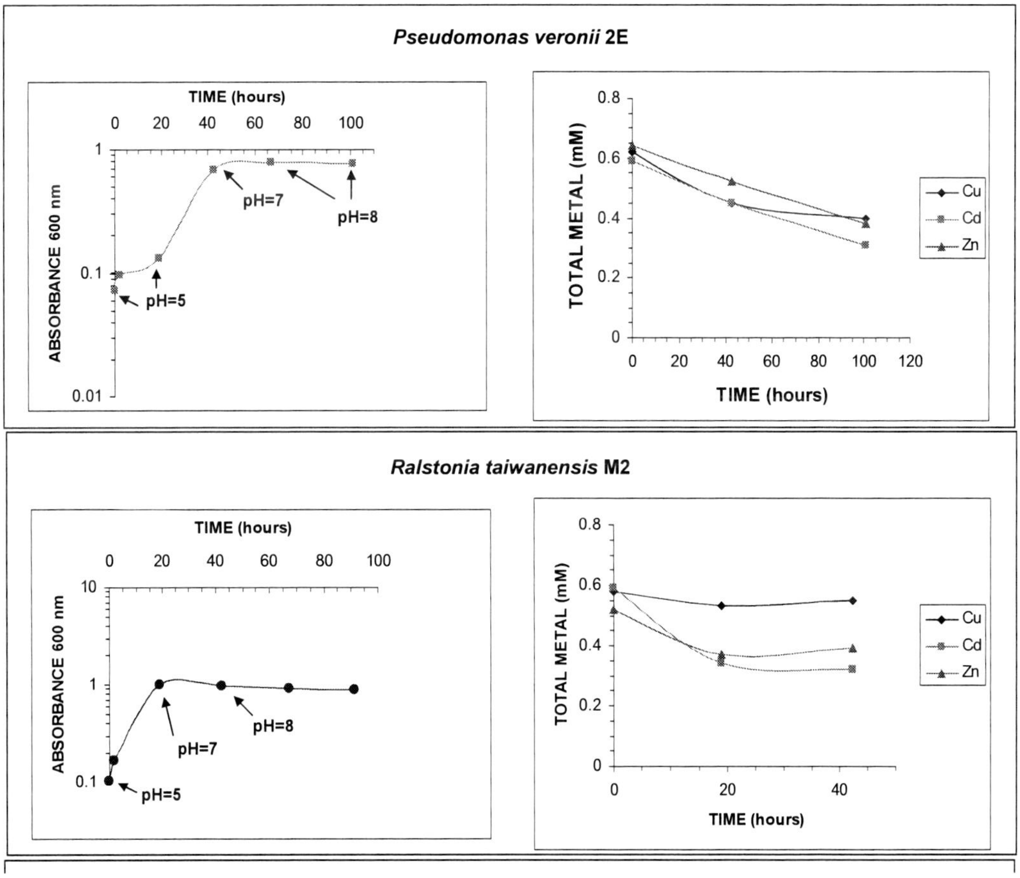

Figure 10. (Continued)

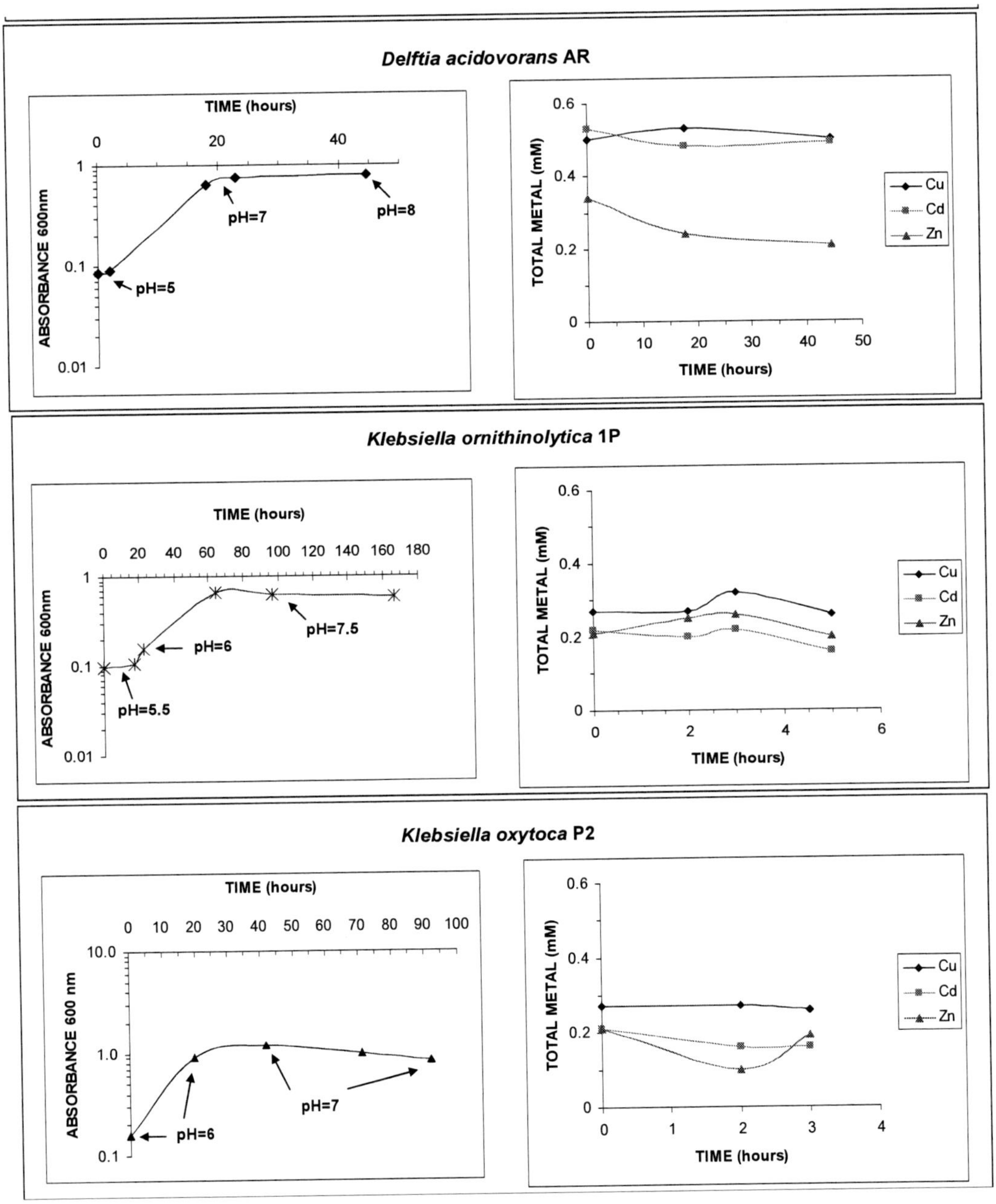

Figure 10. Cu(II), Cd(II) and Zn(II) retention in batch cultures.

Pseudomonas veronii 2E was selected for further studies of Cd(II) biosorption process. This bacteria is a good candidate for a future bioremediation process because of at least three important factors which are: a) high tolerance to cadmium and high biomass yields in a conventional culture medium with low complexing capacity; b) high cadmium biosorption yield at controlled pH = 7.5 in a non complexing mixture and c) good performance to develop films on inert surfaces. This third point is important because allows the utilization of this strain in fixed or fluidized bed reactors. Although *Delftia acidovorans* AR demonstrated a good performance at the assayed pHs, this strain did not show a great ability to develop films or to attach to several matrices.

3.5. Cadmium Biosorption by *Pseudomonas Veronii* 2E

As a first approach, cadmium biosorption kinetics was studied at different incubation times of the biosorption mixture, prepared as described in Section 3.4, with *Pseudomonas veronii* 2E suspension. Buffer HEPES (pH = 7.5) was employed to keep pH constant. The biosorption mixtures were incubated during 30 hours at 32ºC and 200 rpm. Samples were taken at different times and treated as mentioned for cadmium quantification in supernatants. Cd maximal retention (76.8%) was reached in 5 hours and no further changes in Cd(II) concentration were observed during at least 30 hours, under these experimental conditions.

The interaction sorbent - sorbate is evaluated according to two classical models: the Langmuir adsorption model and the Freundlich adsorption model.

The first model assumes that adsorption occurs in a monolayer. There are a finite number of adsorption sites available on a surface, and each one has an equal probability for reaction. Once a particular site is occupied, however, the adsorption probability at that site goes to zero. Furthermore, it is assumed that the adsorbates do not diffuse, so that once a site is occupied it remains unreactive until the adsorbate desorbs from the surface. According to this model, the number of occupied sites, q, is given by:

$$q = q_{max} \, . \, C_f /(K_d + C_f) \qquad (6)$$

or

$$C_f/q = (K_d / q_{max}) + (C_f / q_{max}) \qquad (7)$$

where q_{max} is related to the total number of adsorption sites (theoretical saturation capacity), C_f is Cd (II) final equilibrium concentration in supernatants and K_d is the equilibrium constant for the dissociation of the Cd(II)-surface complex. q was calculated as follows:

$$q = (C_i - C_f) \, . \, V_t / m_t \qquad (8)$$

where C_i is the initial Cd(II) concentration, V_t is the total volume of the biosorption mixture assayed and m_t is the total biosorbent mass as dry weight. Cd(II) equilibrium concentration was determined as previously described.

The empirical Freundlich equation is based on a heterogeneous surface:

$$q = K_F \, . \, C_f^{\,1/n} \qquad (9)$$

where K_F and n are Freundlich constants characteristic of each system, being C_f the Cd(II) final equilibrium concentration in supernatants. K_F and n are indicators of adsorption capacity and intensity, respectively and can be calculated from the linear plot ln q vs. ln C_f. The

Freundlich relation differs from the Langmuir in that not all sites on the surface are considered equal but rather than adsorption becomes progressively, more difficult as more and more adsorbate accumulates. It is assumed that once the surface is covered, additional adsorbed species can still be accommodated, so no maximal monolayer adsorption is predicted. In other words, the Freundlich model does not provide information of the monolayer adsorption capacity as the Langmuir model does (vanLoon et al., 2000).

Cadmium biosorption isotherm was measured. For that purpose, previously optimized conditions (pH = 7.5, HEPES, 32°C, 200 rpm, incubation time 24 hours) were applied. *Pseudomonas veronii* 2E cells were exposed to Cd(II) concentrations, in the range 0.050 to 2mM. In each supernatant, Cd(II) equilibrium concentration was measured after incubation.

Fitting or linearization of equation (7) showed that cadmium biosorption mediated by *Pseudomonas veronii* 2E is in agreement with Langmuir adsorption model. Adsorption parameters obtained by data linearization are K_d = 0.173mM (19.4 mg/L) and q_{max} = 0.476 mmol/g cell dry weight (53.3 mg/ g cell dry weight). K_d (dissociation constant) is the inverse of the affinity constant of binding sites (K_{aff}= 5.776mM^{-1}) and q_{max} is the maximal amount of Cd per unit weight of cell to form a complete monolayer on the surface (theoretical saturation capacity). Comparison of these results with previous reported data suggest that K_d is in good agreement with the value obtained for the bacterium *Bacillus cereus* (20 mg/L, Qi et al., 2006) and for *Rhizopus arrhizus* (0.082mM, Yin et al., 1999). Meanwhile, q_{max} = 0.476 mmol/g cell dry weight is similar to *Pseudomonas aeruginosa* PU21 (40.80 to 57.37 mg/g, Chang et al., 1997), *Bacillus cereus* or *Rhizopus arrhizus* (44.4 mg/g, Qi et al., 2006; 0.56 mmol/g, Yin et al., 1999). As additional data, similar Cu(II) and Zn(II) Langmuir constants from *Pseudomonas putida* CZ1 were obtained (Chen et al., 2005). In this case, q_{max} = 0.471 mmol/g for Cu(II) and a q_{max} = 0.419 mmol/g for Zn(II) were calculated by fitting Langmuir isotherms.

So, *Pseudomonas veronii* 2E shows a Cd(II) adsorption capacity comparable to other microorganisms which have already been tested for bioremedial treatments in aqueous systems.

3.6. Electrophoretic Mobility Measurements and Cd(II) Biosorption in Bacteria: Cellular Surface Interactions

The interface formed between the outer cell envelope and the extracellular environment plays an important role in the bacterial physiology. The outer cell surface mediates exchange and adhesive processes, while also influencing interactions with immunological factors and participating in cell growth and division (Wilson et al., 2001). The components of the surface are macromolecules containing carboxylate, phosphate and amino functional groups which can be ionized as a function of the environmental pH, conferring electrostatic charge to the cell periphery. Electrostatic charge is then a parameter to be considered because of its influence in the polarity and the hydrophilicity, fundamental properties for the cell functions (Hayashi et al., 2003; Tsuneda et al., 2004; Eboigbodin et al., 2006). Net cell surface charge can be assessed on the basis of zeta potential, which is the electric potential of the interfacial region between the bacterial surface and the aqueous environment (Figure 11). Zeta potential (ζ) represents the effective location of the solid-liquid interface and can be estimated by electrophoretic mobility measurements by the Smoluchowski equation:

$$\zeta = (\eta \cdot \mu) / (\varepsilon_0 \cdot \varepsilon) \quad (10)$$

where η is the viscosity of the medium, μ is the electrophoretic mobility, ε_0 is the permittivity of vacuum and ε is the dielectric constant of the medium.

Electrophoretic mobility is an intensive property which indicates the ability of particles to move when an electrical field is applied. Since adsorption is a surface phenomenon, charges on cellular surfaces may be affected when different cations in a bacterial suspension are present, provided that these cations are able to interact with cells.

Determination of electrophoretical mobility of bacteria represents a challenge since experimental conditions need to be carefully established. Ionic strength, pH regulation, cells concentration were taken into account before testing cadmium-cell interaction and its influence in electrophoretic mobility values. As Cd(II) biosorption occurred at pH = 7.5, this parameter was regulated with 10mM HEPES; ionic strength was adjusted with 10mM KCl. The bacterial electrophoretic mobility was evaluated in presence of Cd(II) 0 to 0.5mM. For that purpose, a suspension containing approximately 5 mg dry weight/L of *Pseudomonas veronii* 2E was incubated during 3 hours at 32°C in 10 mL final volume. In all cases, Zeta potential was determined by scattering of laser light in a Brookhaven 90-plus zetameter, as the mean value of seven measurements of the electrophoretic mobility. Significant changes on Zeta potential to more positive values were observed, as shown in Figure 12. As a control experiment, electrophoretic mobility was determined as function of Cd(II) concentration, under conditions where the biosorption phenomenon does not take place. In this case, pH was regulated at pH = 6.5 (10mM HEPES) and ionic strength was again adjusted with 10mM KCl. No significative changes in the mobility were observed (Figure 12). These results mean that the outer negative charge on bacterial surface decreases directly as consequence of Cd(II) interactions.

4. Conclusion

The exploitation of locally available microbiological material and the use of cheap nutrient sources for the cells have been applied for metal removal. Cd(II) bioavailable fractions in the effluent to be treated, as well as in culture media and systems with known chemical composition, were estimated from the determination of the complexing capacity. Isolation of autochthonous strains from contaminated sites and characterization of their behavior towards metals were described. The results summarized in this chapter are a good example of the information needed when the further design step of a wastewater treatment process is considered. This design must take into account the choice of adequate microorganisms and a nutritional source. Knowledge on the bioavailability of metals in the effluent to be treated is relevant, also how bioavailability is affected when nutrients are added if living cells are involved in the process. The bacteria described in this work are suspected to be robust enough to explore also a biological treatment for organic wastes involving either the same strains or a combination of mixed cultures. This exploration may include metal-complex degradation with the consequent changes in heavy metal speciation in the aquatic system.

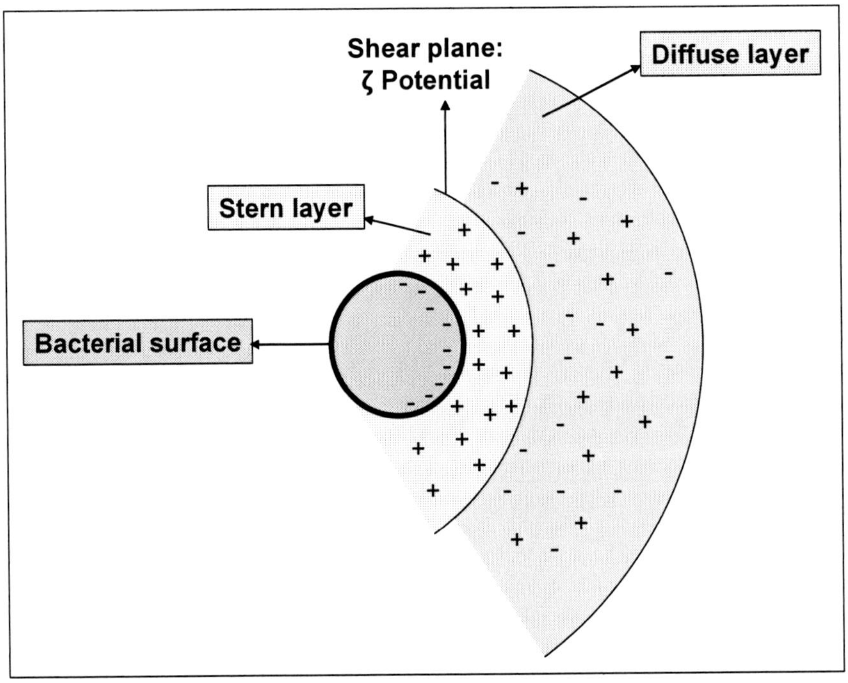

Figure 11. Solvent layers in bacterial cell surroundings.

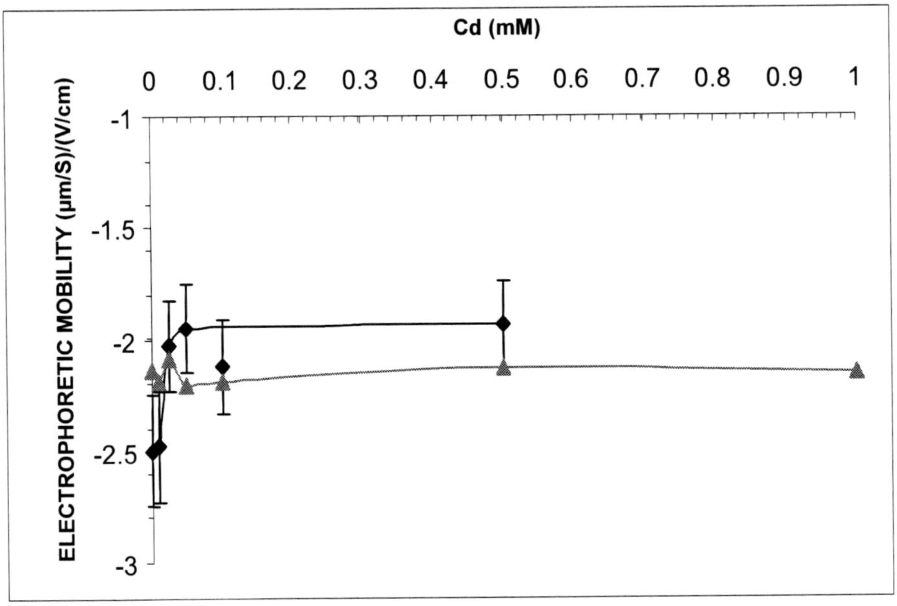

Figure 12. Electrophoretic mobility of *Pseudomonas veronii 2E* in presence of Cd(II) at pH = 7.5 (◆) and pH = 6.5 (▲).

Acknowledgements

This work was supported by the Universidad Nacional de General Sarmiento.

We are grateful to M. Alejandra Daniel for experimental collaboration, to Dr. Marcela Ferrero (PROIMI, Tucumán, Argentina) for the 16S r-RNA identification of P2 strain and to Dr. Roberto Candal (INQUIMAE, Buenos Aires, Argentina) for his help on electrophoretic mobility experiments.

References

Adams, N. and Kramer, J. (1999). Silver speciation in wastewater effluents, surface waters, and pore water. *Environ. Toxicol. Chem.*, 18, 2667-2673.

Alonso Alvarez, E., Callejón Mochón, M., Jiménez Sanchez, J.C. and Ternero Rodriguez, M. (1998). A voltammetry procedure with medium exchange for the speciation of heavy metals in wastewater of a sewage treatment plant. *Electroanalysis,* 10, 917-920.

Anderson, D.M. and Morel, F.M.M. (1978). Copper sensitivity of *Gony-aulax tamarensis.* Limnol. *Oceanogr.*, 23, 283-295.

Belli, S.L. and Zirino, A. (1993). Behavior and calibration of the copper(II) ion-selective electrode in high chloride media and marine waters. *Anal. Chem.*, 65, 2583-2589.

Bergman, U. and Ralph, K.M. (1983). Complexation and toxicity of copper and free metal biossay technique. *Water Res.*, 17, 1697-1703.

Brand, L.E., Sunda, W.G. and Guillard, R.R.L. (1986). Reduction of marine phytoplankton reproduction rates by copper and cadmium. *J. Exp. Mar. Biol. Ecol.*, 96, 225-250.

Bruland, K.W. (1992). Complexation of cadmium by natural organic ligands in the central *North Pacific. Limnol. Oceanogr.*, 37(5), 1008-1017.

Bruland, K.W., Rue, E.L., Donat, J.R., Skrabal, S.A. and Moffett, J.W. (2000). Intercomparison of voltammetric techniques to determine the chemical speciation of dissolved copper in a coastal seawater sample. *Anal. Chim. Acta,* 405, 99-113.

Buffle J. (1988). *Complexation reactions in aquatic systems. An analytical approach.* Ellis Horwood Ltd. Chichester.

Campbell, P.G.C. (1995). Interactions between trace metals and aquatic organisms: a critique to the free-ion activity model. In: *Metal Speciation and Bioavailability in aquatic systems.* A.Tessier and D.R. Turner (ed). John Wiley and Sons, New York, USA, 45-102.

Campos, M.L.A.M. and van den Berg, C.M.G. (1994). Determination of copper complexation in seawater by cathodic stripping voltammetry and ligand competition with salicylaldoxime. *Anal. Chim. Acta,* 284, 481-496.

Capodaglio, G., Coale, K.W. and Bruland, K.W. (1990). Lead speciation in surface waters of the eastern North Pacific. *Mar. Chem.*, 29, 221-233.

Ceretti, H.M., Vullo, D.L., Zalts, A. and Ramírez S.A. (2006). Cadmium complexation in culture media. Electroanalysis 18(5), 493-498.

Chang, J.S., Raw, R. and Chang, C.C. (1997). Biosorption of lead, copper and cadmium by biomass of *Pseudomonas aeruginosa* PU21. *Water Res.* 31(7), 1651-1658.

Chen, X.C., Wang, Y.P., Lin, Q., Shi, J.Y., Wu, W.X. and Chen, Y.X.(2005). Biosorption of copper(II) and zinc(II) from aqueous solution by *Pseudomonas putida* CZ1. *Colloids Surf B Biointerfaces,* 46, 101-107.

Coale, K. H. and Bruland, K.W. (1988). Copper complexation in the northeast Pacific. *Limnol. Oceanogr.,* 33, 1084-1101.

Coale, K. H. and Bruland, K.W. (1990). Spatial and temporal variability in copper complexation in the North Pacific. *Deep-Sea Res.,* 37, 317-336.

Cobelo-García, A. and Prego, R. (2004). Chemical speciation of dissolved copper, lead and zinc in a ria coastal system: the role of the suspended sediments. *Anal. Chim. Acta,* 524, 109-114.

Cobelo-García, A., Prego, R. and Nieto, O. (2003). Chemical speciation of dissolved lead in polluted environments. A case study: the Pontevedra Ria (NW Spain). *Ciencias Marinas,* 29(4), 377-388.

Davey, E.W., Morgan, M.J. and Erickson, S.J. (1973). A biological measurement of copper complexation capacity of Sea Water. Limnol. *Oceanogr.,* 18, 993-997.

Decree 999/92, Environmental Secretary (*Decreto 999/92, Secretaria de Medio Ambiente, República Argentina*).

Demir, A., and Arisoy, M. (2007). *Biological and chemical removal of Cr(VI) from waste water: cost and benefit analysis,* J. Hazard. Mater., doi: 10.1016/j.jhazmat.2006.12.076.

Donat, J. R., Statham, P.J. and Bruland, K.W. (1986). An evaluation of a C-18 solid phase extraction technique for isolating metal-organic complexes from central North Pacific Ocean waters. *Mar. Chem.,* 18, 85-99.

Donat, J., Lao, K.A. and Bruland, K. W. (1994). Speciation of dissolved copper and nickel in South San Francisco Bay; a multi-method approach. *Anal. Chim. Acta,* 284, 547-571.

Donat, J.R. and van den Berg, C. (1992). A new cathodic stripping voltammetry method for determining organic copper complexation in seawater. *Mar. Chem.*, 38, 69-90.

Duffus, J.H. (2002). "Heavy metals" – a meaningless term?, IUPAC technical report, *Pure Appl. Chem.,* 74(5), 793-807.

Eboigbodin, K.E., Newton, J.R.A., Routh, A.F. and Biggs, C.A. (2006). Bacterial quorum sensing and cell surface electrokinetic properties. *Appl. Microbiol. Biotechnol.,* 73, 669-675.

Gadd, G.M. (2000). Bioremedial potential of microbial mechanisms of metal mobilization and immobilization. Curr. Opin. Biotechnol., 11, 271-279.

Gardner, M. and van Veen, E. (2004). Comparability of copper complexation capacity determination by adsorption by chelating resin column and cathodic stripping voltammetry. *Anal. Chim. Acta,* 501, 113-117.

Gardner, M., Dixon, E. and Comber, S. (2000). Copper complexation in English rivers. *Chem. Spec. Bioavail.,* 12, 1-8.

Gonzalez-Gil, G., Jansen, S., Zandvoort, M.H. and van Leewen, H.P. (2003). Effect of yeast extract on speciation and bioavailability of nickel and cobalt in anaerobic bioreactors. *Biotechnol. Bioeng.,* 82, 134-142.

Good, N.E., Winget, G.D., Winter, W., Connolly, T., Izawa, S. and Singh, R.M.M. (1966). Hydrogen Ion Buffers for Biological Research. *Biochemistry,* 5, 467-477.

Guéguen, C., Koukal, B., Dominik, J. and Pardos, M. (2003). Competition between alga (*Pseudokirchneriella subcapitata*), humic substances and EDTA for Cd and Zn control in the algal assay procedure (AAP) medium. *Chemosphere,* 53, 927-934.

Hanson, A.K. and Quinn, J.G. (1993). The distribution of dissolved and organically complexed copper and nickel in the middle Atlantic Bight. *Canadian Journal of Fisheries and Aquatic Sciences,* 20, 151-161.

Hayashi, H., Seiki, H., Tsuneda, S., Hirata, A. and Sasaki, H. (2003). Influence of growth phase on bacterial cell electrokinetic characteristics examined by soft particle electrophoresis theory. *J. Colloid Interface Sci.,* 264, 565-568.

Itabashi, H., Shigeta, Y., Kawamoto, H. and Akaiwa, H. (2000). Simultaneous determination of the complexing capacity and conditional stablility constants of soluble copper(II) complexes in natural water samples by using a chelate extraction technique. *Analytical Sciences, 16*, 1179-1182.

Johnson, D.B. (2006). Biohydrometallurgy and the environment: Intimate and important interplay, *Hydrometallurgy,* 83, 153-166.

Johnson, J., Harper, E.M., Lifset, R. and Graedel T.E. (2007). Dining at the periodic table: metals concentrations as they relate to recycling. *Environ. Sci. Technol.,* 41, 1759-1765.

Jurado-Gonzalez, J.A., Galindo-Riaño, M.D. and Garcia-Vargas, M. (2003). Experimental design in the development of a new method for the sensitive determination of cadmium in seawater by adsorptive cathodic stripping voltammmetry. *Anal. Chim. Acta,* 487, 229-241.

Kirkham, M.B. (2006). Cadmium in plants on polluted soils: effects of soil factors, hyperaccumulation and amendments, *Geoderma,* 137, 19-32.

Kozelka, P.B. and Bruland, K.W. (1998). Chemical speciation of dissolved Cu, Zn, Cd, Pb in Narragansett Bay, R*hode Island. Mar. Chem.,* 60, 267-282.

Kunkel, R. and Manahan, S.E. (1973). Atomic absorption analysis of strong heavy metal chelating agents in water and waste water. *Anal. Chem.,* 45, 1465-1468.

Kurniawan, T.A., Chan, G.Y.S., Lo W. and Babel, S. (2006). Physico-chemical treatment techniques for wastewater laden with heavy metals. *Chem. Eng. J.,* 118, 83-98.

Lohan, M.C., Aguilar-Islas, A.M., Franks, R.P. and Bruland, K.W. (2005). Determination of iron and copper in seawater at pH 1.7 with a new commercially available chelating resin, NTA Superflow. *Anal. Chim. Acta,* 530, 121-129.

Manouchehri, N. and Bermond, A. (2006). Study of trace metal partitioning between soil-EDTA extracts and Chelex-100 resin. *Anal. Chim. Acta,* 557, 337-343.

Martell, A.E. and Smith, R.M. (2001). *NIST Critically Selected Stability Constants of Metal Complexes* Database 46 version 6.0.

Martino, M., Turner, A. and Millward, G.E. (2003). Influence of Organic Complexation on the Adsorption Kinetics of Nickel in River Waters. *Environ. Sci. Technol.* 37, 2383-2388.

Mash, H.E., Chin, Y-P., Sigg, L., Hari, R. and Xue, H. (2003). Complexation of Copper by Zwitterionic Aminosulfonic (Good) Buffers. *Anal. Chem.,* 75, 671-677.

Meylan, S., Odzak, N., Behra, R. and Sigg, L. (2004). Speciation of copper and zinc in natural freshwater: comparison of voltammetric measurements, diffusive gradients in thin films (DGT) and chemical equilibrium models. *Anal. Chim. Acta,* 510, 91-100.

Mills, G.L. and Quinn, J.G. (1981). Isolation of dissolved organic matter and copper-organic complexes from estuarine waters using reverse-phase liquid chromatography. *Mar. Chem.,* 10, 93-102.

Moffet, J. W. (1995). Temporal and spatial variability of copper complexation by strong chelators in the Sargasso Sea. *Deep-Sea Research,* 42, 1273-1295.

Moore, M.R. (2004). A commentary on the impacts of metals and metalloids in the environment upon the metabolism of drugs and chemicals. *Toxicol. Lett.,* 148, 153-158.

Morel, F.M.M. (1983). Trace metals and microorganisms. In: *Principles of aquatic chemistry.* Morel F.M.M. (ed). Wiley Interscience, New York, NY, 300-308.

Mota, A.M. and Gonsalves, M.L.S. (1996). Direct method of speciation of heavy metals in natural waters. In: *Element speciation in bioinorganic chemistry.* Caroli, S. (ed). Wiley, chapter 2.

Murphy, M. (ed) 1998. *Metal finishing. 66th Guidebook and Directory Issue.* 95, 1, ISSN 0026-0576, NY.

Mylon, S.E., Twining, B.S., Fisher, N.S. and Benoit, G. (2003). Relating the speciation of Cd, Cu and Pb in two Connecticut rivers with their uptake in algae. *Environ. Sci. Technol.* 37, 1261-1267.

Nriagu, J. (1997). *Industrial activity and metal emissions, In: Industrial Ecology and Global Change,* Socolow, R., Andrews, C., Berkhout F. and Thomas, V. (ed). Cambridge University Press, 277-285.

Nuester, J. and van den Berg, C.M.G. (2005). Determination of metal speciation by reverse titrations. *Anal. Chem.,* 77, 11-19.

Pagnanelli, F., Mainelli, S. and Toro, L. (2005). Optimisation and validation of mechanistic models for heavy metal bio-sorption onto a natural biomass. *Hydrometallurgy,* 80, 107-125.

Pesavento, M. and Alberti, G. (2000[b]). Determination of the complexing properties of drinking waters towards copper(II) and aluminium(III) by ligand titration. *Wat. Res.,* 34, 4482-4492.

Pesavento, M. and Biesuz, R. (1995). Simultaneous determination of total and free metal ion concentration in solution by sorption on iminodiacetic resin. *Anal. Chem.,* 67, 3558-3563.

Pesavento, M., Alberti, G. and Profumo, A. (2000[a]). Determination of the metal complexing capacity of aqueous solutions containing ligands by titration in the presence of complexing resins. *Anal. Chim. Acta,* 405, 309-319.

Qi, H.Y., Hu, Q., Dou, M.N. and Zeng, J.H. (2006). Biosorption of Cd^{2+} by a hyperresistant newly isolated *Bacillus cereus* strain. *11th International Symposium on Microbial Ecology,* Vienna, Austria, abstract page 79.

Rand, G.M. (ed). (1995). *Fundamentals of aquatic toxicology: effects, environmental fate and risk assessment.* Taylor and Francis, Washington, D.C.

Raspor, B. and Branica, M. (1975). Polarographic reduction of Cd(II)-nitrilotriacetic acid chelate in chloride solutions of pH about 8. *J. Electroanal. Chem. and Interfacial Electrochem.,* 59, 99-109.

Raspor, B. and Branica, M. (1973). Polarographic study of the cadmium-ethylendiamine-tetraacetate chelate at pH about 8. Electroanal. *Chem. and Interfacial Electrochem.*, 45, 79-88.

Rodriguez Presa, M.J., Catoggio, J.A., Posadas, D. and Tuccer, R.I. (1998). Determination of the complexation capacity of waters. Measurements on model systems and natural waters employing copper as the titrating ion. *Environ. Technol.*, 19, 45-54.

Rue, E.L. and Bruland K.W. (1995). Complexation of iron (III) by natural organic ligands in the Central North Pacific as determined by a new competitive ligand equilibration/adsorptive cathodic stripping voltammetric method. *Mar. Chem.*, 50, 117-138.

Ruzic, I. (1982). Theoretical aspects of the direct titration of natural waters and its information yield for trace metal speciation. *Anal. Chim. Acta*, 140, 99-113.

Scally, S., Davison, W. and Zhang, H. (2003). In situ measurements of dissociation kinetics and labilities of metal complexes in solution using DGT. *Environ. Sci. Technol.*, 37, 1379-1384.

Scarponi, G., Capodaglio, G., Barbante, G. and Cescon, P. in Elemental Speciation in Bioinorganic Chemistry. Caroli, S. (ed.), *J. Wiley and Sons Inc.* (1996), 363-392.

Semple, K.T., Doick, K.J., Burauel, P., Craven, A., Harms, H. (2004). Defining bioavailability and bioaccessibility of contaminated soil and sediment is complicated. Environ. *Sci. Technol.*, 228A-231A.

Shuman, M.S. and Woodward, G.P. (1973). Chemical constants of metal complexes from a complexometric titration followed with anodic stripping voltammetry. *Anal. Chem.*, 45, 2032-2035.

Skrabal, S., Donat, J. and Burdige, D. (2000). Pore water distribution of dissolved copper and copper-complexing ligands in estuarine and coastal marine sediments. Geochem. Cosmochim. *Acta*, 64, 1843-1857.

Soares, H., Conde, P., Almeida, A. and Vasconcelos, M. (1999). Evaluation of n-substituted aminosulfonic acid pH buffers with a morpholinic ring for cadmium and lead speciation studies by electroanalytical techniques. *Anal. Chim. Acta,* 394, 325-335.

Sunda, W. G. and Hanson, A.K. (1987). Measurement of free cupric ion concentration in seawater by a ligand competition technique involving copper sorption onto C18 Sep Pak cartridges. *Limnol. Oceanogr.*, 32, 537-551.

Templeton, D.M, Ariese, F., Cornelis, R., Danielsson, L. G., Muntau, H., Van Leeuwen, H. and Lobiński, R. (2000). Guidelines for terms related to chemical speciation and fractionation of elements. Definitions, structural aspects, and methodological approaches. *Pure Appl. Chem.*, 72(8), 1453-1470.

Thomas, V, and Spiro, T. (1997). Emissions and exposure to metals: cadmium and lead, In: *Industrial Ecology and Global Change,* Socolow, R., Andrews, C., Berkhout, F. and Thomas, V. (ed). Cambridge University Press, 297-318.

Town, R.M. and van Leeuwen, H.P. (2002). Effects of adsorption in stripping chronopotentiometric metal speciation analysis. *J. Electroanal. Chem.*, 523, 1-15.

Toze, S. (2006). Reuse of effluent water – benefits and risks. *Agricultural Water Management,* 80, 147-159.

Tsuneda, S. Aikawa, H., Hayashi, H. and Hirata, A. (2004). Significance of cell electrokinetic properties determined by soft-particle analysis in bacterial adhesion onto a solid phase. *J. Colloid Interface Sci.*, 279, 410-417.

Turner, D.R. (1986). *Biological availability at trace elements, in Biogeochemical processes at the land-sea boundary.* Laserre P. and Martin L.M. (ed). Elsevier, Amsterdam, The Netherlands, 191-200.

Tuschall, J.R. and Brezonik, P.L. (1981). Evaluation of the copper anodic stripping voltammetry complexometric titration for complexing capacities and conditional stability constants. *Anal. Chem.*, 53, 1986-1989.

Valls, M. and de Lorenzo, V. (2002). Exploiting the genetic and biochemical capacities of bacteria for the remediation of heavy metal pollution. *FEMS Microbiol. Rev.*, 26, 327-338.

van de Velde, K., Ferrari, C., Barbante, C., Moret, I., Bellomi, T., Hong, S. and Boutron, C. (1999). A 200 year record of atmospheric Co, Cr, Mo and Sb in high altitude alpine firn and ice. *Environ. Sci. Technol.*, 33, 3495-3500.

van den Berg, C. M. G. (1984). Determination of the complexing capacity and conditional stability constants of complexes of copper(II) with natural organic ligands in seawater by cathodic stripping voltammetry of copper-catechol complex ions. *Mar. Chem.*, 15, 1-18.

van den Berg, C.M.G. and Donat, J.R. (1992). Determination and data evaluation of copper complexation by organic ligands in seawater using cathodic stripping voltammetry at varying detection windows. *Anal. Chim. Acta,* 257, 281-291.

van den Berg, C.M.G. (1982[a]). Determination of copper complexation with natural organic ligands in seawater by equilibration with MnO_2 I. *Theory. Mar. Chem.*, 11, 307-322.

van den Berg, C.M.G. (1982[b]). Determination of copper complexation with natural organic ligands in seawater by equilibration with MnO_2 II. Experimental procedures and application to surface seawater. *Mar. Chem.*, 11, 323-342.

van den Berg, C.M.G., Nimmo, M., Daly, P. and Turner, D.R. (1990). Effects of the detection window on the determination of organic copper speciation in estuarine waters. *Anal. Chim. Acta,* 232, 149-159.

van Leeuwen, H., Cleven, R. and Buffle, J. (1989). Voltammetric techniques for complexation measurements in natural aquatic media. *Pure Appl. Chem.*, 61(2), 255-274.

van Leeuwen, H.P. and Town, R.M. (2002). Stripping chronopotntiometry for metal ion speciation analysis at a microelectrode. *J. Electroanal. Chem.*, 523, 16-25.

vanLoon, G.W. and Duffy, S.J. (2000). *Environmental chemistry of colloids and surfaces, Chapter 14 in Environmental Chemistry: A global perspective,* Oxford University Press, Oxford UK, 291-294.

van Veen, E., Burton, N., Comber, S. and Gardner, M. (2002[a]). Speciation of copper in sewage effluents and its toxicity to *Daphnia magna*. Environ. *Toxic. Chem.*, 21, 275-280.

van Veen, E., Comber, S. and Gardner, M. (2002[b]). Interlaboratory comparability of copper complexation capacity determination in natural waters. *J. Environ. Monit.*, 4, 116-120.

vanVeen, E., Gardner, M. and Comber, S. (2001). Temporal variation of copper and zinc complexation capacity in the Humber estuary. *J. Environ. Monit.*, 3(3), 322-323.

Vasconcelos, M.T.S.D., Asenha, M.A.G.O. and Almeida, C.M.R. (1998). Copper(II) complexation properties and surfactant activity of 3-[N,N-Bis(2-hydroxyethyl)amino]-2-

hydroxypropanesulfonic acid and N-(2-hydroxyethyl)piperazine-N′-2-hydroxypropane sulfonic acid pH buffers which may affect trace metal speciation in vitro studies. *Anal. Biochem.,* 265, 193-201.

Volesky, B. (2001). Detoxification of metal-bearing effluents: biosorption for the next century. *Hydrometallurgy,* 59, 203-216.

Vullo, D.L., Ceretti, H.M., Hughes, E.A., Ramírez, S. and Zalts, A. (2005). Indigenous Heavy Metal Multiresistant Microbiota of Las Catonas Stream. *Environ. Monit. Assess.,* 105, 81-97.

Wilson, W.W., Wade, M.M., Holman, S.C. and Champlin, F.R. (2001). Status of methods for assessing bacterial cell surface charge properties based on zeta potential measurements. *J. Microbiol. Methods,* 43, 153-164.

Wong, C.S.C., Li, X. and Thornton, I. (2006). Urban environmental geochemistry of trace metals. *Environ. Pollut.,* 142, 1-16.

Worms, I., Simon, D.F., Hassles, C.S. and Wilkinson, K.J. (2006). Bioavailability of trace metals to aquatic microorganisms: importance of chemical, biological and physical processes on biouptake. *Biochimie,* 88, 1721-1731.

Wu, J. and Luther (III), G.W. (1995). Complexation of Fe(III) by natural organic ligands in the Northwest Atlantic Ocean by a competitive ligand equilibration method and a kinetic approach. *Mar. Chem.,* 50, 159-177.

Xue, H. and Sigg, L. (2002). A Review of Competitive Ligand-Exchange-Voltammetric Methods for Speciation of Trace Metals in Freshwater. *Acs Symposium Series,* 811, 336-370. Washington, DC; American Chemical Society.

Xue, H. and Sigg, L. (1998). Cadmium speciation and complexation by natural organic ligands in fresh water. *Anal. Chim. Acta,* 363, 249-259.

Xue, H. and Sigg, L. (1999). Comparison of the complexation of Cu and Cd by humic or fulvic acids and by ligands observed in lake waters. *Aquatic Geochemistry,* 5, 313-335.

Xue, H., Jansen, S., Prasch, A. and Sigg, L. (2001). Nickel speciation and complexation kinetics in freshwater by ligand exchange and DPCSV. *Environ. Sci. Technol.,* 35(3), 539-546.

Yin, P., Yu, Q., Jin, B. and Ling, Z. (1999). Biosorption removal of cadmium from aqueous solution by using pretreated fungal biomass cultured from starch wastewater. *Water Res.,* 33(8), 1960-1963.

Yu, Q., Kandegedara, A., Xu, Y. and Rorabacher, D.B. (1997). Avoiding interferences from Good's buffers: a contiguous series of noncomplexing tertiary amine buffers covering the entire range of pH 3–11. *Anal. Biochem.,* 253, 50-56.

Zirino, A., De Marco, D.J., van der Weele, D.A. and Belli, S.L. (1998). Direct measurement of copper(II) (aq) in seawater at pH 8 with the jalpaite ion-selective electrode. *Mar. Chem.,* 61, 173-184.

In: Progress in Environmental Microbiology
Editor: Myung-Bo Kim, pp. 151-168
ISBN: 978-60021-940-5

Chapter IV

Community Level Physiological Profiles as Influenced by Soil Management. Critical Considerations about their Interpretation

Elena del Valle Gomez[a] and Olga Susana Correa[b]

[a]Cátedra de Microbiología Agricola, Facultad de Ciencias Agrarias, Universidad Nacional de Rosario, Campo Experimental J. Villarino, 2123, Zavalla, Argentina

[b]Cátedra de Microbiología, Facultad de Agronomia, Universidad de Buenos Aires, Av. San Martin 4453, 1417, Buenos Aires, Argentina

Abstract

Management practices have a significant impact on soil organism populations and their activity. Thus, microbiological properties have been considered as interesting attributes which are highly sensitive to changes produced by soil management. To be considered sustainable, management practices should promote fertility and productivity in the long term and preserve soil microbial diversity.

Soil microbial community dynamics is directly involved with ecosystem functions since it regulates fundamental processes such as organic matter decomposition, nutrient cycling or xenobiotic degradation. Changes in the composition or activity of microbial communities may have short or long-term effects in the ecosystem functioning. Mainly due to this fact, an increasing interest has arisen in the study of microbial communities in the recent years. Different approaches have been developed, based on physiological traits, biochemical analysis of cell constituents or genetic procedures.

By applying the commercial Biolog GN microplates originally designed for Gram negative isolate identification, Garland and Mills (1991) developed a rapid community-level technique based on sole carbon source oxidation for estimating microbial community diversity. Each GN microplate consists of 95 wells with a carbon source incorporated to a basal medium as sole C-source and one control without any C-source; the redox dye tetrazolium violet colorimetrically indicates the utilization of the carbon substrates.

Since then, this approach which was later called community level physiological profile (CLPP) has been increasingly applied and proved to be useful to detect relative changes in microbial communities.

However, as all methods available up to now for describing microbial communities as regards structure and/or function, CLPP has advantages and limitations. Selective enrichment resulting from the high substrate concentrations employed and relatively long incubation periods have to be taken into account in order to prevent over-interpretation of the information obtained.

In this chapter we summarize some of our findings working on CLPP from Biolog GN™ and from the later developed Biolog EcoPlate™ assay (which allows testing a more reduced number of substrates and a larger number of replicates), in field and greenhouse experiments. The suitability of CLPP to be used for monitoring soil quality is also analyzed.

Critical considerations quoted in recent literature as regards interpretation of CLPP are discussed, and also alternative approaches directly measuring oxygen consumption or carbon dioxide production due to the carbon substrate catabolism more recently developed in order to provide a more accurate and ecologically relevant overview on soil functional microbial diversity.

Introduction

Soil is a dynamic system of major importance not only for food and fiber production, but also for ecosystem global balance and functioning which determine the sustainability of life on earth (Gregorich et al., 1994). The thin layer of soil that covers the earth surface represents the difference between life survival and extinction, and makes soil a resource as vital as water, though unlike water, soil is not renewable in a human time scale (Lal, 1994). Soils are constituted by mineral particles, stable and reactive organic matter and numerous organisms; they developed over long periods of time under the influence of climate, topography, parent material, and living organisms. Soil physical and chemical attributes and the interchanges occurring between the solid, liquid and gaseous phases influence and are influenced in turn by the biological activity.

Soil quality determines agricultural sustainability and environmental quality. Beside of being a habitat for the biota and a genetic reservoir, the ecological functions of soils include biomass production and the ability to transform compounds for protecting the environment, groundwater and food chain from pollution (Doran et al., 1996).

Inappropriate soil managements constitute a serious threat for agricultural sustainability. Quality of several soils has significantly declined since forests and prairies were replaced by agriculture. At the beginning, the increasing production of monocultures and incremented use of fertilizers and pesticides resulted in higher crop yields. However, soil organic mater loss, erosion and groundwater pollution have incremented as a consequence of continual agriculture intensification.

The critical role of soil in determining the sustainability of ecosystems has generated an increasing interest in the assessment of the quality of soil. Since soil is one of the most complex habitats on earth, the evaluation of soil quality requires the consideration of physical, chemical and biological aspects.

Yet before the concept of soil quality became widespread, several physical and chemical properties have been used as soil quality indicators. The physical matrix of the soil is the ambit where all chemical and biological processes occur. Among physical attributes texture, aggregate stability, bulk density, water holding capacity, infiltration may be cited. Chemical attributes are involved in nutrient absorption and plant and soil organism population growth. Organic matter plays an important role in soil structure by improving aggregate stability, besides being a nutrient reservoir. Organic carbon, total nitrogen, C/N ratio, pH and cationic interchange capacity (CIC) are some of soil chemical properties.

Biological properties have been less used as soil quality indicators, partially due to the lack of standardized proceedings in many methods and the difficulty to obtain reproducible measurements (Stenberg, 1999). However, some microbiological properties such as microbial biomass, respiration and metabolic quotient have been often quoted in literature as good indicators of soil management effects.

Microorganisms as Soil Quality Indicators

Soil biological components are fundamental to the development of ecologically relevant processes such as organic residue decomposition, nutrient cycling, humus synthesis, improvement of soil structure and xenobiotic decomposition. Microbial activity dominates the degradation of soil organic substrates and it is of major concern for ecosystem functioning. Essential parts of carbon, nitrogen, phosphorus and sulfur global cycles take place in soils by microbial action.

Management practices have a significant impact on soil organism populations and their activity (Roper and Gupta, 1995), as it is shown in Fig. 1 which summarizes the functions of soil microorganisms and their relationship with management practices. Microorganisms are largely sensitive to perturbations, so that modifications in soil biota could precede detectable changes in physical and chemical properties providing an early sign of soil improvement or warning of soil deterioration (Pankhurst et al., 1995; Dick et al., 1996). Thus, microbiological properties have been considered as interesting attributes highly sensitive to changes produced by soil management, and pointed out as suitable indicators complementing physical and chemical properties in soil quality evaluation (Elliot, 1997; Stenberg, 1999).

Integrative Microbial Measurements: Microbial Biomass, Respiration and Metabolic Quotient

Soil organic matter level modifies slowly; however, some of its fractions like microbial biomass may be susceptible to be altered under the influence of different managements (Omay et al., 1997). Microbial biomass, the living component of soil organic matter involved in nutrient transformations and storage, is an active pool of organic carbon which turnover time is less than a year, making it more rapidly responsive to conditions that alter soil organic matter levels (Rice et al. 1996; Melero et al., 2006; Reyes-Reyes et al., 2007).

Soil microbial respiration (SMR) reflects the global activity of aerobic microorganisms and is directly linked to the degradation of soil organic carbon and the release of nutrients (Franchini et al., 2007; Marinari et al., 2007). However, looking only at absolute respiration rates may not be adequate for the interpretation of SMR measurements and it is necessary to relate them to the specific functions the microorganisms perform. For example, the release of nutrients may be positive in terms of crop production but in the long term, it may represent organic matter loss. So, SMR measurements must be related to carbon input in order to estimate if a system is losing or gaining carbon (Parkin et al., 1996).

The metabolic quotient (qCO_2) is an estimation of the microbial metabolic efficiency, which relate microbial respiration per biomass unit and reflects microorganism maintenance energy. Stress conditions may cause an increment in the qCO_2, while a low qCO_2 may be indicative that more carbon is available for biomass production (Anderson, 2003; Franchini et al., 2007; Zhang et al., 2007).

Microbial biomass, SMR and qCO_2 have proven to be responsive to soil management and pollution. However, though they provide meaningful information they are black-box approaches that do not allow looking deeper at soil microbial community complexity (Schloter et al., 2003). Since management practices should promote fertility and productivity in the long term to be considered sustainable, preserving soil microbial structural and functional diversity seems to be of major importance (Giller et al., 1997), and thus, we also should focus on the study of soil microbial community.

Soil Microbial Community Diversity Assessment Procedures

Soil microbial diversity is huge, as it was shown by Torsvik et al. (1990), who reported that one single gram of soil contains thousands of bacterial species. Microbial communities have the ability to adapt to changing environmental conditions by different ways such as modification of individual activity, increasing reproduction of species with favorable abilities or by developing new capabilities via horizontal gene transfer. Due to their high surface to volume ratio, microorganisms have close relations with the surrounding environment, which make them highly responsive to changes and environmental stress (Winding et al., 2005). New methods developing in microbial ecology provide a useful tool to achieve a better understanding about soil microbial composition and dynamics, however, microbial diversity and its role in ecosystem functioning are still partly unraveled (Muyzer, 1999).

Difficulty in the extraction of soil microorganisms, selectivity in growth media that prevents the cultivation of most of the microorganisms and underestimate the real population and difficulties in the classification based on morphological traits are some of the limitations that have restricted the study of soil microbial diversity (Garland, 1997). Methods like biochemical analysis of cell components and genetic approaches have been developed in the last few years; they offer promising perspectives since they do not depend on cultivation, though they also have technical and cost limitations that restrict their use (Garland and Lehman, 1999).

Biochemical methods allow for microbial community structure characterization based on certain cell constituents. Phospholipids and their associated fatty acids from the membranes of living cells are useful biomarkers allowing for the differentiation of subgroups of organisms (Palojärvi et al., 1997). Phospholipids are thought to represent a component of living cells because they break down rapidly when the cells die and do not persist long enough to interact with soil colloids. Though phospholipid fatty acids (PLFA) are found in the membranes of almost all microorganisms, the relative amount varies among different organisms. Signature PLFA exists that are able to differentiate major taxonomic groups within a microbial community (Gram (-) and Gram (+) bacteria, actinobacteria and fungi). Therefore, phospholipid profiles are useful to quantify the presence and relative abundance of in soil microbial communities. Fatty acids are extracted directly from soil, methylated and analyzed by gas chromatography. The method has been extensively used to study microbial community composition and population changes due to agricultural management practices and to xenobiotics. However, PLFA analysis is time-consuming and toxic chemicals are required. Also, since the same fatty acid may be present in more than a single group, this approach has low to intermediate resolution and can be used to estimate only gross changes in community structure (Wander et al., 1995; Nannipieri, 2003). In addition, when using fungal spores to study the potential fungal diversity a high number (130 to 150 spores) of them are needed and this may ignore less predominant species in the microbial population (Kirk et al., 2004).

Molecular methods developed to describe the microbial community structure provide information about the relative similarity and genetic complexity in microbial community diversity. However, those methods are not simple to interpret when applying to soil, partially due to humic compound interference that makes difficult to recover nucleic acids from soil (Garland and Lehman 1999; Nannipieri, 2003). Most of these methods are based either in direct nucleic acid analysis or in multiplying the number of copies from a gene by PCR-amplification to make its detection easier. Re-association kinetics DNA:DNA (Torsvik et al., 1990), nucleic acid hybridization (Buckley et al., 1998), fluorescent probe in situ hybridization (Christensen et al., 1999), microarrays (Smalla et al., 2001) and metagenomic sequence analysis (Handelsman, 2004) are the most used methods directly analyzing nucleic acids. On the other hand, rDNA gene separation by electrophoresis in dissociating gradient gel (DGGE; Muyzer et al., 1993), terminal restriction fragment length polymorphism (T-RFLP; Liu et al., 1997), single strand conformational polymorphism (SSCP; Schiwieger and Tebbe, 1998), ribosomal intergenic space analysis (RISA; Ranjard et al., 2000) and the analysis of clone library sequences of 16 rRNA gene (Olsen et al., 1986) are among the molecular approaches that depend on previous PCR.

The DGGE, originally developed to identify punctual mutations in medical research and then adapted by Muyzer et al. (1993) for the study or soil microbial communities, is the molecular method more applied for their characterization (Nakatsu, 2007). There are primers for the three domains, Bacteria, Eucarya y Archaea, and more specific primers can be chosen for the study of bacterial and fungal populations more frequent in soils (i.e., actinomycetes, ammonia oxidizer □ bacteria, nitrogen fixers, cyanobacteria, mycorrhiza, etc.). The DGGE is a useful method for the analysis of a great number of samples that is often the case in soil microbial ecology. The DNA bands separate according to their sequence and the genetic profiles obtained are representatives of the community structure. The number of bands is an approximation to the number of species, while the intensity of each band is linked to the relative abundance. However, different bands may come from the same microorganism and DNA from more than a microorganism may be present in the same band. As all the methods in which PCR is previously performed, the profiles are representative of the obtained PCR products which, in turn, could not be representing the real proportion of species in the original community.

A physiological approach originally designed for Gram negative isolate identification, the commercial Biolog GN microplates, was first applied by Garland and Mills (1991) for estimating microbial community diversity in soil, water and rhizosphere samples. This approach was later called community level physiological profile (CLPP) (Lehman et al., 1997). In the present chapter, we will focus mainly on this approach and its potential responsiveness to soil management, and also we will summarize results from some of our own experiences. Critical considerations about CLPP interpretation and alternative approaches attempting to overcome its restrictions will be presented.

Community Level Physiological Profiles

This rapid community-level assay based on sole carbon source oxidation consists of 95 wells with a carbon source incorporated to a patented basal medium as sole C-source and one control without any C-source; the redox dye tetrazolium violet colorimetrically indicates the utilization of the carbon substrates. Among the carbon substrates in the plates there are polymers, carbohydrates, carboxylic acids, amino acids, amines and amides, and different commercial plates are available for bacterial and fungal studies. A later assay, the Biolog EcoPlate^{TM}, consisting of 31 carbon sources which allow testing three replicates per plate, was developed with a reduced set of carbon substrates commonly present in soil proposed by Insam (1997).

Aliquots of soil suspensions in sterile saline (0.85% NaCl) or deionized water (Franklin et al., 2001) are inoculated in each well and incubated and color development in the wells is recorded as optical density with a plate reader. A pre-incubation period (18-24 h) before inoculation may be performed to allow microbial utilization of any soluble organic carbon derived from the soil that could interfere in the sole carbon substrate response (Dick et al., 1996; Gomez et al., 2004).

The simpler way to analyze the response in the microplate is looking at univariate measurements such as average well color development (AWCD), richness of metabolized substrates and evenness, or some index that combines both richness and evenness like the Shannon Weaver Index. But in order to capture all the potential differences in the CLPP, multivariate analysis provide a pattern of relative similarity between samples based in the use of each carbon substrate (Garland, 1997).

Since Garland and Mills first used this approach, it has been increasingly applied.

Several findings have proved the assay to be useful to detect relative changes in microbial communities from different environmental and soil samples. Zak et al., 1994 classified samples from one transect with different types of vegetation. Bossio and Scow (1995) found that microbial communities were affected in their response pattern in Biolog GN by carbon input and flooding in rice crop experiments. Campbell et al. (1997) found that soil microbial communities from different grassland sites discriminated better using plant root exudates than those carbon sources present in Biolog GN. Lupwayi et al. (1998) could distinguish samples on the basis of tillage and previous crop. Sharma et al. (1998) found that functional diversity increased in soil incubated with maize litter amendment. Both the Biolog and the phospholipids fatty acid analysis showed successional patterns in microbial communities over time when a silt loam soil was amended with simple and complex compounds and crop residue (Schutter and Dick, 2001). More recently, Bending et al. (2004) included Biolog GN in their study of interrelationships between microbial properties as indicators of soil quality; Avidano et al. (2005) used Biolog GN to assess soil health in a polluted site from Italy. The effect of green manuring on soil microbial community structure and function was investigated by Elfstrand et al. (2007) applying PLFA and Biolog EcoPlate, respectively. Also both approaches were used by Williams and Rice (2007) to analyze how long-term enhancement of soil water availability influences microbial communities. The above mentioned are only some of the works from the profuse literature on CLPP. Following we summarize our own experiences utilizing CLPP from BIOLOG® GN plates and BIOLOG ® EcoPlate in field and in controlled condition experiments.

CLPP from a Land Use Intensification Gradient

In Entre Ríos Province (Argentina), like in other sites all over the world, agriculture intensification has promoted clearing of native vegetation, in many cases in an indiscriminate way. The study presented following was addressed to explore the consistency of CLPP to show patterns associated to land use intensification and the suitability of the CLPP to be used for monitoring soil quality in relation to other soil physical and chemical properties traditionally used to evaluate soil quality.

The experiment was carried out in Vertic Argiudolls (31°30′S latitude; 59°45′W longitude), with increased time elapsed since clearing of the xerophytic and herbaceous native vegetation: NV (native vegetation); NP (cleared in 1982, moldboard plowed for 8 years and cropped with corn and soybean; then under naturalized prairie); CT (cleared in 1972, moldboard plowed and cropped with corn and soybean); and NT (cleared in 1958, moldboard plowed till 1994; then cropped with corn and soybean under no till). Composite samples were collected from the surface layer (0–7.5 cm) from each one of the sites in three consecutive years. More detailed information about the sites under study and the sampling can be found in Gomez et al. (2000) and Gomez et al. (2004). Biolog GN microplates were inoculated with aliquots from soil suspensions previously pre-incubated and then incubated at 25 °C, and data from optical density at 54 h of incubation were analyzed. The average activity in the plate (AWCD), Richness (R), as the number of oxidized C substrates, and the Shannon–Weaver index (*H*) (i.e., the richness and evenness of response) were calculated and analyzed by ANOVA, and Principal components (PCA) was performed to analyze the relative use of the 95 carbon sources.

The average activity in the plate (AWCD) and R was significantly different among NV, NP, CT and NT, and showed always the highest values in samples from NV and the lowest values in NT, in all the three sampling years. Table 1 shows the AWCD and the Shannon–Weaver index (*H*), which was always significantly higher in NV and significantly lower in NT with respect to the other sites. The naturalized prairie and CT did not differ in *H* in the first sampling date, but it was significantly higher in NP compared to CT in the other two years (Gomez et al., 2004). Proportional reduction in R with respect to the 95 carbon substrates (maximum richness equal to 100%) present in BIOLOG GN plates is shown in Fig.2.

The PCA based on the intensity in the utilization of the 95 carbon sources showed a consistent separation of samples from the native condition and the sites with increasing time since clearing and different further management. The first and second principal components explained 62.8, 73.1 and 68.3% in each one of the three sampling years, respectively (Gomez et al., 2004). The distribution of the samples in the bi-dimensional space in all the sampling dates is summarized in Fig 3.

Consistent distinctive CLPP patterns were found from sites with increasing time elapsed since clearing of the native vegetation and different further managements over three consecutive years, evidenced by the significant difference in AWCD, R and *H*, and by the high proportion of the information originally contained in the 95-dimensional data that was displayed in a two-dimensional space.

The CLPP were compared with other soil physical and chemical properties traditionally used to evaluate soil quality, like the variation between dry and wet average aggregate diameter (ΔAAD) after sieving (which is indicative of structural stability), organic carbon (OC) and total nitrogen (tN). The CLPP differentiated each one of the sites in the land use intensification gradient, while ΔAAD did not differentiated NP from NV and CT, while OC and tN could not distinguish between the two agricultural sites (Gomez et al., 2004). The CLPP resulted to be more sensitive to distinguish the sites under study than ΔAAD, OC and tN, as it is reflected in Fig. 4.

CLPP as Influenced by Organic Amendment Application

Continuous soil removal and intensive use of pesticides and fertilizers may cause deterioration of soil physical, chemical and biological properties in conventional horticultural cropping. Organic amendment application is an alternative practice that helps improving soil condition and acts as a source of carbon and other nutrients, favor microbial biodiversity and activity, and improve soil structure.

The aim of the work described below was to evaluate the short-term response of microbial community functional diversity to organic amendment incorporation, in a horticultural soil with a long history of conventional use.

The experimental site, a Vertic Argiudoll located in Zavalla, Argentina (32°43′27″ S; 60°55′18″ W) had been conventionally managed with horticultural crops for many years. Plots in a complete randomized block design with three replicates were amended with vermicompost from source-separated household solid waste (HSW), vermicompost from horse and rabbit manure (HRM) and chicken manure (CM) that were incorporated at rates of 10 and 20 Mg ha^{-1} on a dry base (255, 188 and 428 kg C Mg^{-1} HSW, HRM and CM, respectively); control plots without any amendment (C) were also included in the experiment. Six months after amendment incorporation, composite samples were collected, inoculated in BIOLOG® EcoPlate and analyzed as it is detailed in Gomez et al. (2006).

Significant differences among treatments were found in the average well-color development (AWCD), richness (R) and Shannon–Weaver index (H) calculated from the use of the different carbon sources after the amendment incorporation as it is shown in Table 2. The rate of 20 Mg ha^{-1} always showed values significantly higher of AWCD and H compared to the unamended plots, and only the HSW treatment showed significant differences when comparing the rates of 10 and 20 Mg ha^{-1} (Table 2). A PCA performed on data from each of the amendment application rates showed that plots amended with 10 Mg ha^{-1} and 20 Mg ha^{-1} grouped together and differentiated from the control and one undisturbed adjacent site included in the analysis (data not shown). A regression analysis was performed between richness and Shannon–Weaver index with soil organic carbon; the regression coefficients were 0.77 and 0.72, respectively, suggesting that greater carbon availability due to amendment incorporation could be explaining the increase in microbial community functional potential (Gomez et al., 2006).

Short-term changes in the microbial functional diversity in response to amendment applications could be detected by the CLPP resulting from the use of carbon substrates present in BIOLOG EcoPlate.

CLPP of Root and Leaf Microbial Communities as Affected by Plant Genotype and PGPR Inoculants

Plant genotype determines microbial community composition on roots and leaves. Agricultural management practices, such as applying plant growth promoting rhizobacteria (PGPR) inoculants, could induce alterations on natural microbial communities settled on plant surfaces. These changes would be responsible at least in part of plant growth, health and yield (Simon et al., 2001; Lindow and Brandl, 2003).

The purpose of the following study was to characterize changes in microbial communities associated to tomato (*Lycopersicon esculentum*) roots and leaves as induced by plant genotype and the application of *Azospirillum brasilense*. Culture-dependent and independent methods were used (CLPP and DGGE, respectively).

The experiment was carried out in Buenos Aires (34° 38′ S, 58° 28′ W), Argentina. Seeds of two tomato varieties, fresh-market tomato and cherry tomato [*L. esculentum* Mill. var. *cerasiforme* (Dunal) A. Gray) hybrid Supersweet100 (Rogers, CA, USA)] and fresh-market tomato [(*L. esculentum* var. *esculentum* Mill.) cultivar ACE 55 (Asgrow Seed Co. Mendon, MI, USA)] were inoculated with *A. brasilense* strain BNM65; controls without inoculation were included in the experiment. Plants were randomly placed in a plant growth chamber and grown at 25°C with 12-h light/12-h dark cycles. Sixty days after sowing plants were collected, root and leaf extracts were prepared and aliquots inoculated in BIOLOG® EcoPlate. Denaturing gradient gel electrophoresis (DGGE) of PCR-amplified 16S rRNA gene was performed with DNA prepared from the same plant extracts as reported in Correa et al. (2007). Principal component analysis (PCA) was used to examine the relationship among treatments, while DGGE banding patterns were assessed by cluster analysis using the Pearson correlation coefficient (r), and similarities among profiles were depicted with a dendrogram using the UPGMA method (Sneath and Sokal, 1973).

PCA from CLPP data based on the intensity in the utilization of 31 carbon sources showed a clear differentiation between the two tomato varieties, either in roots as in leaves and also between non-inoculated and inoculated plants. The first and second principal components explained 56.5 and 51.7% of data variance for phyllosphere and rhizoplane microbial community, respectively. Tomato genotype was the main factor determining the CLPP of the phyllosphere bacterial community (Fig. 5A). The inoculation with *Azospirillum* induced changes in the phyllosphere of cherry tomato (Fig. 5A) and in the rhizoplane of both tomato varieties (data not shown).

The DGGE allowed differentiating the phyllosphere microbial community of cherry and fresh-market tomato (Fig. 5B). The bacterial community fingerprints obtained from fresh-market tomato leaves non-inoculated with *Azospirillum* (P FMT NI) was much more diverse than that obtained from cherry tomato leaves (CHT NI); only a 20% of similarity between them was observed. Inoculation with *Azospirillum* induced more changes in the less diverse phyllosphere of cherry tomato (Fig. 5B, compare CHT NI and CHT I) than in those more diverse of fresh-market tomato (Fig. 5B, FMT NI and FMT I). Modifications in rhizoplane bacterial communities of cherry tomato, induced by inoculation with *A. brasilense*, also could be detected both by CLPP and DGGE (Correa et al., 2007).

Using both combined approaches allowed us to establish that it was the plant genotype the determining factor of the structure and function of bacterial community in tomato phyllosphere and also in the changes induced, either in root or leaves, by inoculation with a PGPR.

In other experiment, the CLPP was used to study the impact of a PGPR-plant fungal pathogen bio-controller; *Bacillus amyloliquefaciens* strain BNM122, on soybean rhizosphere bacterial community (Fernández Ferrari, 2006). This bacterial strain was isolated in our laboratory from suppressive soils of soybean crop and showed wide antifungal activity against phytopathogens (Souto et al., 2004). The use of biological agents to control plant disease is a friendly agronomical practice that allows reducing doses of fungicide avoiding ecosystem pollution and protecting animals and human beings health. Before a biological agent can be commercially used, assays evaluating the effect on not target microorganisms must be conducted. In Argentina there is an increasing use of soybean seed inoculated with *Bradyrhizobium japonicum,* in order to obtain nitrogen from biological fixation is a usual practice. In addition, the low phosphorus levels of many of our soils make soybean dependent on mycorhization. The introduction of BNM122 as an inoculant on soybean seeds should not have negative impact on these two microbial-plant symbioses.

Our objective was to evaluate the impact of the bio-controller BNM122 on the native microbial community from the rhizosphere, the nodulation by *B. japonicum* and the mycorhization of soybean plants and to compare it with the effect of synthetic fungicides (Thiram+ Carbendazim). Soybean seeds were inoculated with *B. japonicum* E109 and BNM122 or treated with a mix of chemical fungicides (thiram + carbendazim) just before sowing them in 4 L plastic-pots, filled with soil from an agricultural experimental field. Controls with seeds only inoculated with *B. japonicum* were included. Pots were arranged in a glasshouse in a completely randomized design with five replicates per treatment. Plants grew under natural temperature and light conditions. At R6 stage (full seed) soybean plants were collected, rhizosphere extract prepared and loaded in BIOLOG® Ecoplates. Microplates were incubated for 72 h at 25 °C. Principal component analysis (PCA) was used to examine the relationship among treatments.

As in the study described before, the CLPP allowed evidencing alterations in microbial communities from the rhizosphere of plants inoculated with BNM122 and treated with a chemical fungicide compared with control plants (Fig. 6). However, significant differences in the CLPP were only evident between plants treated with the chemical fungicides and control plants. On the other hand, both symbioses were negatively affected by the use of chemical fungicide, while the control plants and those inoculated with the bio-controller agent showed no significant differences either in nodulation or in mycorhization (Fernández Ferrari, 2006). In this study it was demonstrated that the chemical fungicide produced greater alterations in microorganisms from the rhizosphere than the use of a bio-control agent with antifungal activity, and the CLPP was a sensitive approach that allowed showing those differences.

Critical Considerations about Biolog-CLPP. Alternative CLPP Approaches

The CLPP obtained from Biolog microplates has been widely used and proved to be valuable in detecting temporal and spatial differences in microbial community functional potential. However, such approach has been also questioned, arguing that it provides a biased view, since it is dependent on inocula density and cultivation, and it can not be unraveled what fraction of the microbial community is responsible for the obtained pattern of substrate oxidation (Garland, 1997; Konopka et al., 1998). Smalla et al. (1998), applied molecular approaches, DGGE and temperature gradient gel electrophoresis (TGGE) in the original inocula and in individual wells from incubated Biolog microplates; they could determine that they were the fast-growing bacteria those responding in the Biolog assay. Selective enrichment occurs in the wells due to the high substrate concentrations and long incubation periods from more than 48 hours (Garland et al., 2003).

Alternative methods addressed to obtain more ecologically relevant soil CLPP had been developed in the last years. Degens and Harris (1997) developed an approach based on substrate-induced respiration (SIR) differences, which does not depend from extraction and cultivation of soil microorganisms. Soil samples are directly incubated in hermetic vials with solutions of single organic substrates, for short periods of time (4 hours) and respiration is assessed measuring CO_2 evolved by gas chromatography. This method has proved to be useful for characterizing soils from different agricultural managements (Degens and Harris, 1997, Graham and Haynes, 2005). However, according to our own experience, the approach is very laborious and time-consuming, which restricts the number of treatments and replicates to be analyzed. Previous studies must be performed in order to determine which substrates and concentrations can be expected to allow the maximum response in the minimum time. A great variability exists among replicates from a same treatment and yet in seudo-replicates from the same sample (Casas, 2006). A high concentration of substrate is still required in this technique in relation to soil sample size, which may generate enrichment conditions, and also makes it necessary to adjust the pH of substrates with strongly acidic reaction in the solution. This last becomes an additional factor of variability among substrates that can not be attributed to the substrate *per se* but to the buffer solution used in pH adjustment (Casas, personal communication). Though this method closely approaches to "in situ" activity, the reasons mentioned before possibly have prevented a more widespread use of it.

Also based on substrate-induced respiration, but combining it with Biolog assay, Campbell et al. (2003) applied a method called MicroResp to measure the CO_2 production from the catabolism of different carbon sources added to each well of a microplate, and a radioactive carbon substrate method. The MicroResp system consists of a microplate with deep wells where the soil samples and the added substrates are placed and a top microplate where the evolved CO_2 is detected. The CO_2 evolved is trapped by absorption in an alkali and can be determined by colorimetric detection trough a pH indicator dye or by radioactivity scintillation counting. Though this method allow for physiological fingerprinting of soil microbial communities with the use of very low substrate concentrations, its use is restricted mainly by cost and, in the case of radio-labeled substrates, the difficulty to manage radioactive wastes.

The BD Oxygen Biosensor System (BD Oxy) is a fluorescence-based microplate for detecting dissolved oxygen. This approach was first used by Garland et al. (2003) to assess CLPP from rhizosphere and salt marsh litter samples, acclimated and unacclimated to specific substrates. The BD Oxy plate consists of an O_2-sensitive fluorophore (a ruthenium dye) absorbed in a silicon matrix that is permeable to oxygen, and allows defining the substrates to be used or the manipulation of factors such as inorganic nutrient levels. Since the fluorescence is quenched by the presence of O_2, the consumption of O_2 by respiration in the sample will increase the fluorescent signal that is read in a fluorometer. Fluorescence reaches a maximum value and then declines as O_2 diffusion exceeds O_2 consumption. The low substrate concentrations, 10 to 100 times less than those employed in the Biolog assay and in the SIR method, and the short incubation times needed to reach the peak (less than 24 hours) prevent selective enrichment (Garland et al., 2003).

The BD Oxy approach was used by Gomez et al. (unpublished data) to evaluate the catabolism of different carbon sources as influenced by tillage, fertilization, and mineral supplementation. The BD Oxy CLPP allowed differentiating no till from chisel plowing; as regards mineral supplementation two groups could be distinguished: nitrogen and nitrogen plus phosphorus supplements showed the highest values of fluorescence for most of the carbon substrates tested, compared to phosphorus supplemented and unsupplemented treatments.

Conclusion

Substrates in Biolog microplates provide a wide set of compounds that enable to estimate relative potential metabolic versatility. Distinctive patterns of carbon source utilization by soil microbial communities can be obtained from Biolog system, suggesting that Biolog-CLPP could be a useful tool for soil quality monitoring, particularly given its relative simplicity, and in fact, some monitoring programs have included the assay (Winding et al., 2005).

However, CLPP from Biolog has biases and limitations as well as other methods for studying microbial communities. It provides information about a portion of the community, that which is able to grow under the assay conditions, thus, it should not be emphasized on particular substrates catabolism, since they can not be related to "in situ" utilization. Selective enrichment resulting from the high substrate concentrations employed and relatively long incubation periods also must be taken into account in order to prevent over-interpretation of the information obtained.

Other approaches directly measuring respiration products, like BD Oxy, offers a promising perspective. Nevertheless, they still need exhaustive testing in order to detect strength and weakness.

References

Anderson, T.H., 2003. Mirobial eco-physiological indicators to assess soil quality. Agriculture, *Ecosystems and Environment* 98, 285-293.

Avidano, L., Gamalero, E., Cossa, G.P., Carraro, E., 2005. Characterization of soil health in an Italian polluted site by using microorganisms as bioindicators. *Applied Soil Ecology* 30, 21-33.

Bending, G.D., Turner, M.K., Rayns, F., Marx, M.C., Wood, M., 2004. Microbial and biochemical soil quality indicators and their potential for differentiating areas under contrasting agricultural management regimes. *Soil Biology and Biochemistry* 36, 1785-1792.

Bossio, D.A., Scow, K.M., 1995. Impact of carbon and flooding on the metabolic diversity of microbial communities in soils. *Applied and Environmental Microbiology* 61, 4043–4050.

Buckley, D.H., Graber, J.R., Schmidt, T.M., 1998. Phylogenetic analysis of nonthermophilic members of the kingdom Crenarchaeota and their diversity and abundance in soils. *Applied and Environmental Microbiology* 64, 4333-4339.

Campbell, C.D., Grayston, S.J., Hirst, D.J., 1997. Use of rhizosphere carbon sources in sole carbon source tests to discriminate soil microbial communities. *Journal of Microbiological Methods* 30, 33-41.

Campbell, C.D., Chapman, S.J., Cameron, C.M., Davidson, M.S., Potts, J.M., 2003. A rapid microtiter plate method to measure carbon dioxide evolved from carbon substrate amendments so as to determine the physiological profiles of soil microbial communities by using whole soil. *Applied and Environmental Microbiology* 69, 3593-3599.

Casas, C., 2006. Respuesta de la microflora edáfica a la presencia del endofito Neothyphodium sp. en plantas de Lolium multiflorum. Graduate thesis. Facultad de Agronomía. Universidad de Buenos Aires, pp 52.

Christensen, H., Hansen, M, Sorensen, J., 1999. Counting and size classification of active soil bacteria by fluorescence in situ hybridization with an rRNA oligonucleotide probe. *Applied and Environmental Microbiology* 65, 1753-1761.

Correa, O.S., Romero, A.M., Montecchia, M.S., Soria, M.A., 2007. Tomato genotype and Azospirillum inoculation modulate the changes in microbial communities associated with roots and leaves. *Journal of Applied Microbiology* 102, 781-786.

Degens, B.P., Harris, J.A., 1997. Development of a physiological approach to measuring the catabolic diversity of soil microbial communities. *Soil Biology and Biochemistry* 29, 1309-1320.

Dick, R.P., Breakwell, D.P., Turco, R.F., 1996. Soil enzyme activities and biodiversity measurements as integrative microbiological indicators. In: Doran, J.W., Jones, A.J. (Eds.). *Methods for Assessing Soil Quality,* 49 SSSA Special Publication, Madison, WI, p. 247-271.

Doran, J.W., Parkin, T.B., 1996. Quantitative indicators of soil quality: a minimum data set. In: Doran, J.W., Jones, A.J. (Eds.). *Methods for Assessing Soil Quality,* 49 SSSA Special Publication, Madison, WI, p. 25-37.

Elfstrand, S., Hedlund, K., Mårtensson, A., 2007. *Soil enzyme activities, microbial community composition and function after 47 years of continuous green manuring.* Applied Soil Ecology 35, 610-621.

Elliot, E.T., 1997. Rationale for developing bioindicators of soil health. In: Pankhurst, C.E., Doube, B.M., Gupta, V.V.S.R. (Eds.). *Biological Indicators of Soil Health.* CAB International, Wallingford, UK, p. 49-78.

Fernández Ferrari, M.C., 2006. Efecto de la inoculación de un agente de control biológico sobre la comunidad microbiana de la rizosfera y sobre el crecimiento y desarrollo de plantas de soja (Glycine max L. Merr). *Graduate thesis. Facultad de Agronomía.* Universidad de Buenos Aires, pp 40.

Franchini, J.C., Crispino, C.C., Souza, R.A., Torres, E., Hungría, M., 2007. Microbiological parameters as indicators of soil quality under various soil management and crop rotation systems in southern Brazil. *Soil and Tillage Research* 92, 18-29.

Franklin, R.B., Garland, J.L., Bolster, C.H., Mills, A.L., 2001. Impact of dilution on microbial community structure and functional potential: comparison of numerical simulations and batch culture experiments. *Applied and Environmental Microbiology* 67, 702–712.

Garland, J., 1997. Analysis and interpretation of community-level physiological profiles in microbial ecology. *FEMS Microbiology Ecology* 24, 289-300.

Garland, J., Mills, A., 1991. Classification and characterization of heterotrophic microbial communities on the basis or patterns of community level sole carbon source utilization. *Applied and Environmental Microbiology* 57, 2351-2359.

Garland, J.L., Lehman, R.M., 1999. Dilution/extinction of community phenotypic characters to estimate relative structural diversity in mixed communities. *FEMS Microbiology Ecology* 30, 333-343.

Garland, J.L., Roberts, M.S., Levine, L.H., Mills, A.L., 2003. Community-level physiological profiling performed with an oxygen-sensitive fluorophore in a microtiter plate. *Applied and Environmental Microbiology* 69, 2994-2998.

Giller, K.E., Beare, M.H., Lavelle, P., Izac, A.M.N., Swift, M.J., 1997. Agricultural intensification, soil biodiversity and agroecosystem function. *Applied Soil Ecology* 6, 3-16.

Gomez, E., Bisaro, V., Conti, M., 2000. Potencial C-source utilization patterns of bacterial communities as influenced by clearing and land use in a vertic soil of Argentina. *Applied Soil Ecology* 15, 273–281.

Gomez, E., Garland, J., Conti, M., 2004. Reproducibility in the response of soil bacterial community-level physiological profiles from a land use intensification gradient, *Applied Soil Ecology* 26, 21–30.

Gomez, E., Ferreras, L., Toresani, S., 2006. Soil bacterial functional diversity as influenced by organic amendment application. *Bioresource Technology* 97, 1484-1489.

Gregorich, E.G., Carter, M.R., Angers, D.A., Monreal, C.M., Ellert, B.H., 1994. Towards a minimum data set to assess soil organic matter quality in agricultural soils. *Canadian Journal of Soil Science* 74, 367-385.

Graham, M.H., Haynes, R.J., 2005. Catabolic diversity of soil microbial communities under sugarcane and other land uses estimated by Biolog and substrate-induced respiration methods. *Applied Soil Ecology* 29, 155-156.

Handelsman, J., 2004. Metagenomics: Application of genomics to uncultured microorganisms. *Microbiology and Molecular Biology Reviews* 68, 669-685.

Insam H., 1997. Substrate utilization tests in microbial ecology. A preface to the special issue of the Journal of Microbiological Methods. *Journal of Microbiological Methods* 30, 1-2.

Konopka, A., Oliver, L., Turco, R.F., 1998. The use of carbon substrate utilization patterns in environmental and ecological microbiology. *Microbial Ecology* 35, 103-115.

Kirk, J.L., Beaudette, L.A., Hart, M., Moutoglis, P., Klironomos, J.N., Lee, H., Trevors, J.T., 2004. Methods of studying soil microbial diversity. *Journal of Microbial Methods* 58, 169-188.

Lal, R., 1994. Sustainable land use systems and soil resilience. In: Greenland, D.J., Szabolcs, I. (Eds.). *Soil Resilience and Sustainable Land Use.* CAB International, Wallingford, UK, p. 41-67.

Lehman, M.R., Colwell, F.S., Garland, J.L., 1997. Physiological profiling of indigenous aquatic microbial communities to determine toxic effects of metals. *Environmental Toxicology and Chemistry* 16, 2232-2241.

Lindow, S.E., Brandl, M.T., 2003. Microbiology of the Phyllosphere. *Applied and Environmental Microbiology* 69, 1875-1883.

Liu, W-T., Marsh, T.L., Cheng, H., Forney, L.J., 1997. Characterization of microbial diversity by determining terminal restriction fragment length polymorphisms of genes enconding 16S rRNA. *Applied and Environmental Microbiology* 63, 4516-4522.

Lupwayi, N.Z., Rice, W.A. and Clayton, G.W., 1998. Soil microbial diversity and community structure under wheat as influenced by tillage and crop rotation. *Soil Biology and Biochemistry* 30, 1733–1741.

Marinari, S., Masciandaro, G., Ceccanti, B., Grego, S., 2007. Evolution of soil organic matter changes using pyrolysis and metabolic indices: A comparison between organic and mineral fertilization. *Bioresource Technology* 98, 2495-2502.

Melero S., Ruiz Porras, J.C., Herencia, J.F., Madejon, E., 2006. Chemical and biochemical properties in a silty loam soil under conventional and organic management. *Soil and Tillage Research* 90, 162-170.

Muyzer, G., 1999. DGGE/TGGE a method for identifying genes from natural ecosystems. *Current Opinion in Microbiology* 2, 317–322.

Muyzer, G., de Wall, E.C., Uitterlinden, A.G., 1993. Profiling of complex microbial populations by denaturing gradient gel electrophoresis analysis of polymerase chain reaction-amplified genes coding for 16S rRNA. *Applied and Environmental Microbiology* 59, 695-700.

Nakatsu, C.H., 2007. Soil Microbial community analysis using denaturing gradient gel electrophoresis. *Soil Science American Journal* 71, 562-567.

Nannipieri, P., 2003. Microbial diversity and soil functions. *European Journal of Soil Science* 54, 655-670.

Nannipieri, P., Ascher, J., Ceccherini, M.T., Landi, L., Pietramellara, G., Renella, G., 2003. Microbial diversity and soil functions. *European Journal of Soil Science* 54, 655-670.

Olsen, G.J., Lane, D.J., Giovannoni, S.J., Pace, N.R., Stahl, D.A., 1986. Microbial ecology and evolution: A ribosomal RNA approach. *Annual Review of Microbiology* 40, 337-365.

Omay, A.B., Rice, C.W., Maddux, L.D., Gordon, W.B. 1997. Changes in soil microbial and chemical properties under long-term crop rotation and fertilization. *Soil Science Society of America Journal* 61, 1672-1678.

Palojärvi, A., Sharma, S., Rangger, A., von Lützow, M., Insam, H., 1997. Comparison of Biolog and phospholipid fatty acid patterns to detect changes in microbial community. In: Insam, H., Rangger, A. (Eds.). *Microbial Communities: Functional versus Structural Approaches.* Springer-Verlag Berlin Heidelberg, p. 37-48.

Pankhurst, C.E., Hawke, B.G., McDonald, H.J., Kirby, C.A., Buckerfield, J.C., Michelsen, P., O'Brien, K.A., Gupta, V.V.S.R., Doube, B.M., 1995. Evaluation of soil biological properties as potential bioindicators of soil health. *Australian Journal of Experimental Agriculture* 35, 1015-1028.

Parkin, T., Doran, J., Franco-Vizcaíno, E., 1996. Field and laboratory tests of soil respiration. In: Doran, J., Jones, A. (Eds.), *Methods for Assessing Soil Quality.* SSSA Special Publication 49, SSSA Madison, WI, pp. 231–245.

Ranjard, L., Poly, F., Combrisson, J., Richaume, A., Gourbiere, F., Thioulouse, J., Nazaret, S., 2000. Heterogeneous cell diversity and genetic structure of bacterial pools associated with various soil microenvironments as determined by enumeration and DNA fingerprinting approach (RISA). *Microbial Ecology* 39, 263-272.

Reyes-Reyes, B.G., Alcántara-Hernández, R., Rodríguez, V., Olalde-Portual, V., Dendooven, L., 2007. Microbial biomass in a semi arid soil of the central highlands of Mexico cultivated with maize or under natural vegetation. *European Journal of Soil Biology,* in press.

Rice, C.W., Moorman, T.B., Beare, M., 1996. Role of microbial biomass carbon and nitrogen in soil quality. In: Doran, J., Jones, A. (Eds.), *Methods for Assessing Soil Quality.* SSSA Special Publication 49, SSSA Madison, WI, pp. 203-215.

Roper, M.M., Gupta, V.V.S.R., 1995. Management practices and soil biota. *Australian Journal of Soil Research* 33, 321-339.

Schloter, M., Dilly, O., Munch, J.C., 2003. Indicators for evaluating soil quality. *Agriculture, Ecosystems and Environment* 98, 255-262.

Schiwieger, F., Tebbe, C.C., 1998. A new approach to utilize PCR-single-strand-conformation polymorphism for 16S rRNA gene-based microbial community analysis. *Applied and Environmental Microbiology* 64, 4870-4876.

Schutter, M., Dick, R., 2001. Shifts in substrate utilization potential and structure of soil microbial communities in response to carbon substrates. *Soil Biology and Biochemistry* 30, 1481-1491.

Sharma, S., Rangger, A., von Lützow, M., Insam, H., 1998. Functional diversity of soil bacterial communities increases after maize litter amendment. *European Journal of Soil Biology* 34, 53-60.

Simon, H.M., Smith, K.P., Dodsworth, J.A., Guenthner, B., Handelsman, J., Goodman, R.M., 2001. Influence of tomato genotype on growth of inoculated and native bacteria in the spermosphere. *Applied and Environmental Microbiology* 67, 514-20.

Small, J., Call, D.R., Brockman, F.J. Straub, T.M., Chandler, D.P., 2001. Direct detection of 16S rRNA in soil extracts by using oligonucleotide microarrays. *Applied and Environmental Microbiology* 67, 4708-4716.

Smalla, K., Wachtendorf, U., Heuer, H., Liu, W.T. and Forney, L., 1998. Analysis of Biolog GN substrate utilization patterns by microbial communities. *Applied and Environmental Microbiology* 64, 1220–1225.

Sneath, P.H., Sokal, R.R., 1973. *Numerical Taxonomy,* W.H. Freeman and Co., San Francisco.

Souto, G.I., Correa, O.S., Montecchia, M.S., Kerber, N.L., Pucheu, N.L., Bachur, M., García, A.F., 2004. Genetic and functional characterization of a Bacillus sp. strain excreting surfactin and antifungal metabolites partially identified as iturin-like compounds. *Jouranl of Applied Microbiology* 97, 1247-1256.

Stenberg, B., 1999. Monitoring soil quality of arable land: microbiological indicators. Acta Agriculture Scandinavian, Section. B. *Soil and Plant Science* 49, 1-24.

Torsvik, V., Goksoyr, J., Daae, F.L., 1990. High diversity in DNA of soil bacteria. *Applied and Environmental Microbiology* 56, 782-787.

Wander, M.M., Hedrick, D.S., Kaufman, D., Traina, S.J., Stinner, B.R., Kehrmeyer, S.R., White, D.C., 1995. The functional significance of the microbial biomass in organic and conventionally managed soils. *Plant and Soil* 170, 87-97.

Williams, M.A., Rice, C.W., 2007. Seven years of enhanced water availability influences the physiological, structural, and functional attributes of a soil microbial community. *Applied Soil Ecology* 35, 535-545.

Winding, A., Hund-Rinke, K., Rutgers, M., 2005. The use of microorganisms in ecological soil classification and assessment concepts. *Ecotoxicology and Environmental Safety* 62, 230-248.

Zak, J., Willig, M., Moorhead, D., Wildman, H., 1994. Functional diversity of microbial communities: a quantitative approach. *Soil Biology and Biochemistry* 26, 1101–1108.

Zelles, L., 1999. Fatty acid patterns of phospholipids and lipopolysaccharides in the characterization of microbial communities in soil: a review. *Biology and Fertility of soils,* 29, 111-129.

Zhang, J., Song, Ch., Yang, W., 2007. Effects of cultivation on soil microbiological properties in a freshwater marsh soil in Northeast China. *Soil and Tillage Research* 93, 231-235.

In: Progress in Environmental Microbiology
Editor: Myung-Bo Kim, pp. 169-186

ISBN: 978-60021-940-5

Chapter V

Endogenic and Anthropogenic Adsorption of Cu and Zn onto the Non-Residual and Residual Components in the Surficial Sediments (Natural Surface Coating Samples)

Y. Li[1,2*], *X. L. Wang*[2], *X. Y. Du*[2], *T. Wang*[2]

[1]Energy and Environmental Research Centre, North China Electric Power University, Beijing 102206, China

[2]College of Environment and Resources, Jilin University, Changchun 130012, China

Abstract

The selective extraction procedure was further developed and perfected newly in the present study in order to understand more mechanisms of heavy metals adsorption onto the surficial sediments (SSs) and a kind of special surficial sediments: natural surface coating samples (NSCSs). Then the extraction of endogenic Cu and Zn in company with the extraction of Fe, Mn oxides and organic materials (OMs) and adsorption of anthropogenic (added) Cu and Zn onto the solid particles before and after the extractions were investigated using the selective extraction- (metal adsorption)-statistical analysis methods. The results indicate that 0.1 mol/L $NH_2OH{\cdot}HCl$ + 0.1 mol/L HNO_3 has great potential for removing Mn oxides, $(NH_4)_2C_2O_4$ (0.2 mol/L) + $H_2C_2O_4$ (pH 3.0) for extracting both Fe and Mn hydrous oxides, and H_2O_2 (30%) for selectively extracting OMs. And the statistical analyses of experimental results suggest that both for endogenic and anthropogenic heavy metals, the adsorption capacities of NSCSs were higher than those of SSs, implying that the role of the NSCSs in the transformation and cycling of

* Corresponding Author. Tel.: 86-10-5197-1241, Fax: 86-10-5197-1241, Email: liyuxx@jlu.edu.cn

heavy metals in aquatic environments was more important than that of SSs. For the two sorts of metal pollutants, the relative adsorption roles of non-residual and residual fractions of solid particles were different from each other, i.e. non-residual fraction contributed more roles in anthropogenic Cu adsorption to the solid particles, and the contributions of non-residual and residual fractions to the anthropogenic Zn adsorption were comparative; but for endogenic pollutants, the contribution of non-residual fraction was significant more than that of residual fraction for Zn, and the role of non-residual fraction was similar to that of residual fraction for Cu. However, the relative roles of Fe, Mn oxides and OMs in the non-residual fraction of the particles were similar to each other for endogenic and anthropogenic metals, and the greatest contribution to metals adsorption on a molar basis was from Mn oxides. Metals adsorption capacities of Mn oxides exceeded those of Fe oxides by one order of magnitude, fewer roles were found attributing to adsorption by OMs. These results reveal that Mn oxides in the non-residual fractions were the most important component in controlling heavy metals in aquatic environments.

Keywords: surficial sediments; natural surface coating samples; Mn oxides; Fe oxides; organic materials; selective extraction; adsorption; Cu and Zn

Introduction

Natural surface coatings (NSCs) consist of various microorganisms and develop on various surfaces and under various conditions (Flemming, 1995). They form on surfaces of all materials in rivers, lakes, and wetlands, such as rocks and sediments, and account for a wide range of microbial organisms on earth (Costerton et al., 1987). The behavior, transport, and ultimate fate of trace metals in aquatic environments have prompted great interest due to the toxicity and bioaccumulation potential of heavy metals. It has been found that solid particles, such as surficial sediments (SSs), suspended particulates, and the natural surface coatings: kind of special surficial sediments play crucial roles in the cycling and bioavailability of heavy metals (Vuceta and Morgan, 1978; Santschi et al., 1997; Lion et al., 1988; Nelson et al., 1999; Dong et al., 2002, 2003a, b). Within the SSs (NSCs), the three most important geochemical components were identified as Fe, Mn oxides and organic materials (OMs) (Lion et al., 1982, 1988; Young and Harvey, 1992; Perret et al., 2000).

Studies on the capacities and mechanisms of heavy metals adsorption onto the solid particulates by using the selective extraction technique were mainly carried out on the NSCs developed on glass slides fixed on polypropylene racks for approximately 2 weeks at the depth of 30 cm in natural waters (Nelson et al., 1994; Dong et al., 2000, 2001a, 2001b, 2002, 2003a, 2003b). The results indicated that trace metal adsorption to NSCs is expected to be governed by Mn and Fe oxides, with smaller roles attributed to the adsorption by OMs and Al oxides in NSCs (Nelson et al., 1994, 1999; Dong et al., 2000, 2001a). However, less research was reported on the natural surface coating samples (NSCSs) grown on the surface of river shingles (Farag et al., 1998; Dong et al., 2003c), especially less investigation was focused on the adsorption mechanisms of heavy metals onto the NSCSs. The NSCSs should be more representative than the NSCs for studying heavy metals transportation in aquatic environments.

There are some differences between NSCSs and NSCs, for instance, the growing surroundings, such as velocity of flow, intensity of sunshine, temperature and pH of water, type of matrix, and species of microorganism and so on. From this view, the NSCSs are more similar to SSs, and more common than NSCs. On the other hand, the differences in growing environments result in the differences in composition. The previous studies showed that Mn existed in residual fraction of NSCs was below the detection limit, but that in SSs (NSCSs) was up to 45% (10-25%), respectively; and Fe presented in residual fraction of NSCs was only 3% with Fe up to 81% (45-76%) in SSs (NSCSs). The similar tendency was also investigated for the content of OMs, i.e. the amount of OMs in SSs (NSCSs) (1050 and 1593 μmol/g respectively for SSs and NSCSs) was much lower than that in NSCs (8019 μmol/g).

The purpose of the present study is to establish and further improve a selective chemical extraction technique based on the previous studies to selectively remove Fe, Mn oxides and OMs from SSs (NSCSs), and to estimate the relative contributions from the non-residual and residual components in the SSs (NSCSs) through using the selective extraction- (metal adsorption)-statistical analysis methods. These results will help us to further understand the adsorption mechanism of Cu and Zn by SSs (NSCSs) in aquatic environments.

Methodology

Collection of SSs (NSCSs)

The NSCSs were directly collected from the surfaces of river shingles. Shingles in the near shore of river were taken out of the water, and the NSCSs attached to the shingles were scraped into plastic containers containing minimal mineral salts (MMS) solution (Li et al., 2006a) using a plastic scoop. The SSs (approximately 5 cm in depth from the surface) were collected with a plastic scoop and stored in polyethylene bags. Prior to collection of SSs (NSCSs), the plastic containers and apparatus were pre-cleaned with detergent, soaked for 24 h in soap solution, acid washed for 24 h in 6:1 (v/v) H_2O/HNO_3, and rinsed in distilled-deionized water (ddH_2O), followed by a second 24 h acid wash and a final rinse in ddH_2O. Then the SSs (NSCSs) were transported to the laboratory and dumped into more voluminous plastic containers in which contain minimal mineral salts (MMS) solution. The samples were pretreated through sieving to remove the debris and settling to separate the sand and gravel. The treated samples were then preserved at 4 °C in MMS solution.

OMs expressed as total organic carbon (TOC) in the SSs (NSCSs) were determined by 'oxidative combustion-infrared analysis' method and measured by a TOC-analyzer (TOC-V_{CPH}, Shimadzu, Japan); pseudo-total amounts of Fe and Mn were digested by HCl-HNO_3-$HClO_4$, and the metals in supernatant were measured by a WYX-9004 flame atomic absorption spectrometer (FAAS) equipped with a SML-III graphite furnace (Shenyang, China). Distilled-deionized water (ddH_2O) was used through this study. All glassware and other containers were thoroughly cleaned and finally rinsed with ddH_2O prior to use.

Selective Extraction of SSs (NSCSs)

SSs (NSCSs) sample of 0.1g was extracted in 20-ml of extraction reagent in 100-ml polyethylene centrifugal tubes using the extraction techniques modified from John and Latifatu (2004) and Dong et al. (2002 and 2003b). The extraction techniques used by John and Latifatu (2004) and Dong et al. (2002 and 2003b), especially used by Dong et al. (2002 and 2003b) have been already optimized for extracting soils and natural surface coatings, so we directly used the extraction techniques as the initial experiment conditions.

The initial experiments with hydroxylamine extraction method for selective removal of Mn oxides suggested that extraction with 0.02 mol/L $NH_2OH{\cdot}HCl$ + 0.01 mol/L HNO_3 for 20 min as used by Dong et al. (2002 and 2003b) did not affect on Mn oxides extraction with lower extraction efficiency. This result was mostly due to the significant difference between SSs (NSCSs) and NSCs in components, for example, Fe and Mn existed in residual fraction in SSs (NSCSs) were more than those in NSCs. Thus, we modified this extraction procedure by improving the $NH_2OH{\cdot}HCl$ and HNO_3 concentrations both to 0.1 mol/L, shortening the extraction time to 30 min. Preliminary experiments with 0.3 mol/L $Na_2S_2O_4$ (pH 6.0) (Dong et al., 2002 and 2003b) to extract both Mn and Fe oxides indicated that the determination of Mn and Fe in supernatant by FAAS was disturbed significantly by $S_2O_4^{2-}$. Thus, we eliminated the sodium dithionite reagent to extract Mn and Fe oxides and adopted 0.2 mol/L $(NH_4)_2C_2O_4$ (pH 3.0) (John and Latifatu, 2004) to extract Mn and Fe oxides. Extraction with 30% H_2O_2 heating at 40 °C in water batch for 48h was conducted to remove OMs, as compared with the use of 1.5 ml H_2O_2 (60%) heated on a hot plate by John and Latifatu (2004), to reduce the simultaneous extraction amounts of Mn and Fe by reducing the H_2O_2 concentration to 30% and the temperature to 40 °C. All reagents in the extraction processes were in trace metal grades.

Previously, several authors suggested that Fe and Mn in the SSs (NSCSs) could be defined as two fractions by using sequential extraction procedures, i.e. non-residual fraction (also called total extractable fractions, including five binding phases of exchangeable, bound to carbonates, to Mn-and Fe- oxides, and to OMs) and residual one. In this study, the total extractable amounts of Fe and Mn in the surficial sediments were determined by using the modified sequential extraction procedure (MSEP) (Tessier et al., 1979; Belzile et al., 1989; Yu et al., 2001). Furthermore, some studies indicated that determining total contents of heavy metals in SSs was insufficient to assess the environmental impact of heavy metals on the environment or organism (Ma and Rao, 1997; Mester et al., 1998; Barona et al., 1999); they suggested that metals associated with the residual fraction in soils or surface sediment samples should be highly unavailable to environments. Therefore, the Fe and Mn oxides in the non-residual fraction of SSs (NSCSs) were defined as the total Fe and Mn oxides for the convenience of calculating the efficiency of selective extraction for Fe and Mn oxides in the SSs (NSCSs).

Cu and Zn Adsorption to SSs (NSCSs)

Adsorption of Cu and Zn onto the unaltered SSs (NSCSs) and the SSs (NSCSs) extracted by three different extraction reagents were measured in duplicate in chemically defined solutions under eight different Cu and Zn concentrations (15.5, 31.0, 62.0, 93.0, 124.0, 155.0, 232.5, and 310.0 μmol/L). The solutions were prepared by diluting 15.5 mmol/L $Cu(NO_3)_2$ and $Zn(NO_3)_2$ reference solutions through a minimal mineral salt (MMS) solution, and adjusting pH at 6.0 ± 0.1 using 0.01 mol/L HNO_3 and NaOH (25 ± 1 °C). SSs (NSCSs) of 0.1 g were weighted into a 100-ml polyethylene tubes, and then 50-ml solutions with eight different Cu and Zn concentrations were added into the tubes. The suspensions were stirred continuously for 24 h and filtered through 0.45-μm Millipore filters. Before filtering, the pH was re-adjusted to the initial level and the changes in pH after 24 h were found to be minor (all these adjustments were only within ± 0.5 pH). Cu and Zn in the initial and equilibrium solutions were measured by FAAS. The amounts of adsorbed Cu and Zn were calculated as the difference between the amount added initially and that remaining in the equilibrium solutions. In order to compare with the endogenic pollutants, the mixed metals (Cu and Zn) adsorption from an equimolar Cu-Zn solution to the SSs (NSCSs) before and after each treatment was taken place instead of single metal adsorption.

Statistical Analyses

Adsorption of Cu and Zn to the SSs (NSCSs) was analyzed by the Langmuir adsorption isotherm, which was expressed as follows:

$$\Gamma = \frac{\Gamma_{\max} K[\mathrm{M}^{2+}]}{1 + K[\mathrm{M}^{2+}]} \tag{1}$$

where Γ is the adsorption of M^{2+} by the SSs (NSCSs) per unit weight (μmol/g), $\Gamma_{\max}$ is the maximum adsorption of M^{2+} by the SSs (NSCSs) per unit weight (μmol/g), K is the Langmuir equilibrium coefficient (l/μmol), and $[M^{2+}]$ is the concentration of free Cu or Zn in the supernatant of the solution (μmol/L).

As described in the previous studies (Dong et al., 2002, 2003a and 2003b; Turner et al., 2004), it is possible that each extraction reagent could remove additional components from the particles in addition to the target materials. As the first approximation, the amounts of Cu and Zn adsorbed to Fe/Mn oxides, OMs and residues in the SSs (NSCSs) could be estimated, assuming that the decrease in the adsorption amounts was due to the partial removal of Fe/Mn oxides and OMs. This assumption requires that each extraction technique should not remove any significant amount of other surfacial components and that components of the SSs (NSCSs) are independent variables (i.e. the presence of one component would not affect the precipitation of the others and vice versa). The results from each adsorption to the SSs (NSCSs) extracted with a given reagent, together with the adsorption to the unaltered SSs (NSCSs), could therefore be expressed as follows (Dong et al., 2003a):

$$\Gamma = a \times M_{Fe} + b \times M_{Mn} + c \times M_{OMs} + d \times M_{Res} \quad (2)$$

where, Γ represents the amount of M^{2+} adsorbed by the unaltered particles and the particles digested with each extraction reagent; a, b, c and d are percentages of Fe, Mn, OMs, and residues in the SSs (NSCSs) (unextracted or extracted), respectively; M_{Fe}, M_{Mn}, M_{OMs} and M_{Res} are the amounts of M^{2+} adsorbed to the SSs (NSCSs) due to the presence of Fe oxides, Mn oxides, OMs and residues in the SSs (NSCSs), respectively (they indirectly represent the amounts of M^{2+} bound to the Fe oxides, Mn oxides, OMs, and residues in the SSs (NSCSs)). Finally, the relative contributions of Fe oxides ($\Gamma_{\text{mmol M/g Fe}}$), Mn oxides ($\Gamma_{\text{mmol M/g Mn}}$), OMs ($\Gamma_{\text{mmol M/g TOC}}$) and residues ($\Gamma_{\text{mmol M/g Res}}$) to M^{2+} adsorption on a molar basis could be estimated by using M_{Fe}, M_{Mn}, M_{OMs}, and M_{Res} divided by the total extractable amounts of Fe oxides, Mn oxides, OMs, and residues in the SSs (NSCSs), respectively.

Meanwhile, if Γ represents the amount of M^{2+} removed by the individual extraction, the relative association of endogenic M^{2+} with Mn and Fe oxides, OMs or residues were also estimated.

Results

Selective Extraction of SSs (NSCSs)

The results of SSs (NSCSs) analyzed by selective extraction procedures described before are summarized in Tables 1 and 2. The removal efficiencies of Mn oxides in the non-residual fractions by $NH_2OH \cdot HCl$ treatment reached 85% (94%) and 63% (72%) for the samples collected in May 2006 and May 2004, respectively, with only 4.6% and 11% (10%) and 3% (1%) of OMs was extracted, implying that for the target components, i.e. Mn oxides and non-target components, i.e. OMs, $NH_2OH \cdot HCl$ was an effective extraction regent for Mn oxides removal. However, for the other non-target component, Fe oxides, the effect of $NH_2OH \cdot HCl$ extraction was greater than that on OMs and the removal of Fe oxides by $NH_2OH \cdot HCl$ extraction was up to more than 11%. The slightly more Fe oxides extracted in the selective removal of Mn oxides could be due to the binding form of Fe in the NSCSs and the considerable part of Fe in the NSCSs were existed as carbonatic phase and bound to Mn oxides as determined by sequential extraction procedure (Table 3). Meanwhile, Turner et al. (2004) reported out that although $NH_2OH \cdot HCl$ could remove almost all of the Mn oxides in the non-residual fractions from the SSs (NSCSs), the little part of Fe existed as carbonatic phase and amorphous form would be also extracted simultaneously. In the present study, Fe present as carbonatic phase and bound to Mn oxides was up to 0.9% (1.7%) of pseudo-total Fe in average, and equivalent to approximately 5% (7%) of the total extractable Fe (Table 3). So, the slightly higher extraction of Fe was ineluctable and should be acceptable.

After treatment with $(NH_4)_2C_2O_4$, the removal of target components Fe and Mn oxides were satisfactory and the extraction efficiencies were respectively up to more than 91%. And that for the non-target component organic materials, the effect of $(NH_4)_2C_2O_4$ was negligible and only less than 9% of OMs extracted. These results imply that $(NH_4)_2C_2O_4$ used as the selective extraction reagent for both Fe and Mn oxides in the non-residual fraction of the SSs (NSCSs) was feasible to the further mechanism investigation of metals adsorption onto the SSs (NSCSs) via combination of selection extraction - (adsorption) - additional model analysis.

In the previous studies on the NSCs, 10% of HNO_3 was always used to remove OMs as well as almost all of the Fe and Mn oxides and did not selectively extract OMs. Here, we applied H_2O_2 as the extraction agent for the OMs removal and the results of SSs (NSCSs) extracted with 30% of H_2O_2 heating at 40 °C in water batch for 48 h are also summarized in Tables 1 and 2. The results showed that the removal efficiencies of OMs reached 70%; the effect of this extraction treatment on Fe oxides was invisible and only 1.5% of Fe oxides were digested. However, the H_2O_2 extraction removed slightly more Mn oxides from the non-residual fractions of the SSs (NSCSs). This could be explained by the study of Chao (1984): when the sediments were extracted by H_2O_2 to digest OMs, the part of reducible Mn oxides would be also removed if the sediments have not been treated by reducing reagent.

Extraction of Cu and Zn in the SSs (NSCSs)

As showed in Table 1, a considerable fraction of the total extractable Zn in the SSs (NSCSs) (35% and 51% respectively for SSs and NSCSs) was solubilized after the solid particles were extracted with $NH_2OH{\cdot}HCl$ to remove Mn oxides. Virtually almost all of the extractable Zn (83% and 88% respectively for SSs and NSCSs) was solubilized after extraction with $(NH_4)_2C_2O_4$ to remove Mn and Fe oxides. Meanwhile, only 13% (15%) of extractable Zn was solubilized in company with the removal of OMs. The solubilization of the total extractable Cu in the SSs was in the same tendency as that of Zn. But the major fraction of extractable Cu in NSCSs (94%) was solubilized after the NSCSs were treated with H_2O_2 to remove OMs, and small part of extractable Cu (21% and 25%) was solubilized after extraction with $NH_2OH{\cdot}HCl$ and $(NH_4)_2C_2O_4$.

Adsorption of Cu and Zn onto the SSs (NSCSs)

The experimental and fitted Cu and Zn adsorption levels onto unextracted and extracted SSs (NSCSs) are shown in Table 2. It is indicated that Cu and Zn adsorptions to the SSs (NSCSs) followed Langmuir adsorption isotherm before and after extraction of SSs (NSCSs). The maximum adsorption Γ could be used for estimation of the adsorption capacity of the solid particles. The maximum adsorption of Cu was about 2 orders of magnitude greater than that of Zn before and after the SSs (NSCSs) were treated by chemical reagents. This implies that Cu was more strongly retained by SSs (NSCSs) than Zn. The preferential retention of Cu than Zn by the SSs (NSCSs) was in agreement with the results reported in the literature for not only Cu and Zn, but also the other metals adsorbed to different adsorbents, such as goethite and hematite (Schwertmann and Taylor, 1989), aluminum hydroxides (Hsu, 1989), humic substances (Schnitzer, 1969), oxisols, ultisols and alfisols (Gomez et al., 2001). The adsorption capacity of NSCSs (taking the untreated NSCSs as an example) was about 10%

larger than that of SSs for Cu and 16% for Zn, suggesting that NSCSs were more important than SSs for the transformation and cycling of trace heavy metals in aquatic environments.

The effects from the removal of Fe/Mn oxides and OMs on Cu adsorption were all significant, especially the removals of combined Fe and Mn oxides. A similar tendency was also observed for Zn adsorption (Table 2). When OMs were removed from the SSs (NSCSs), Γ_{max} of Cu decrease by 21% (26%) based on the unaltered SSs (NSCSs) system; in comparison, Γ_{max} of Zn decrease by 21% (20%) as compared to the unextracted SSs (NSCSs). The effect of Mn oxides removal was similar to that of OMs removal, in terms of reducing Γ_{max} levels of Cu and Zn. As to the removals of combined Fe and Mn oxides, the Γ_{max} levels of Cu and Zn drastically decrease by 53% (56%) and 47% (36%), respectively, as compared to the untreated SSs (NSCSs).

Discussion

Modification and Improvement of the Selective Extraction Procedure

Although 0.02M $NH_2OH{\cdot}HCl$ +0.01M HNO_3, $Na_2S_2O_4$ (pH6.0, 30min) and 10% HNO_3 (24h) could effectively remove target components (Dong et al., 2001b) in the NSCs developed on glass slides, Fe and Mn in 10% HNO_3 (24h) extraction solution did not represent the total extractable Fe and Mn in the SSs and NSCSs grown on the shingles, because of the obvious difference in Fe/Mn fraction levels between NSCSs and NSCs (Li et al., 2006a, b). So the first modification based on the selective extraction procedure applied to the NSCs was to take the place of the 10% HNO_3 with the sequential extraction for determining the total extractable Fe and Mn in the SSs (NSCSs). The results indicate that the extraction efficiencies of Mn by $NH_2OH{\cdot}HCl$, or those of Fe and Mn by $(NH_4)_2C_2O_4$ using the modified selective extraction procedure for NSCSs were slightly lower than those for NSCs using the previous selective extraction procedure (Dong et al., 2001b). This was mainly due to the obvious difference in Fe/Mn fraction levels between SSs (NSCSs) and NSCs (Li et al., 2006a, b).

As described above, in order to avoid the effects of $S_2O_4^{2-}$ on the determination of Fe and Mn by FAAS, $(NH_4)_2C_2O_4$ was employed instead of $Na_2S_2O_4$. The extraction results showed well selective. On the other hand, in the previous studies, the OMs were all extracted together with Fe and Mn (Dong et al., 2001b), and there was no report on the selective extraction of OMs. In this study, we applied 30% H_2O_2 heating at 4 °C in the water batch to selectively remove OMs and the results show that the extraction efficiencies of OMs by H_2O_2 were more than 70%, to some extent lower than those obtained by 10% HNO_3 (95.6% for OMs, and 100% for Fe and Mn), but the effects of H_2O_2 treatment on Fe and Mn were respectively less than 2% and 36%, generally reaching the purpose of selectively extraction of OMs and advancing the progress of extraction technique.

Table 1. Assessment of removal of Fe, Mn oxides and OMs as well as Cu and Zn in the SSs (NSCSs) by selective extractions [†]

		Pseudo-total amount		Total extractable amount		$NH_2OH{\cdot}HCl$ extraction		$(NH_4)C_2O_4$-$H_2C_2O_4$ extraction		H_2O_2 extraction	
		SSs	NSCSs	SSs	NSCSs	SSs	NSCSs	SSs	NSCSs	SSs	NSCSs
Fe [a]	µmol Fe/g	623.9± 8.5	680.1 ± 10.8	275.1 ± 7.3	305.2 ± 3.1	74.1 ± 5.3	52.5 ± 0.8	252.9 ± 2.3	293.5 ± 16.9	4.00 ± 0.2	1.6 ± 0.1
	Extraction efficiency (%)	–	–	100	100	26.9	17.2	91.9	96.1	1.5	0.5
Mn [a]	µmol Mn/g	11.9 ± 0.3	23.1 ± 0.4	8.8 ± 0.3	20.3 ± 0.4	7.5 ± 0.4	19.1 ± 0.5	8.4 ± 0.1	19.4 ± 0.4	1.9 ± 0.6	4.2 ± 0.1
	Extraction efficiency (%)	–	–	100	100	85.1	94.3	95.1	95.7	21.2	20.7
TOC [b]	µmol TOC/g	2077.4 ± 29.2	2789.7 ± 1.8	2077.4 ± 29.2	2789.7 ± 1.8	1857.8 ± 43.1	2506.5 ± 52.9	1969.5 ± 16.9	2549.9 ± 45.1	473.4 ±5.1	352.2 ± 17.3
	Extraction efficiency (%)	–	–	100	100	10.6	10.2	5.2	8.6	77.2	87.4
Cu [a]	nmol Cu/g	597.9 ± 4.7	769.7 ± 4.6	240.8 ± 5.3	394.8 ± 6.1	157.8 ± 2.5	83.8 ± 1.4	224.2 ± 7.3	99.7 ± 5.4	43.2 ± 1.3	370.9 ± 1.3
	Extraction efficiency (%)	–	–	100	100	65.5	21.2	93.1	25.3	17.9	93.9
Zn [a]	nmol Zn/g	1854.9 ± 80.0	2295.0 ± 70.9	1507.6 ± 51.5	1674.7 ± 54.5	525.3 ± 16.8	845.2 ± 23.3	1244.2 ± 18.6	1478.3 ± 67.6	200.5 ± 16.8	250.5 ± 11.8
	Extraction efficiency (%)	–	–	100	100	34.9	50.5	82.5	88.3	13.3	14.9

[†] SSs and NSCSs were collected in May 2006; [a] Mean (n=5) ± SD and the concentrations were the extracted amount; [b] Mean (n=5) ± SD and the concentration was the TOC amount of the remaining, digested NSCSs.

Table 2. Assessment of removal of Fe, Mn oxides and OMs in the SSs (NSCSs) by selective extractions and adsorption of Cu and Zn to the SSs (NSCSs) before and after extraction treatments[†]

		Pseudo-total amount		Total extractable amount		$NH_2OH \cdot HCl$ extraction		$(NH_4)C_2O_4$-$H_2C_2O_4$ extraction		H_2O_2 extraction	
		SSs	NSCSs	SSs	NSCSs	SSs	NSCSs	SSs	NSCSs	SSs	NSCSs
Fe[a]	μmol Fe/g	437.7± 6.7	502.5± 10.9	83.7± 2.2	123.7± 8.4	18.5± 0.6	13.3± 0.7	81.7± 3.9	118.5± 2.8	0.3± 0.02	0.7± 0.06
	Extraction efficiency (%)	–	–	100	100	22.1	10.8	97.6	95.8	0.3	0.6
Mn[a]	μmol Mn/g	7.3±0.3	11.3±0.3	4.0±0.2	8.6±0.4	2.5±0.03	6.2±0.2	3.7±0.1	8.25±0.4	1.3±0.3	3.1±0.2
	Extraction efficiency (%)	–	–	100	100	63.2	72.1	91.3	95.8	33.0	36.1
TOC[b]	μmol TOC/g	1049.9± 110.5	1592.9± 107.4	1049.9± 110.5	1592.9± 107.4	1014.3± 103.1	1573.0± 103.1	1005.7± 85.8	1588.1± 125.4	313.8± 12.2	428.9± 43.2
	Extraction efficiency (%)	–	–	100	100	3.4	1.3	4.2	0.3	70.1	73.1
Cu[c]	Γ_{max} (μmol Cu/g)			69.8±4.9	76.7±5.0	59.9±5.5	65.6±7.6	32.8±5.3	34.1±7.2	54.8±5.4	56.8±7.1
	Coefficient (r)			0.9527	0.9672	0.9863	0.9607	0.9269	0.9721	0.9631	0.9821
Zn[c]	Γ_{max} (μmol Zn/g)			23.8±1.8	27.6±1.8	19.0±3.0	24.5±4.0	12.6±2.9	17.8±2.4	18.7±2.9	22.2±3.3
	Coefficient (r)			0.9811	0.9840	0.9687	0.9902	0.9763	0.9794	0.8289	0.9328

[†] SSs and NSCSs were collected in May 2004; [a] Mean (n=5) ± SD and the concentrations were the extracted amount; [b] Mean (n=5) ± SD and the concentration was the TOC amount of the remaining, digested NSCSs; [c] Mean (n=2) ± AD (average deviation) and the concentration was the adsorbed amount.

Table 3. Fractions of Fe and Mn in SSs (NSCSs) determined by the modified sequential extraction procedure [†]

Fractions	Mn (%) [a]		Fe (%) [a]	
	NSCSs	SSs	NSCSs	SSs
Exchangeable	1.9	0.6	0.0	0.0
Bound to carbonates	9.2	4.8	0.1	0.1
Bound to Mn oxides	48.3	26.5	1.6	0.8
Bound to Fe oxides	13.2	16.9	20.3	16.2
Bound to OMs	2.8	5.7	1.8	2.1
Residues	24.6	45.4	76.2	80.9

[†] SSs and NSCSs were collected in May 2004.

[a] Mean (n=3).

Table 4. Estimation of Cu and Zn adsorption by Fe, Mn oxides, OMs and residues in SSs (NSCSs)

Adsorption capacity		Endogenic metals [a]				Anthropogenic metals [b]			
		Cu		Zn		Cu		Zn	
		SSs	NSCSs	SSs	NSCSs	SSs	NSCSs	SSs	NSCSs
Metal adsorption to the particles through Fe oxides, M_{Fe} / (nmol M/g)		78.7	27.7	1081.0	790.5	34.8×10^3	34.6×10^3	6.4×10^3	7.1×10^3
Metal adsorption capacity of Fe oxides	Γ_{Fe} / (μmol M/g Fe)	5.1	1.6	70.4	46.4	7.4×10^3	5.0×10^3	1.4×10^3	1.0×10^3
	Γ_{Fe} / (mmol M/mol Fe)	0.3	0.09	3.9	2.6	0.4×10^3	0.3×10^3	0.08×10^3	0.06×10^3
Metal adsorption to the particles through Mn oxides, M_{Mn} / (nmol M/g)		159.1	39.1	254.1	740.8	2.5×10^3	9.8×10^3	5.1×10^3	3.1×10^3
Metal adsorption capacity of Mn oxides	Γ_{Mn}/ (μmol M/g Mn)	328.8	35.0	525.2	663.7	11.5×10^3	20.7×10^3	23.1×10^3	6.6×10^3
	Γ_{Mn}/ (mmol M/mol Mn)	18.1	1.9	28.9	36.5	0.6×10^3	1.1×10^3	1.3×10^3	0.4×10^3
Metal adsorption to the particles through OMs, M_{OMs} / (nmol M/g)		10.9	415.1	169.7	106.6	20.1×10^3	22.1×10^3	4.8×10^3	5.8×10^3
Metal adsorption capacity of OMs	Γ_{OMs} / (μmol M/g TOC)	0.4	12.4	6.8	3.2	1.6×10^3	1.2×10^3	0.4×10^3	0.3×10^3
	Γ_{OMs} / (mmol M/mol TOC)	0.01	0.15	0.08	0.04	0.02×10^3	0.01×10^3	0.005×10^3	0.004×10^3
Estimated metal adsorption to the particles through non-residues, M_{Non} / (nmol M/g)		248.6	481.8	1504.8	1637.9	57.4×10^3	66.5×10^3	16.3×10^3	16.0×10^3
Detected metal adsorption to the particles through non-residues, M_{Non} / (nmol M/g)		240.8	394.8	1507.6	1674.7	57.4×10^3	66.5×10^3	16.3×10^3	16.0×10^3
Detected metal adsorption to the particles through residues, M_{Res} / (nmol M/g)		357.1	374.9	347.1	620.3	12.5×10^3	10.2×10^3	7.4×10^3	11.4×10^3
Metal adsorption capacity of residues, Γ_{Res} / (μmol M/g residues)		0.4	0.4	0.4	0.7	0.01×10^3	0.01×10^3	0.008×10^3	0.01×10^3

[a] SSs and NSCSs were collected in May 2006.
[b] SSs and NSCSs were collected in May 2004.

Scavenging Contribution of SSs (NSCSs) Components

Based on Table 1, Cu_{Fe}, Cu_{Mn}, Cu_{OMs} and Cu_{Res} were obtained by solving four equations based on the separate results of the four extractions. The resulting values reflect the relative association of Cu with Mn oxides, Fe oxides, OMs and residues. The Cu adsorption capacities of Mn oxides (Γ_{Mn}), Fe oxides (Γ_{Fe}), OMs (Γ_{OMs}) and residues (Γ_{Res}) were also estimated by using Cu_{Fe}, Cu_{Mn}, Cu_{OMs} and Cu_{Res} divided by the total extractable amounts of Mn oxides, Fe oxides and OMs in the non-residual fraction and the total content of residues in the SSs (NSCSs), respectively (Table 4). Similar results for Zn are also given in Table 4.

The calculated values of Cu_{Fe}, Cu_{Mn}, Cu_{OMs}, and Cu_{Res} indicate that the contribution to total Cu binding to the SSs (NSCSs) from the residues was similar to that from the non-residual components. The estimated Cu adsorption capacities of Mn oxides, Fe oxides and OMs to the total extractable Cu binding to the SSs were different from those to the NSCSs. For instance, the greatest contribution for Cu binding to the SSs was from Mn oxides, followed by Fe oxides; the smallest was from the OMs; but the relative contributions to total Cu adsorption to the NSCSs were ordered as: OMs » Mn oxides > Fe oxides. The calculated values for Zn show that the contribution to total Zn binding to the SSs (NSCSs) from the residues was significantly less than that from non-residues, and the estimated Zn adsorption capacities of Mn oxides, Fe oxides and OMs to the total extractable Cu binding to the SSs and NSCSs showed a similar tendency of Fe oxides > Mn oxides > OMs. But it should be noted that although the relative contribution of each non-residual component was different from Cu and Zn, or SSs and NSCSs, the metal (Fe/Mn) oxides could contribute more to Cu and Zn adsorptions; less significant roles are indicated for OMs in affecting Cu and Zn adsorptions excepting for Cu binding to NSCSs. The more importance of metal oxides (than OMs) for Zn adsorption was in agreement with the results of Fujiyoshi et al. (1994) who stated that OMs was not of primary importance for $^{65}Zn(II)$ adsorption to the scale sample, whereas $^{65}Zn(II)$ scavenging was predominantly controlled by hydrous Fe(III) oxides in the scale.

Cu and Zn binding to residues ($\Gamma_{mmol\ M/g\ Res}$) on a unit mass basis was obviously lower than those of the other three components ($\Gamma_{mmol\ M/g\ Fe}$, $\Gamma_{mmol\ M/g\ Mn}$ and $\Gamma_{mmol\ M/g\ TOC}$). This suggests that the adsorption capacity of residues was negligible compared to those of Fe/Mn oxides and OMs. Thus, in order to compare the adsorption properties of the other three most important geochemical components (Fe, Mn oxides, and OMs) on a molar basis, $\Gamma_{mmol\ M/g\ Fe}$, $\Gamma_{mmol\ M/g\ Mn}$ and $\Gamma_{mmol\ M/g\ TOC}$ were transformed into the forms of $\Gamma_{mmol\ M/mol\ Fe}$, $\Gamma_{mmol\ M/mol\ Mn}$ and $\Gamma_{mmol\ M/mol\ TOC}$ (Table 4). Cu adsorption to Mn oxides on a molar basis was almost two orders of magnitude greater than that to Fe oxides and approximately three orders of magnitude greater than that to OMs. While Zn binding to Mn oxides was about one order of magnitude higher than that to Fe oxides and approximately two orders of magnitude higher than that to OMs. This result suggests that Fe/Mn oxides were more important than OMs for Cu and Zn scavenging at lower metal concentrations. In the other words, the greatest contribution to total extractable Cu and Zn binding to the SSs (NSCSs) (on a molar basis) was from Mn oxides in the non-residual fraction of the SSs (NSCSs).

Adsorption Contribution of SSs (NSCSs) Components

Based on Table 2, the estimated contributions of SSs (NSCSs) components to the total Cu and Zn adsorption by unextracted SSs (NSCSs) (M_{Fe}, M_{Mn}, M_{OMs}, and M_{Res}) based on relative concentration of each component (a, b, c, or d) and the natural of the component (Γ_{Mn}, Γ_{Fe}, Γ_{OMs} and Γ_{Res}) are shown in Table 4 and Figures 1 and 2. Non-residues contributed more in the SSs (NSCSs) than residues. For non-residual components, Fe oxides contributed most next was OMs and the lowest was from Mn oxides.

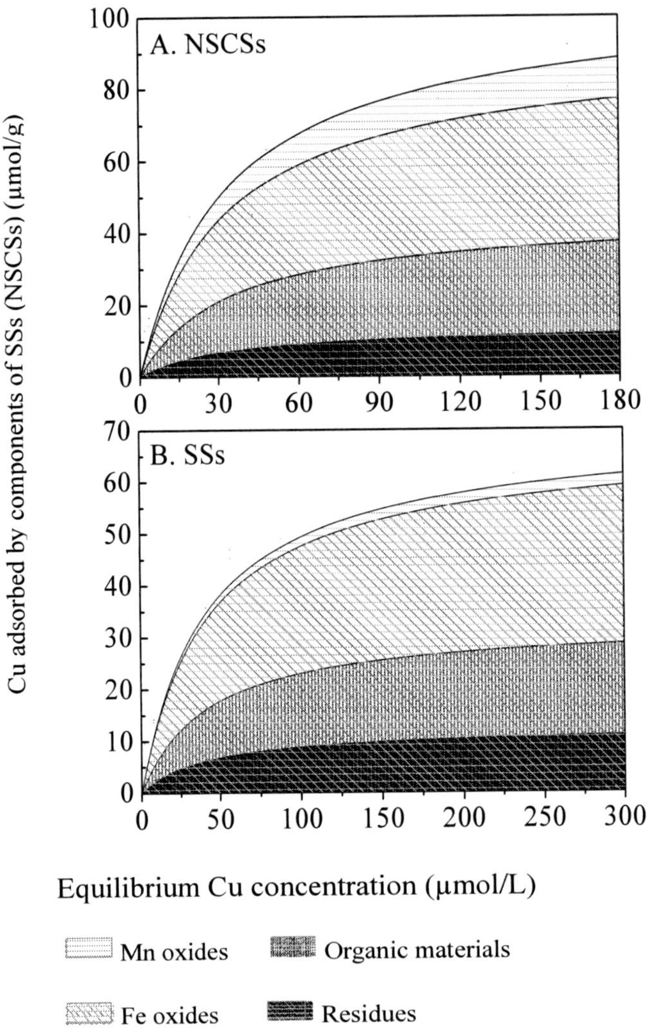

Figure 1. Estimated Cu adsorption to components of untreated NSCSs (A) and SSs (B) based on nonlinear least-squares fitting of Cu adsorption isotherm data.

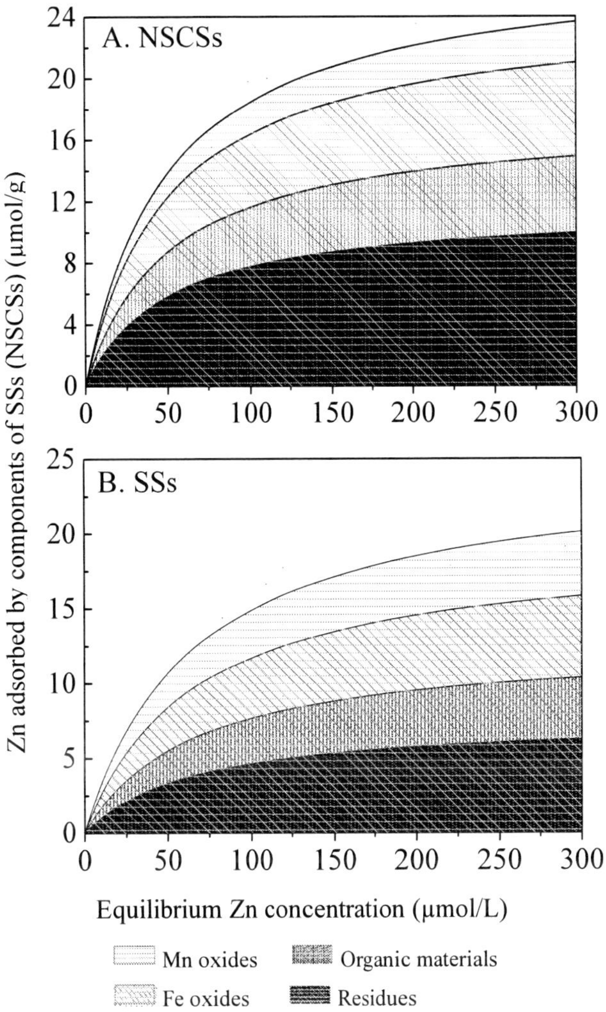

Figure 2. Estimated Zn adsorption to components of untreated NSCSs (A) and SSs (B) based on nonlinear least-squares fitting of Zn adsorption isotherm data.

As described above, the adsorption contribution of each component was determined not only by the natural of the component but also by relative concentration of each component in the SSs (NSCSs) compared with other components. Therefore, to compare the adsorption property of these components without the effect of relative concentration, adsorption ability was compared on a unit mass basis as discussing in the study of endogenic Cu and Zn. For non-residual components, the Cu or Zn adsorption ability of components followed the order Mn oxides > Fe oxides > OMs. The adsorption ability of OMs was nearly 2 orders of magnitude greater than that of residues for SSs (NSCSs), implying that compared to non-residual components, the Cu or Zn adsorption ability of residues was almost invisible. Then, as described above, in order to compare the adsorption ability of non-residual components on a molar basis, $\Gamma_{\text{mmol M/g Fe}}$, $\Gamma_{\text{mmol M/g Mn}}$ and $\Gamma_{\text{mmol M/g TOC}}$ were transformed into the forms of $\Gamma_{\text{mmol M/mol Fe}}$, $\Gamma_{\text{mmol M/mol Mn}}$ and $\Gamma_{\text{mmol M/mol TOC}}$ (Table 4). For Cu, the adsorption ability of Mn oxides was slightly greater than that of Fe oxides and about one order of magnitude greater than that of OMs; but for Zn, the adsorption ability of Mn oxides was almost on order of magnitude greater than that of Fe oxides and nearly two orders of magnitude higher than that of OMs. This result suggests that Fe/Mn oxides were more important than OMs for Cu and Zn adsorption at higher metal concentrations. In the other words, the greatest contribution to total extractable Cu and Zn binding to the SSs (NSCSs) (on a molar basis) was from Mn oxides in the non-residual fraction of the SSs (NSCSs).

Comparison of Adsorption Ability of SSs and NSCSs and their Components

Considering the results as showed in Table 4 and Figures 1 and 2, it could be found that there were some differences and also some similarities between the endogenic and anthropogenic pollutants in adsorption ability of SSs (NSCSs) and their components. Here, endogenic pollutants represent a long process of the interaction between the adsorbent (SSs and NSCSs) and adsorbate (Cu and Zn), and the enrichment of Cu and Zn at lower concentrations went along with the growing of SSs (NSCSs); but anthropogenic pollutants represent a short process of adsorption, and Cu and Zn binding to the SSs (NSCSs) at higher concentration taken place within 24 h.

For the two kinds of solid particles, the adsorption capacity of NSCSs was greater than that of SSs. The endogenic Cu (Zn) in the NSCSs was about 29% (24%) higher than that in SSs and the anthropogenic Cu (Zn) adsorption to the NSCSs was nearly 10% (16%) higher than those in SSs. This result agrees closely with that obtained from the other trace metals adsorption to SSs and NSCs or NSCSs (Li et al., 2005; Guo et al., 2006), implying that NSCSs were more important than SSs for the transformation and cycling of trace metals in aquatic environment. The result also strongly suggests that with the interaction time increased, the higher adsorption ability of NSCSs was more remarkable.

For the two kinds of trace metals, the amount of endogenic Zn was almost three orders of magnitude greater than that of endogenic Cu; but the adsorption of anthropogenic Cu in contrary was nearly three orders of magnitude greater than that of Zn. This in part could be due to the initial Cu/Zn ratio in the solution, and in the laboratory experiments the initial Cu/Zn ration was 1 and in the river water the initial Cu/Zn (Cu and Zn in the river water were respectively 0.11 ppm and 0.61 ppm) ratio was about 0.2. The higher adsorption of anthropogenic Cu was mainly due to the difference in characters of Cu and Zn, such as the higher standard electrode potential of Cu, the lower covalent radius of Cu (Dong et al., 2003d). The higher amount of endogenic Zn was dominantly due to the higher concentration of Zn in the river water.

For the endogenic and anthropogenic pollutants, the relative adsorption roles of non-residual and residual fractions of SSs (NSCSs) were different from each other. Non-residual fraction contributed more roles in anthropogenic Cu adsorption to the SSs (NSCSs), and the contributions of non-residual and residual fractions to the anthropogenic Zn adsorption were comparative; but for endogenic pollutants, the contribution of non-residual fraction was significant more than that of residual fraction for Zn, and the role of non-residual fraction was similar to that of residual fraction for Cu. However, the adsorption abilities of Mn oxides, Fe oxides and organic materials in the non-residual fraction of particles were similar to each other for endogenic and anthropogenic pollutant. Metals adsorption capacities of Mn oxides exceeded those of Fe oxides by one order of magnitude, fewer roles were found attributing to adsorption by organic materials (OMs). These results imply that Mn oxides in the non-residual fractions were the most important component in controlling heavy metals in aquatic environments.

Conclusion

In conclusion, the following may be drawn based on the results of this study:

(1) A selective extraction technique was modified and improved to be suitable for the separation of Mn oxides, Fe oxides and OMs in non-residual fractions of the SSs (NSCSs). The target components were removed with efficiencies above 63%, and the non-target materials with levels up to 36% were also extracted.
(2) Anthropogenic Cu and Zn adsorption onto the SSs (NSCSs) and their components fitted adequately well to Langmuir isotherms, and Cu adsorption was approximately 3 times greater than that of Zn. But endogenic Zn enrichment in the SSs (NSCSs) was nearly 3 times greater than that of Cu.
(3) Adsorption abilities of anthropogenic Cu and Zn onto the NSCSs were respectively about 10% and 16% greater than those onto SSs, and the endogenic Cu (Zn) enrichment in the NSCSs was about 29% (24%) higher than that in SSs.
(4) The adsorption ability of Mn oxides on a molar basis exceeded that of Fe oxides by approximately an order of magnitude. The value of metal adsorption ability of OMs was smallest among the three components in the non-residual fraction, and the estimated adsorption capacity of the residues to metal adsorption was insignificant.

Acknowledgements

This research was supported by the Ministry of Science and Technology of China ("973" Project no. 2004CB3418501). Support for Li Yu was provided by the scientific start-up fund from North China Electric Power University, China.

References

Chao, T. T. *J Geochem Explor* 1984, *20*, 101-135.

Costerton, J. W.; Chen, K. J.; Geesey, G. G.; Ladd, T. I.; Nickel, J. C.; Dasguta, M.; Marrie, T. J. *Annu Rev Microbiol* 1987, *41*, 435-464.

Dong, D. M.; Derry, L. A.; Lion, L. W. *Water Res* 2003a, *37*, 1662-1666.

Dong, D. M.; Hua, X. Y.; Li, Y.; Li, Z. H. *Environ Pollut* 2002, *119*, 317-321.

Dong, D. M.; Hua, X. Y.; Li, Y.; Zhang, J. J.; Yan, D. X. *Environ Sci Technol* 2003b, *37*, 4106-4112.

Dong, D. M.; Li, Y.; Hua, X. Y. *Microchem J* 2001a, *70*, 25-33.

Dong, D. M.; Li, Y.; Hua, X. Y.; Zhang, J. J.; Yang, F. *Chinese J Environ Sci* 2003c, *24*, 131-134.

Dong, D. M.; Li, Y.; Zhang, B. Y.; Hua, X. Y.; Yue, B. H. *Microchem J* 2001b, 69, *89-94*.

Dong, D. M; Li, Y.; Zhang, J. J.; Hua, X. Y. *Chemosphere* 2003d, *51*, 369-373.

Dong, D. M.; Nelson, Y. M.; Lion, L. W.; Shuler, M. L.; Ghiorse, W. C. *Water Res* 2000, *34*, 427-436.

Flemming, H.-C. *Water Sci Technol* 1995, *32*, 27-33.

Fujiyoshi, R.; Okamoto, T.; Katayama, M. *Appl Radiat Isotopes* 1994, *45*, 165-170.

Gomez, P. C.; Fontes, M. P. F.; Silva, A. G.; Mendora, E. S.; Netto, A. R. *Soil Sci Soc Am J* 2001, *65*, 1115-1121.

Guo, S. H.; Wang, X. L.; Li, Y.; Chen, J. J.; Yang, J. C. *J Environ Sci* 2006, *18*, 1193-1198.

Hsu, P.H. *Aluminum oxides and oxyhydroxides;* Minerals in soil environments; ASA and SSSA: Madison, WI, 1989; pp 331-378.

John, O. A.; Latifatu, A. O. *Geoderma* 2004, *119*, 85-95.

Li, Y.; Chen, J. J.; Wang, X. L.; Dong, D. M.; Guo, S. H. *Chem J Chinese U* 2006a, *27*, 627-631.

Li, Y.; Wang, X. L.; Wang, Y.; Dong, D. M.; Zhang, H. P.; Li, Q. S.; Li, X. C. *J Environ Sci* 2005, *17*, 126-129.

Li, Y.; Yang, F.; Dong, D. M.; Lu, Y. Z.; Guo, S. H. *Chemosphere* 2006b, *62*, 1709-1717.

Lion, L. W.; Altmann, R. S.; Leckle, J. O. *Environ Sci Technol* 1982, *16*, 660-666.

Lion, L. W.; Shuler, M. L.; Ghiorse, W. C. *CRC Critical Reviews in Environmental Control* 1988, *17*, 273-306.

Nelson, Y. M.; Lion, L. W.; Shuler, M. L.; Ghiorse, W. C. *Limnol Oceanogr* 1999, *44*, 1715-1729.

Nelson, Y. M.; Lo, W.; Lion, L. W.; Shuler, M. L.; Ghiorse, W. C. *Water Res* 1994, *29*, 1934-1944.

Perret, D.; Gaillard, J. F.; Dominik, J.; Atteia, O. *Environ Sci Technol* 2000, *34*, 3540-3546.

Santschi, P. H.; Lenhart, J. J.; Honeyman, B. D. *Mar Chem* 1997, *58*, 99-125.

Schnitzer, M. *Soil Sci Soc Am P* 1969, *33*, 75-81.

Schwertmann, U.; Taylor, R. M. *Iron oxides;* Minerals in soil environments; ASA and SSSA: Madison, WI, 1989; pp 379-438.

Turner, A.; Millward, G. E.; Roux, S. M. L. Mar Chem 2004, 88, 179-192.

Vuceta, J.; Morgan, J. J. Environ Sci Technol 1978, 12, 1302-1308.

Young, L. B.; Harvey, H. H. Geochim Cosmochim Acta 1992, 56, 1175-1186.

In: Progress in Environmental Microbiology
Editor: Myung-Bo Kim, pp. 187-202

ISBN: 978-60021-940-5

Chapter VI

Colonisation of Water Systems in the Built Environment of Northern Germany by *Legionella spp.* and *Pseudomonas spp.*

B. P. Zietz [*] ***and H. Dunkelberg***
Medical Institute of General Hygiene and Environmental Health,
University of Göttingen, Germany

Abstract

Pneumonia with *Legionella spp.* presents a public health challenge especially because fatal outcomes still remain frequent. *Pseudomonas aeruginosa* is a significant source of hospital-acquired pneumonia and can also cause devastating chronic infections in compromised hosts, for example respiratory infections in cystic fibrosis patients. The aim of our studies was to describe the abundance and epidemiology of *Legionellaceae* in the man-made environment. In total, water systems of 70 different buildings in the German town of Göttingen (Lower Saxony) were examined for the presence of *Legionella* in two sampling cycles. Of these 22 (31%) had the bacterium in at least one water sample. *Legionella pneumophila* serogroups 1, 3, 4, 5 and 6 could be identified in the water samples. Most of the buildings were colonized solely by one distinct strain, as proven by PCR typing. Some buildings contained more than one PCR type or even more than one serogroup. Additionally the colonization of greenhouse misting systems with *Legionella spp.* and *Pseudomonas spp.* was studied in 20 different greenhouse misting systems located in Northern Germany. In total 80 water samples were collected. Each system was tested on two different occasions. Water was drawn at a central tap and at the

[*] Correspondence concerning this article should be addressed to Dr. B. P. Zietz, MPH; Medical Institute of General Hygiene and Environmental Health, University of Göttingen, Lenglerner Str. 75, D-37079 Göttingen, Germany. Tel.: +49 551 5007886-1; fax: +49 551 5007886-3. E-mail: bjoern.zietz@nlga.niedersachsen.de

outlet of spray nozzles. Sampled greenhouses were used to cultivate various plants and trees for commercial, recreational or scientific reasons, some of them in tropical conditions. *Legionella spp.* was detected in 10% of the systems (two systems), but only in low numbers. *Pseudomonas spp.* was recovered from 70% of the greenhouse watering systems (fourteen systems), occasionally at counts greater than 10,000 CFU/100 ml. Each colonized greenhouse had one or several individual strains of *Legionella* and *Pseudomonas* that could not be detected in any other system. This was demonstrated by a random amplified polymorphic DNA typing method. The possible health hazard caused by these water systems for both genera of bacteria is evaluated and discussed.

Keywords: *Legionella*, *Pseudomonas*, drinking water, hot water supply, greenhouse misting systems, molecular typing, PCR, RAPD

Introduction

Legionellae are gram-negative bacteria (rods) that require special culture conditions (Fields et al., 2002). They were first recognized as the causative agent of Legionnaire's disease in 1977 following an epidemic of acute pneumonia at an American Legion convention in Philadelphia (Fraser et al., 1977; McDade et al, 1977). The bacterium has been recovered from a wide range of man-made water systems including hot water supplies, cooling towers and whirlpool spas (Breiman and Butler 1998). Surveys of lakes, ponds, streams, and soils have indicated that this bacterium is also a common inhabitant of natural waters (Fliermans, 1996).

Bacteria of the genus *Pseudomonas* are gram-negative rods with polar flagella (Timmis, 2002). Similar to *Legionella* they are common environmental bacteria (Costerton et al., 1999; Stover et al., 2000). They can be detected in groundwater and drinking water systems (Schoenen et al., 1986; Codony et al., 2002; Banning et al., 2003). *Pseudomonads* and especially *Pseudomonas aeruginosa* are significant nosocomial pathogens, which for example cause wound infections and bacteraemia in burn victims or urinary tract infections in catheterized patients (Stover et al., 2000). They are also a source of hospital-acquired pneumonia, particularly in patients on respirators (Garau and Gomez, 2003). A chronic respiratory infection with *P. aeruginosa* is typical for patients with cystic fibrosis (Gibson et al., 2003). Additionally outbreaks of *Pseudomonas* dermatitis / folliculitis and otitis externa associated with swimming pool and whirlpool spa use have been described (CDC, 2000; Ratnam et al., 1986).

Materials and Methods

Legionellaceae in Warm Water Supplies of Buildings

In a first sampling period, water was collected from 70 private and public buildings between February and September 1999 in the Göttingen area, in Germany. In private buildings water was taken in the bath room from the hot-water taps. Water samples from public buildings were obtained by turning on the hot-water taps (many from showers heads) and taking the first water. A second sample was taken when the water reached the highest temperature. In the first sampling period, in total 129 samples were collected (Zietz et al., 2001). In a second sampling period, 196 samples from the same buildings (including additional follow-up samples of colonized buildings) were collected between September 1999 and November 2000 (Brengelmann et al., 2003).

All sampled buildings were supplied by the water plant of Göttingen (Stadtwerke Göttingen). The distribution system consists of three main pressure zones according to different elevations of the city and several small higher or more distant zones. Three facilities that combine water from a transport pipe from the Harz Mountains (about 80%) and local wells supplied water to the lowest zone (Schumacher et al., 1988). Out of this zone, water is pumped up to the other zones.

Water samples (1 l volume) were filtered through 0.45-µm-pore-size cellulose-mixed ester filters with a diameter of 50 mm (Schleicher and Schuell, Dassel, Germany) using a vacuum pump. Then 10 ml of a KCl/ HCl-buffer (0.2 M, adjusted to pH 2.2) was poured onto the filter and removed again after 5 minutes. The filters were placed on a MWY agar plate (Oxoid, Wesel, Germany) and incubated at 37°C in a humidified atmosphere (plastic bag) for seven days and examined daily. Additionally 1 ml of water was added to 1 ml of a KCl/ HCl-buffer and after 5 minutes 0.5 ml of the solution was used to inoculate the surface of the MWY agar. This was done in duplicate. Colonies that morphologically matched *Legionella* colonies were subcultured onto blood and MWY agar. Representative colonies (1-2) of those that failed to grow on blood agar were examined by direct fluorescent antibody technique. Isolates were stored at -70°C (Microbank, Mast Diagnostica).

RAPD-Polymerase chain reaction: To identify different strains of *Legionella* we used three different primers to amplify DNA fragments in crude bacterial lysates to generate banding profiles (Wiese et al., 2004; Zietz et al., 2002). Used primers were ERIC2 (5'-AAG TAA GTG ACT GGG GTG AGC G-3') (van Belkum et al., 1993) and a combination of Lpm-1 (5'-GGT GAC TGC GGC TGT TAT GG-3') and Lpm-2 (5'-GGC CAA TAG GTC CGC CAA CG-3') (Jaulhac et al., 1992). ERIC2 is an enterobacterial repetitive intergenic consensus motif. Lpm-1 and Lpm-2 are part of the macrophage infectivity potentiator (mip) gene of *Legionella*. Gels were stained by adding ethidium bromide to the agarose gel and bandings were visualized under ultraviolet light. Details of the methods can be found in Zietz et al. (2001).

Legionella spp. and *Pseudomonas spp.* in Greenhouse Misting Systems

Water was collected from private and public greenhouse misting systems between June and September 2003 in Lower Saxony and Hessia, Germany. Sampled greenhouses were used to cultivate various plants and trees for commercial, recreational or scientific reasons, some of them in tropical conditions. Some of the greenhouses were open to the public. In total 20 different greenhouse misting systems were tested on two different occasions. Each time, water was drawn at the outlet of spray nozzles and at a central tap of the internal misting system. Therefore in total 80 water samples were collected. Water was not flushed except the first water stagnated in the pipes connecting the tap with the central (circulating) system (amount estimated). Greenhouse misting systems were supplied by public drinking water systems, by rain water cisterns or by private wells. In one case water was supplied from a pond (Zietz et al., 2006).

To detect *Legionella spp.*, water samples (100 and 20 ml) were treated as described above with the exception that MWY agar was replaced by GVPC agar. Per sample 5-8 colonies of those that failed to grow on blood agar were examined using a *Legionella spp.* latex test kit.

For the detection of *Pseudomonas spp.*, samples of 100 and 20 ml water were vacuum filtered through 0.45-µm-pore-size cellulose-mixed ester filters (Schleicher and Schuell) and placed on Cetrimide agar plates (Merck, Darmstadt, Germany). Additionally, 0.5 ml of water was directly inoculated on the surface of the Cetrimide agar in duplicate. Colonies were examined under UV light for fluorescence and representative suspicious colonies were subcultured onto blood agar. Finally, following a Gram stain diagnosis and a positive oxidase test, the species of the isolated colonies were determined using BBL Crystal Enteric/NF System (Becton Dickinson).

For every sample, numbers of total heterotrophic plate counts per millilitre at 20 and 36°C were determined. The parameters pH, conductivity, total hardness and different ions were determined from an additional water sample at every greenhouse sampling date. For details of the methods please refer to Zietz et al., 2006.

For typing isolates of *Legionella* and *Pseudomonas* the same RAPD-Polymerase chain reaction method was used as described above. Primer 272 (5′-AGC GGG CCA A-3′) (Mahenthiralingam et al., 1996; Ruimy et al., 2001) was used to generate banding profiles for *Pseudomonas spp.* isolates.

Results

Legionellaceae in Warm Water Supplies of Buildings

Of the 70 buildings examined in the first sampling period 18 (26%) had the bacterium in at least one water sample. *Legionella pneumophila* serogroups 1, 4, 5 and 6 could be identified in the water samples (Zietz et al., 2001). In the second period 17 buildings (24%) had positive results in the culture (Brengelmann et al., 2003). Of these, 4 buildings were newly detected to have cultureable *Legionellaceae* compared with the first sampling period. In addition to the serogroups found before, serogroup 3 was detected for the first time (Table A). No other species than *L. pneumophila* was found. In total 31% of the buildings had the bacterium in at least one water sample

Table A. Detected *Legionellaceae* in different types of buildings

Type of building	Number of tested buildings	First sampling period			Second sampling period		
		Buildings with at least one sample containing *Legionellaceae*		Found serogroups *	Buildings with at least one sample containing *Legionellaceae*		Found serogroups *
Sports halls and swimming baths	24	5	(21%)	1, 4, 5, 6	4	(17%)	1, 4, 6
University buildings	19	5	(26%)	1, 4, 6	3	(16%)	1, 6
Hospitals and old people's homes	8	5	(63%)	1, 4, 6	7	(88%)	1, 3, 4, 5
Halls of residence	4	2	(50%)	1, 5	2	(50%)	1, 4
Hotels	4	0	(0%)	-	0	(0%)	-
Other buildings	11	1	(9%)	1	1	(9%)	4
Total	70	18	(26%)	1, 4, 5, 6	17	(24%)	1, 3, 4, 5, 6

* all isolates *L. pneumophila.*

Table B. The distribution of the maximum found colony forming units (CFU/l) of *Legionella spp.* for all tested buildings

Range of detected colony forming units (CFU/l)	Number of buildings - first sampling period	Number of buildings - second sampling period
Not detectable	52 buildings	53 buildings
$1\text{-}10^2$	7 buildings	5 buildings
$>10^2\text{-}10^3$	1 building	1 building
$>10^3\text{-}10^4$	6 buildings	3 buildings
$>10^4\text{-}10^5$	4 buildings	5 buildings
$> 10^5$	-	3 buildings

The overall trend was that the larger the building's plumbing had been, the more samples were positive for *Legionella* and the more bacteria could be found. The highest detected concentration of bacteria was 78,000 CFU/l in the first period and 150,000 CFU/l in the second period. In the second sampling period 3 buildings were found to have a bacteria concentration above 10^5 CFU/l (Table B).

Most of the buildings were colonized solely by one distinct strain, as proven by PCR. In three cases equal patterns were found in separate buildings. There were two buildings in this study where isolates with different serogroups were found at the same time.

No association of serogroups or identical PCR types and water supply zones was found (Zietz et al., 2001; Brengelmann et al., 2003).

Legionella spp. and *Pseudomonas spp.* in Greenhouse Misting Systems

Legionella spp. were detected in low numbers in two of the greenhouse misting systems (10%). In the first greenhouse (designated K), *L. pneumophila* serogroup 6 (35 CFU per 100 ml, equivalent to 350 CFU/l) were recovered from the peripheral sample (outlet of spray nozzles) of the first sampling period in July 2003. The central system of the second greenhouse R was found to be colonized by *L. pneumophila* serogroup 6 (6 CFU per 100 ml, equivalent to 60 CFU/l) in the second sampling period. The misting system of greenhouse K was supplied by a private well and the system of greenhouse R by a rainwater cistern that was filled up with public drinking water if there was a lack of rainwater (in the hot summer probably filled up several times during the sampling period, but there was no documentation). In total, 13 isolates (all belonging to serogroup 6) have been stored and typed by RAPD. Isolates from each greenhouse had a unique typing pattern and all isolates from the same greenhouse were identical (Zietz et al., 2006).

Pathogenic *Pseudomonas spp.* could be found in 70% of the greenhouse watering systems (14 systems), occasionally in excess of 10^4 CFU per 100 ml. The species *P. aeruginosa*, *P. putida*, *P. fluorescens* and one single isolate of *P. stutzeri* were detected. In three systems, only one of the four samples (2 x central, 2 x peripheral/outlet of spray nozzles) contained *Pseudomonas spp.* In seven systems, *Pseudomonas spp.* were found in two different samples, in two systems in three samples and in one system they were cultured from all four samples.

In total, 59 isolates of *P. aeruginosa* were recovered from the greenhouses, of which 34 different PCR patterns could be identified. Additionally, each of the five *P. aeruginosa* reference strains had an individual pattern. Each greenhouse had one or more individual patterns that could not be detected in any other system. In three cases, a pattern from an isolate of the first sampling cycle could be found again in isolates of the second cycle. *P. fluorescens* was recovered from five different greenhouses, each having one single isolate with its own individual PCR banding pattern. *P. putida* was also detected in five different greenhouses. From one greenhouse, three isolates were recovered and a single isolate from each of the other ones. All greenhouse strains and the reference strain (DSM 291) had their own distinct pattern (Fig. 1). The three isolates of greenhouse M produced an equal banding pattern (isolates PpM1Cb, PpM1P, PpM1Ca) (Zietz et al., 2006).

Total heterotrophic counts at 20 and 36°C varied in a broad range between 0 CFU and $>10^4$ CFU per ml. The average CFUs from peripheral samples were higher than that from central samples. There were frequently significant ($P <0.05$ or $P <0.01$) Spearman rank correlations between *Pseudomonas* numbers and total heterotrophic counts (only in second sampling cycle) or between different types of total heterotrophic counts in the same sampling cycle. The results indicate that microbial enumeration of several species can occur simultaneously, probably related to biofilms (Zietz et al., 2006).

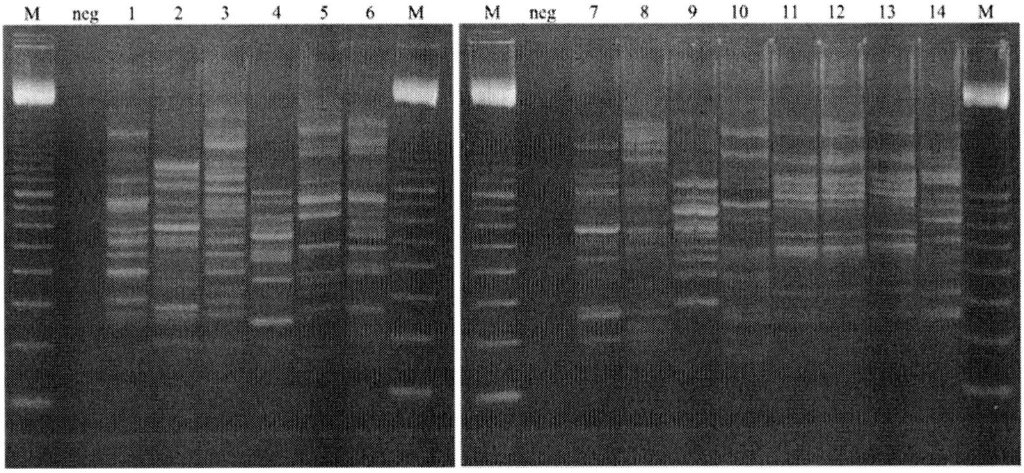

Figure 1. Fingerprint profiles of *P. fluorescens* isolates (left) and *P. putida* isolates (right) from greenhouse misting systems tested with primer sequence 272. Lanes (left to right): molecular weight marker (100 bp), negative control (water blank), *P. fluorescens* reference strain DSM 50090 (lane 1), isolates PfG2C, PfO1C, PfL1P, PfF1C, PfC1C, molecular weight marker. Second part: molecular weight marker (100 bp), negative control, *P. putida* DSM 291 (lane 7), PpL2P, PpC2P, PpN1C, PpM1Cb, PpM1P, PpM1Ca, PpF1C, molecular weight marker.

Discussion

Legionella spp. in Warm Water Supplies

Also in Lower Saxony, Germany, Habicht and Müller (1988) found that 70% of the 103 hospitals and 18% of the 62 hotels investigated were positive for *Legionella*. These findings are similar to our results testing only one supply area. Boschek et al. (1995) were able to culture *Legionellaceae* from warm water of 11 out of 12 sampled hospitals in a German town. A study in south-eastern Germany investigated *Legionella* contamination in 75 different buildings. In total 85% of all public buildings (hotels, scientific facilities, swimming baths) and 65% of all private households in large buildings had the bacteria in their water samples. Most samples contained *Legionella spp.* in the range of 10^3 - 10^5 CFU/l, the highest concentration was 4 x 10^6 CFU/l. Serogroups 1, 3, 6, and 10 were most commonly isolated with frequencies of 22%, 14%, 16% and 18%, respectively. Of the buildings, 31 were colonized by one serogroup and 15 by two or more. Additionally 5 strains (6%) were non-pneumophila species (Lück et al., 1993). A high constancy (regular detection, only a few minor genetical changes) of *Legionella* colonization was found in a hospital connected with a ring pipe warm water system over 7 years (Lück et al., 1995). It has to be taken into account that used methods varied between these studies.

It is important to know that hospitals and old people's homes used by a risk group for infections frequently had high rates of contamination including the Göttingen study.

No association of serogroups or identical PCR types and water supply zones was found in our study in Göttingen. So the exact evolution and origin of these populations remains unclear. But it seems likely that occasionally *Legionella* colonies from environmental sources are transported through the water supply system to the user's installation. In some cases they are able to colonize especially the hot water systems.

Greenhouse Misting Systems

Regarding greenhouse contamination only one related study has been published which examined different gardening farms located in eastern Poland (Lublin province) for the presence of *Legionella spp.* in tap water, soil and air (Stojek and Dutkiewicz, 2002). Of the 20 samples taken from indoor taps in traditional greenhouses used for sprinkling on plants, 25.0% were culture positive for *Legionella spp.* In outdoor taps used for sprinkling on plants, 22.2% of the 36 samples contained *Legionella spp.* No bacteria were detected in the tap-water, soil and air samples collected from modern greenhouses.

Legionella spp. infections (especially *L. longbeachae*) have been associated with gardening and the use of potting soil in Australia, Asia, North America and Europe (Cameron et al., 1991; Anonymous, 2000; Suzuki et al., 2002; den Boer et al., 2007). *Legionella spp.* was also detected within slow sand filters used for fungal plant pathogen suppression in horticultural crops (Calvo-Bado et al., 2003) and composted plant materials (Hughes and Steele, 1994).

Our study on greenhouse misting systems indicates that sources other than hot water installations, cooling towers and whirlpool spas should be observed as possibly critical for a relevant *Legionella* contamination. A possible *Pseudomonas* contamination should also be kept in mind especially when untreated water is used in man-made systems.

Health Evaluation of *Legionella spp.* Contamination

The numbers of colony forming units found in greenhouse misting systems (350 and 60 CFU/l, respectively) appear to be relatively low compared with the numbers of *Legionella* detected in our study of hot water systems in public and private buildings in the city of Göttingen. The city is located in centre of the area studied in the greenhouse investigation. In the 70 public and private buildings, up to 150,000 CFU/l of *Legionella spp.* were recovered from hot water taps. Of the 70 buildings examined, 31% had the bacterium in at least one water sample compared to 10% of greenhouse misting systems (Zietz et al. 2001; Brengelmann et al., 2003).

The conditions in these building hot water systems may have been more favourable for *Legionella spp.* growth than in the greenhouse misting systems. The summer of 2003 (June to September) was rather warm in Germany, generally providing good conditions for the enumeration of *Legionella spp.* in greenhouse water systems. Air temperatures often exceeded 25°C during the sampling period. The temperatures of the water samples in the greenhouse misting systems were usually between 20 and 30°C, although there was sometimes a high turnover of water (statement of the system operators) and probably lower soil temperatures. *Legionellaceae* are known to multiply at temperatures between 25 and 42°C, with an optimal growth temperature of 35°C (Fields et al., 2002).

The US Occupational Safety and Health Administration recommends action level 2 ("immediate cleaning and/or biocide treatment; take prompt steps to prevent employee exposure") for *Legionella* concentrations in domestic water exceeding 10^5 CFU/l and action level 1 ("prompt cleaning and/or biocide treatment of the system") for concentrations exceeding 10^4 CFU/l. For *Legionella* concentrations in humidifiers, both of these action levels are only one tenth of the values stated above (OSHA, 1999)

In Germany, the action level for the renovation of existing tap-water systems is regulated by a DVGW technical norm. The most recent norm states that if *Legionella* concentration exceeds 10^3 CFU/l, a renovation and/or cleaning has to be initiated. Concentrations above 10^5 CFU/l are considered extremely high, which make immediate cleaning and prompt steps for exposure prevention necessary (DVGW, 2004). These numbers are technically orientated and are not based on estimates of infective doses. As one can see, the *Legionella* concentrations found in tap-water of several buildings in Göttingen were in the range where prevention steps are necessary. In greenhouse misting systems detected *Legionella* concentrations (350 CFU/l and 60 CFU/l, respectively) were below all of these action levels, but they indicate that these systems were contaminated.

Evaluation of Possible Health Effects of *Pseudomonas spp.* Contamination

Water of misting systems in our greenhouse study contained *Pseudomonas spp.* in concentrations sometimes higher than 10^4 CFU per 100 ml. High counts were more frequent in central samples and in water collected in the second sampling cycle. In comparison with published data on other water systems, detected numbers of *Pseudomonas spp.* appear high to very high.

For example, the water distribution system of Barcelona and surrounding cities was tested for the presence of members of the *Pseudomonadaceae* family (Ribas et al., 2000). *Pseudomonas spp.* were found in concentrations of up to 74 CFU/100 ml. The authors reported that untreated water contained high numbers of these bacteria. In the Italian city of Bologna shower water of 12 swimming pools contained *P. aeruginosa* in 22 out of 48 samples. Concentrations ranged between 2 and 1500 CFU per 100 ml (Leoni et al., 2001).

Pseudomonas spp. is known to be a common species in the first step of the formation of a biofilm in (model) drinking water distribution systems as well as in later phases of biofilm communities (Martiny et al., 2003; Martiny et al., 2005; Schmeisser et al., 2003). In the biofilm environment they are able to successfully adapt to situations of protozoan grazing (Matz et al., 2004). The genus can be detected throughout the stages of typical water purification systems including filtration, softening, deionization and UV treatment (Penna et al., 2002; Mazzola et al., 2006).

Generally, community-acquired pneumonia (CAP) with *Pseudomonas spp.* in previously healthy adults seems to be relatively rare. Although, *Pseudomonas* CAP should be considered in the differential diagnosis for anyone with a smoking history who presents with rapidly progressive pneumonia (Hatchette et al., 2000). In many cases of *Pseudomonas* pneumonia patients had certain risk factors or comorbidities such as lung diseases. This type of pneumonia trends to have severe courses and outcomes (Lesprit et al., 2003; Ramsdell et al., 2005).

In a case description a community-onset *Pseudomonas* pneumonia was associated with the use of a home humidifier. The 39 year-old-man had a medical history including 25 years of asthma (Harris et al., 1984).

To assess detected *Pseudomonas* concentrations in greenhouse misting systems in relation to results of laboratory animal tests, several factors have to be taken into account. Such factors are for example species differences, susceptibility of exposed persons, differing pathogenicity of strains and differences in ways of exposure. Additionally the amount of aerosol deposited in the lungs and the particle size is important. The particle size is known to be an important determinant of the depth of deposition in the lung (ratio peripheral to central deposition).

In laboratory tests, it is reported that in rats a mortality rate of >50% can be induced by a transtracheal infection with 10^6 CFU of *P. aeruginosa* strain PAO1 (Lesprit et al., 2003). In mice, an intranasal inoculation of either *P. aeruginosa* strain PA103 or strain PA103-29 led to a dose-dependent mortality. While inoculation with 10^4 CFU of strain PA103 did not cause death, mortality rates were between 40% and 80%, following administration of 10^5–10^6 CFU (Schultz et al., 2001). Similar results with the intranasal administration of *P. aeruginosa* strain AC869 were found in mice. The 50% lethal dose with this strain was 2.7 x 10^7 CFU. Treatment with 1.61 x 10^7 CFU resulted in slight morbidity and some mortality, generally within 3–4 days after the microbial dose (George et al., 1991). According to a review by Rusin et al. (1997), the oral infectious dose for *P. aeruginosa* in animal and human test subjects ranges between 10^8 and 10^9 CFU.

The volume of deposited aerosol water in the respiratory tract of a person staying in a greenhouse is supposed to be low. The volume of deposited aerosol and the location in the respiratory tract where the water is deposited is dependent on the aerosol size generated by the system.

Comparing different available laboratory mortality and morbidity tests with our highest *Pseudomonas* counts (>10^4 CFU/100 ml), there is probably a safety margin remaining (although the exact maximum CFU numbers in our study are unclear). However a detailed risk assessment is still pending and requires further data and investigation. For immune-compromised patients, this calculation may be different (Kooguchi et al., 1998; Faure et al., 2004).

The German drinking water commission recommends a repetition of sampling in 100 ml if *Pseudomonas spp.* is detected in a drinking water sample. In case of confirmation the contamination should be quantified and the systems should be assessed for the reasons of contamination. Afterwards measures for removal of the contamination should be undertaken (disinfection, flushing) (Umweltbundesamt, 2002).

The WHO has evaluated health hazards of *P. aeruginosa* associated with recreational waters (WHO, 2006). It recommends that for continuously disinfected pools, operational levels of *P. aeruginosa* should be below 1 CFU per 100 ml. In natural spas operating with no residual disinfectant, *Pseudomonas* concentrations should be below 10 CFU/100 ml. "If high counts are found (>100/100 ml), pool operators should check turbidity, disinfectant residuals and pH, resample, backwash thoroughly, wait one turnover and resample. If high levels of *P. aeruginosa* remain, the pool should be closed and a thorough cleaning and disinfection programme initiated. Hot tubs should be shut down, drained, cleaned and refilled." Routine monitoring of *P. aeruginosa* is recommended for public and semipublic hot tubs and natural spas.

Conclusion

The presented investigations showed that many water systems of buildings in Northern Germany are contaminated with *Legionellaceae*. This is especially true for larger buildings and more complex hot water systems using recirculation of water. Our studies also indicate that aerosolizing greenhouse watering systems may under certain circumstances be a potential source of *Legionella spp.* or *Pseudomonas spp.* infection in gardeners and visitors.

Typing of cultured isolates of *Legionella spp.* and *Pseudomonas spp.* with RAPD PCR showed that this method can be a useful epidemiological tool to investigate a possible infection as a result of a water system contamination with these bacteria. Another main conclusion of our studies is that there exists a great diversity of *Legionella* and *Pseudomonas* strains in man-made water systems as detectable by culture and PCR typing. In epidemiological studies testing water samples for the presence of *Legionella spp.* and *Pseudomonas spp.* several isolates should identified from a sample because a co-colonization of a system with different strains is possible. A non-detection of strains in epidemiological investigations may lead to false conclusions.

Greenhouse misting systems should be part of water management programs that include *Legionella spp.* and *Pseudomonas spp.* monitoring and control. Additionally other previously unconsidered man-made water systems should be systematically assessed in the future to uncover a possible bacterial contamination and infection risk.

Acknowledgements

The authors thank F. Brengelmann, J. Ebert, B. Gerhart, S. Luthin and M. Narbe, for their cooperation in the investigations.

References

Anonymous. Legionnaires' disease associated with potting soil – California, Oregon, and Washington, May–June 2000. *MMWR Morb Mortal Wkly Rep* 2000, 49, 777–778.

Banning, N; Toze, S; Mee, BJ. Persistence of biofilm-associated *Escherichia coli* and *Pseudomonas aeruginosa* in groundwater and treated effluent in a laboratory model system. *Microbiol* 2003, 149, 47-55.

Boschek, HJ; Langer, BK; Exner, M. Vorkommen von Legionellen in Warmwasserversorgungssystemen von Krankenhäusern und Pflegeheimen einer deutschen Großstadt. [Abstract]. *Zbl Hyg Umweltmed* 1995, 197, 314-315.

Breiman, RF; Butler, JC. Legionnaires' disease: Clinical, epidemiological, and Public Health perspectives. *Sem Resp Inf* 1998, 13, 84-89.

Brengelmann, F; Zietz, B; Dunkelberg, H. Nachweis von Legionellen in Gebäuden im Bereich einer kommunalen Trinkwasserversorgung. Aachen: Shaker Verlag; 2003.

Calvo-Bado, LA; Morgan, JA; Sergeant, M; Pettitt, TR; Whipps, JM. Molecular characterization of *Legionella* populations present within slow sand filters used for fungal plant pathogen suppression in horticultural crops. *Appl Environ Microbiol* 2003, 69, 533-541.

Cameron, S; Roder, D; Walker, C; Feldheim, J. Epidemiological characteristics of *Legionella* infection in South Australia: implications for disease control. *Aust N Z J Med* 1991, 21, 65-70.

CDC (Centers for Disease Control and Prevention). *Pseudomonas* dermatitis / folliculitis associated with pools and hot tubs--Colorado and Maine, 1999-2000. *MMWR Morb Mortal Wkly Rep* 2000, 49, 1087-1091.

Codony, F; Morato, J; Ribas, F; Mas, J. Effect of chlorine, biodegradable dissolved organic carbon and suspended bacteria on biofilm development in drinking water systems. *J Basic Microbiol* 2002, 42, 311-319.

Costerton, JW; Stewart, PS; Greenberg, EP. Bacterial biofilms: a common cause of persistent infections. *Science* 1999, 284, 1318-1322.

den Boer, JW; Yzerman, EP; Jansen, R; Bruin, JP; Verhoef, LP; Neve, G; van der Zwaluw, K. Legionnaires' disease and gardening. *Clin Microbiol Infect* 2007, 13, 88-91.

DVGW (Deutsche Vereinigung des Gas- und Wasserfaches e.V.). Trinkwassererwärmungs- und Trinkwasserleitungsanlagen; Technische Maßnahmen zur Verminderung des Legionellenwachstums; Planung, Errichtung, Betrieb und Sanierung von Trinkwasser-Installationen. DVGW Arbeitsblatt W 551. Bonn: DVGW; 2004.

Faure, K; Sawa, T; Ajayi, T; Fujimoto, J; Moriyama, K; Shime, N; Wiener-Kronish, JP. TLR4 signaling is essential for survival in acute lung injury induced by virulent *Pseudomonas aeruginosa* secreting type III secretory toxins. *Respir Res* 2004, 5, 1.

Fields, BS; Benson, RF; Besser, RE. *Legionella* and Legionnaires' disease: 25 years of investigation. *Clin Microbiol Rev* 2002, 15, 506-526.

Fliermans, CB. Ecology of *Legionella*: From data to knowledge with a little wisdom. *Microb Ecol* 1996, 32, 203-228.

Fraser, DW; Tsai, TR; Orenstein, W; Parkin, WE; Beecham, HJ; Sharrar, RG; Harris, J; Mallison, GF; Martin, SM; McDade, JE; Shepard, CC; Brachman, PS. Field

Investigation team. Legionnaires' disease: description of an epidemic of pneumonia. *N Engl J Med* 1977, 297, 1189-97.

Garau, J; Gomez, L. *Pseudomonas aeruginosa* pneumonia. *Curr Opin Infect Dis* 2003, 16, 135-143.

George, SE; Kohan, MJ; Whitehouse, DA; Creason, JP; Kawanishi, CY; Sherwood, RL; Claxton, LD. Distribution, clearance, and mortality of environmental *pseudomonads* in mice upon intranasal exposure. *Appl Environ Microbiol* 1991, 57, 2420-2425.

Gibson, RL; Burns, JL; Ramsey, BW. Pathophysiology and management of pulmonary infections in cystic fibrosis. *Am J Respir Crit Care Med* 2003, 168, 918-951.

Habicht, W; Müller, HE. Occurrence and parameters of frequency of *Legionella* in warm water systems of hospitals and hotels in Lower Saxony. *Zbl Bakt Hyg B* 1988, 186, 79-88.

Harris, AA; Goodman, L; Levin, S. Community-acquired *Pseudomonas aeruginosa* pneumonia associated with the use of a home humidifier. *West J Med* 1984, 141, 521–523.

Hatchette, TF; Gupta, R; Marrie, TJ. *Pseudomonas aeruginosa* community-acquired pneumonia in previously healthy adults: case report and review of the literature. *Clin Infect Dis* 2000, 31, 1349–1356.

Hughes, MS; Steele, TW. Occurrence and distribution of *Legionella* species in composted plant materials. *Appl Environ Microbiol* 1994, 60, 2003-2005.

Jaulhac, B; Nowicki, M; Bornstein, N; Meunier, O; Prevost, G; Piemont, Y; Fleurette, J; Monteil, H. Detection of *Legionella spp.* in bronchoalveolar lavage fluids by DNA amplification. *J Clin Microbiol* 1992, 30, 920-924.

Kooguchi, K; Hashimoto, S; Kobayashi, A; Kitamura, Y; Kudoh, I; Wiener-Kronish, J; Sawa, T. Role of alveolar macrophages in initiation and regulation of inflammation in *Pseudomonas aeruginosa* pneumonia. *Infect Immun* 1998, 66, 3164-3169.

Leoni, E; Legnani, PP; Bucci Sabattini, MA; Righi, F. Prevalence of *Legionella spp.* in swimming pool environment. *Water Res* 2001, 35, 3749-3753.

Lesprit, P; Faurisson, F; Join-Lambert, O; Roudot-Thoraval, F; Foglino, M; Vissuzaine, C; Carbon, C. Role of the quorum-sensing system in experimental pneumonia due to *Pseudomonas aeruginosa* in rats. *Am J Respir Crit Care Med* 2003, 167, 1478-1482.

Lück, PC; Leupold, I; Hlawitschka, M; Helbig, JH; Carmienke, I; Jatzwauk, L; Guderitz, T. Prevalence of *Legionella* species, serogroups, and monoclonal subgroups in hot water systems in south-eastern Germany. *Zentralbl Hyg Umweltmed* 1993, 193, 450-460.

Lück, PC; Schütz, M; Helbig, JH; Jatzwauk, L. Konstanz und Variabilität von *Legionella pneumophila*-Populationen im Wassersystem eines Universitätsklinikums. [Abstract]. *Zbl Hyg Umweltmed* 1995, 197, 317-318.

Mahenthiralingam, E; Campbell, ME; Foster, J; Lam, JS; Speert, DP. Random amplified polymorphic DNA typing of *Pseudomonas aeruginosa* isolates recovered from patients with cystic fibrosis. *J Clin Microbiol* 1996, 34, 1129-1135.

Martiny, AC; Albrechtsen, HJ; Arvin, E; Molin, S. Identification of bacteria in biofilm and bulk water samples from a nonchlorinated model drinking water distribution system: detection of a large nitrite-oxidizing population associated with *Nitrospira spp. Appl Environ Microbiol* 2005, 71, 8611-8617.

Martiny, AC; Jorgensen, TM; Albrechtsen, HJ; Arvin, E; Molin, S. Long-term succession of structure and diversity of a biofilm formed in a model drinking water distribution system. *Appl Environ Microbiol* 2003, 69, 6899-6907.

Matz, C; Bergfeld, T; Rice, SA; Kjelleberg, S. Microcolonies, quorum sensing and cytotoxicity determine the survival of *Pseudomonas aeruginosa* biofilms exposed to protozoan grazing. *Environ Microbiol* 2004, 6, 218-226.

Mazzola, PG; Martins, AM; Penna, TC. Chemical resistance of the gram-negative bacteria to different sanitizers in a water purification system. *BMC Infect Dis* 2006, 6, 131.

McDade, JE; Shepard, CC; Fraser, DW; Tsai, TR; Redus, MA; Dowdle, WR. Legionnaires' disease: isolation of a bacterium and demonstration of its role in other respiratory disease. *N Engl J Med* 1977, 297, 1197-1203.

OSHA (Occupational Safety & Health Administration). (1999) OSHA Technical Manual (TED 1-0.15A), Section III - Chapter 7. Legionnaires' disease (1999, January 20). Washington, DC: Occupational Safety & Health Administration. [cited 2007, May 22nd]. Available from: URL: http://www.osha.gov/dts/osta/otm/otm_iii/otm_iii_7.html

Penna, VT; Martins, SA; Mazzola, PG. Identification of bacteria in drinking and purified water during the monitoring of a typical water purification system. *BMC Public Health* 2002, 2, 13.

Ramsdell, J; Narsavage, GL; Fink, JB; American College of Chest Physicians, Home Care Network Working Group. Management of community-acquired pneumonia in the home: an American College of Chest Physicians clinical position statement. *Chest* 2005, 127, 1752–1763.

Ratnam, S; Hogan, K; March, SB; Butler, RW. Whirlpool-associated folliculitis caused by *Pseudomonas aeruginosa*: report of an outbreak and review. *J Clin Microbiol* 1986, 23, 655-659.

Rello, J; Bodi, M; Mariscal, D; Navarro, M; Diaz, E; Gallego, M; Valles, J. Microbiological testing and outcome of patients with severe community-acquired pneumonia. *Chest* 2003, 123, 174–180.

Ribas, F; Perramon, J; Terradillos, A; Frias, J; Lucena, F. The *Pseudomonas* group as an indicator of potential regrowth in water distribution systems. *J Appl Microbiol* 2000, 88, 704-710.

Ruimy, R; Genauzeau, E; Barnabe, C; Beaulieu, A; Tibayrenc, M; Andremont, A. Genetic diversity of *Pseudomonas aeruginosa* strains isolated from ventilated patients with nosocomial pneumonia, cancer patients with bacteremia, and environmental water. *Infect Immun* 2001, 69, 584–588.

Rusin, PA; Rose, JB; Haas, CN; Gerba, CP. Risk assessment of opportunistic bacterial pathogens in drinking water. *Rev Environ Contam Toxicol* 1997, 152, 57-83.

Schmeisser, C; Stockigt, C; Raasch, C; Wingender, J; Timmis, KN; Wenderoth, DF; Flemming, HC; Liesegang, H; Schmitz, RA; Jaeger, KE; Streit WR. Metagenome survey of biofilms in drinking-water networks. *Appl Environ Microbiol* 2003, 69, 7298-7309.

Schoenen, D; Stoeck, B; Hienzsch, S; Emmel, B. Dekontamination von mit *Pseudomonas aeruginosa* besiedelten Trinkwasserhähnen. *Zentralbl Bakteriol Mikrobiol Hyg [B]* 1986, 182, 551-557.

Schultz, MJ; Rijneveld, AW; Florquin, S; Speelman, P; Van Deventer, SJ; van der Poll, T. Impairment of host defence by exotoxin A in *Pseudomonas aeruginosa* pneumonia in mice. *J Med Microbiol* 2001, 50, 822-827.

Schumacher, PG; Wagner, I; Kuch, A. Die Trinkwasserversorgung von Göttingen mit Mischwasser. Erfahrungen über den Einfluß der Wasserqualität und von Inhibitoren auf Korrosion im Rohrnetz. *gwf Wasser / Abwasser* 1988, 129, 146-152.

Stojek, NM; Dutkiewicz, J. *Legionella* in sprinkling water as a potential occupational risk factor for gardeners. *Ann Agric Environ Med* 2002, 9, 261-264.

Stover, CK; Pham, XQ; Erwin, AL; Mizoguchi, SD; Warrener, P; Hickey, MJ; Brinkman, FS; Hufnagle, WO; Kowalik, DJ; Lagrou, M; Garber, RL; Goltry, L; Tolentino, E; Westbrock-Wadman, S; Yuan, Y; Brody, LL; Coulter, SN; Folger, KR; Kas, A; Larbig, K; Lim, R; Smith, K; Spencer, D; Wong, GK; Wu, Z; Paulsen, IT; Reizer, J; Saier, MH; Hancock, RE; Lory, S; Olson, MV. Complete genome sequence of *Pseudomonas aeruginosa* PA01, an opportunistic pathogen. *Nature* 2000, 406, 959-964.

Suzuki, K; Tachibana, A; Hatakeyama, S; Yamaguchi, K; Tateda, K. Clinical characteristics in 8 sporadic cases of community-acquired *Legionella* pneumonia. *Nihon Kokyuki Gakkai Zasshi* 2002, 40, 282-286.

Timmis KN. *Pseudomonas putida*: a cosmopolitan opportunist par excellence. *Environ Microbiol* 2002, 4, 779-781.

Umweltbundesamt (Federal Environmental Agency). Empfehlung der Trinkwasserkommission zur Risikoeinschätzung, zum Vorkommen und zu Maßnahmen beim Nachweis von *Pseudomonas aeruginosa* in Trinkwassersystemen. Empfehlung des Umweltbundesamtes nach Anhörung der Trinkwasserkommission des Umweltbundesamtes. *Bundesgesundheitsbl* 2002, 45, 187-188.

van Belkum, A; Struelens, M; Quint, W. Typing of *Legionella pneumophila* strains by polymerase chain reaction-mediated DNA fingerprinting. *J Clin Microbiol* 1993, 31, 2198-2200.

WHO (World Health Organization). Guidelines for safe recreational waters. Volume 2 - Swimming pools and similar recreational-water environments. World Health Organization: Geneva; 2006. [cited 2007, May 22nd]. Available from: URL: http://www.who.int/entity/water_sanitation_health/bathing/srwe2full.pdf

Wiese, J; Helbig, JH; Lück, PC; Meyer, HG; Jansen, B; Dunkelberg H. Evaluation of different primers for DNA fingerprinting of *Legionella pneumophila* serogroup 1 strains by polymerase chain reaction. *Int J Med Microbiol* 2004, 294, 401-406.

Zietz, BP; Wiese, J; Brengelmann, F; Dunkelberg, H. Presence of *Legionellaceae* in warm water supplies and identification of strains by polymerase chain reaction. *Epidemiol Infect* 2001, 126, 147-152.

Zietz, BP; Dunkelberg, H; Ebert, J; Narbe, M. Isolation and characterization of *Legionella spp.* and *Pseudomonas spp.* from greenhouse misting systems. *J Appl Microbiol* 2006, 100, 1239-1250. Erratum in: *J Appl Microbiol* 2006, 101, 976.

Zietz, BP; Wiese, J; Lück, PC; Helbig, J; Dunkelberg, H. Epidemiological typing of *Legionella pneumophila* serogroup 5 strains. In: *Legionella*. Marre, R; Aabu Kwaik, Y; Bartlett, C; Cianciotto, N; Fields, BS; Frosch, M; Hacker, J; Lück, PC, editors.

In: Progress in Environmental Microbiology
Editor: Myung-Bo Kim, pp. 203-221
ISBN: 978-60021-940-5

Chapter VII

Improving Fecal Coliform Removal in Maturation Ponds

Nibis Bracho** *and Clark L. Casler
Centro de Investigaciones del Agua, Ciudad Universitaria, Universidad del Zulia, Maracaibo 4001-A, Estado Zulia, Venezuela

Abstract

Maturation ponds are commonly used as a treatment method for improving or polishing effluent from secondary biological processes, activated sludge, trickling filters or facultative ponds. Methods for removing fecal coliforms (FC) have been studied by many authors, and all indicate that sunlight, temperature and retention time are the principal factors that cause FC reduction. Pond geometry affects retention time, and this has been demonstrated on both pilot- and real-scales by several authors. The objective of this chapter is to demonstrate the effect of sunlight exposure time on FC removal in maturation ponds on a pilot- and full-scale basis. The chapter will also explain how to take advantage of this important resource, by improving the geometric configuration of the system, to increase retention time and, therefore, natural disinfection. A case study developed in tertiary treatment with maturation ponds, located after a conventional percolating filter plant, is used as an example. The use of baffles to change pond configuration is the best way to control, handle or manipulate pond hydraulic behavior, and thus provide a series of benefits directly related to sunlight intensity. The two parameters, retention time and sunlight exposure time, constitute the principal binomial for FC removal in maturation ponds as tertiary treatment after conventional wastewater treatment.

* nibisbracho@hotmail.com

Introduction

Maturation ponds are commonly used as a treatment method for improving or polishing effluent from secondary biological processes, activated sludge, trickling filters or facultative ponds (Droste 1997, Mara and Pearson 1998, and Bitton 1999). Reduction of pathogenic organisms is normally demonstrated by fecal indicator bacteria, viruses, and protozoa and less frequently by enumeration of Most Probable Number (MPN) and membrane filtration methods.

Several studies have identified factors involved in bacterial reduction in Waste Stabilization Ponds (WSPs) or maturation ponds, including retention time, exposure to sun or ultraviolet (UV) light, and visible light and temperature. Each of these parameters, in turn, depends on a series of physico-chemical and environmental factors.

Maturation pond design is based on bacterial decay. The first-order rate constant or coefficient (k_T) for fecal coliform (FC) removal is recognized to be highly dependent on temperature (Marais 1974). However, other parameters also play an important role in FC removal. Curtis (1990) and Curtis and Mara (1994) investigated the photo-oxidation process occurring in waste stabilization ponds and concluded "*...light kills faecal coliforms in waste stabilization ponds by an oxygen-mediated exogenous photosensitization that interacts synergistically with elevated pH*". Under controlled laboratory conditions, Alkan *et al.* (1995) found that light intensity -although an important factor for bacterial die-off- did not depend on temperature. Davies-Colley *et al.* (1999) agreed with Curtis (1990) and Alkan *et al.* (1995), that exposure to sunlight is considered to be the most important cause of natural disinfection in WSPs. Mutamara and Puetpaiboon (1997), Lloyd *et al.* (2002), and Bracho *et al.* (2006), showed that FC decreased with increased retention time of water in the ponds. In addition, Bracho (2003) showed in a pilot study, that FC removal did increase by 6% in presence of sunlight. Also, it was demonstrated on a full-scale that increasing retention time increased water exposure time.

Still, Curtis and Mara (1994) believe that parameters affecting FC removal are still very much a matter of debate. On a real scale, it is difficult to isolate and measure each variable that may affect stabilization ponds. For example, sunlight warms the earth's surface, increasing environmental temperature, and consequently pond temperature. Light intensity, dissolved oxygen (DO), pH and temperature combinations are the result of natural phenomena difficult to measure in open ponds directly exposed to the environment. However, pond geometry, plant operation and maintenance may be controlled. Hydraulic studies, accompanied by engineering interventions, may also be employed to improve hydraulics, increase retention time (Lloyd *et al.* 2002, Bracho 2006a, b), and therefore, increase sunlight exposure to remove FC with greater efficiency (Bracho 2003). Bracho (2006a) improved hydraulic behavior in a maturation pond by using baffles, and increased retention time by five hours. It is noteworthy that this intervention improved FC removal by 50%, and used existing facilities, with no need for additional land.

Methodology

Pilot- and full-scale experiments had to be designed and implemented for the present research. These included determining some of the parameters involved in fecal coliform (FC) removal mechanisms. Several strategies were implemented to achieve this objective. Additional time was later dedicated to determining FC removal mechanisms on a pilot- and full-scale basis.

1. The engineering design of the South pond, at the Lidsey-England sewage treatment plant, was modified and changed into a channel pond, by increasing the Length/Width ratio from 9:1 to 79:1.
2. Operational changes were made, such as modifying flow rate both in the original pond (denominated open pond) and in the pond with the engineering intervention (called the channel pond).
3. Bacteriological and hydraulic evaluations were carried out under different operating conditions in the open and channel ponds to compare the advantages of each configuration.

1. Effect of Retention Time, Sunlight and Temperature on FC Removal in a Pilot-Scale Experiment

A two-week experiment designed to show the effect produced by sunlight, retention time and temperature on FC removal, with and without sunlight, was carried out at the Godalming-England sewage treatment plant, in August 2002. The aim was to determine if FC removal was better at the same temperature under light, as opposed to dark conditions.

1.1. Description of the Experiment

Godalming sewage treatment plant has a conventional plant and tertiary maturation ponds. To determine the effect of sunlight exposure time on FC removal, a batch experiment was undertaken in two tanks filled with water from the maturation pond of the Godalming sewage treatment plant. Sunlight was blocked from one of the tanks with plywood and both tanks were covered with plastic sheeting to eliminate wind, and thus guarantee similar conditions for each tank (Figure 1).

Temperature, sunlight intensity, dissolved oxygen (DO), pH, turbidity, and ammonium were measured at the site, and fecal coliforms (FC), suspended solids (SS) and chlorophyll *a* were measured in the laboratory. The batch experiment was run with different retention times, to observe how sunlight affected bacterial removal. The results showed that Chlorophyll *a* concentration, pH, DO and temperature were similar in both treatments (Table 1). Under these working conditions, FC removal efficiency in the treatment exposed to light was 6% greater than the one under dark conditions, which can only be attributed to the presence of light.

Figure 1. Experiment at the Godalming sewage treatment plant (tanks with and without sunlight).

Table 1. Average values of parameters under analysis in the pilot experiment under dark and light conditions, from 01 - 12 August 2002

Parameters	Outlet light conditions	Outlet dark conditions
FC (cfu/100 ml)	5.8×10^3	1.05×10^4
FC removal efficiency (%)	85.53	79.75
SS (mg/l)	2.63	4.51
Chlorophyll (µg /l)	4.4	3.61
Temperature (°C)	19.46	18.07
Oxygen (mg/l)	9.78	8.94
pH	7.42	7.52
NH_4	0.75	0.84
Turbidity (NTU)	2.25	2.44

In the linear regression plot presented for light and dark conditions (Figure 2), the regression for light conditions is above that for dark conditions, with an average distance between them equivalent to 6% efficiency.

Removal efficiency was greater in the tank exposed to sunlight, due to the natural disinfection produced by sunlight (Curtis *et al.* 1992a, b).

Statistical models for the batch experiment were obtained for retention time between 5 and 34 hours and for a temperature range between 17°C and 21°C.

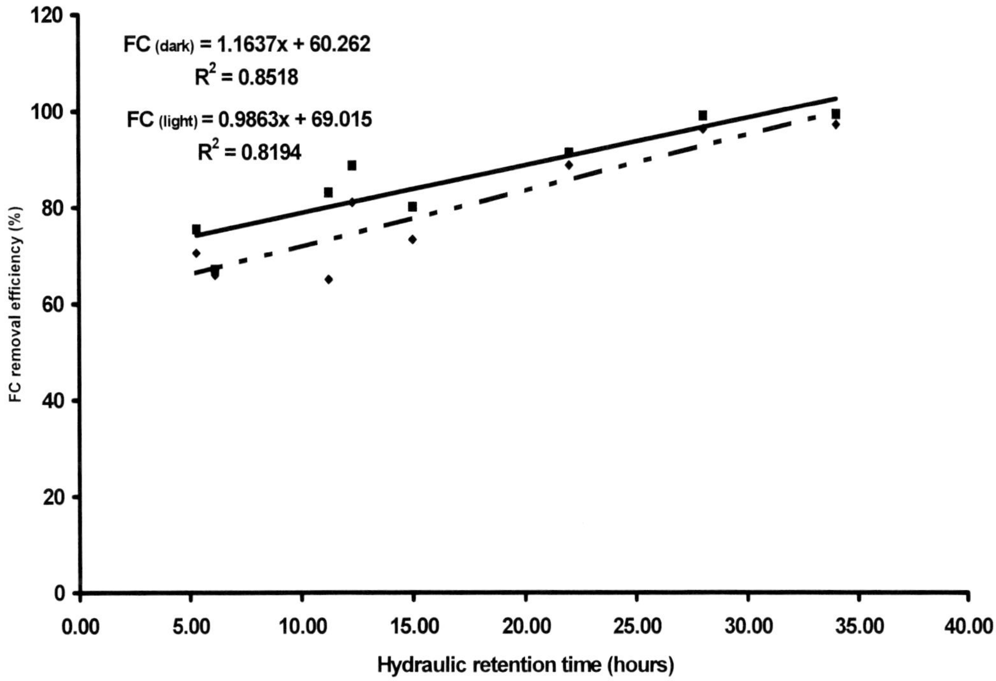

Figure 2. Linear regression between hydraulic retention time and FC removal efficiency in dark and light conditions.

For dark conditions:

$$FC\ (\%) = 60.262 + 1.1637\ (tm) \tag{1}$$

For light conditions:

$$FC\ (\%) = 69.015 + 0.986\ (tm) \tag{2}$$

Where: tm = retention time (h)

It was statistically demonstrated that retention time was the most important variable, attributing to 85% of FC removal. Light intensity, on the other hand, played a secondary role, with 6% more FC removal being due to sunlight exposure. The sun acts as a natural disinfectant, but has a greater effect on FC removal when exposure time of water to sunlight increases; the parameter exposure time is described as retention time.

2. Effect of Retention Time, Sunlight, Temperature, and other Parameters on FC Removal in a Full-scale Experiment

The Center for Environmental Health Engineering (CEHE), at the University of Surrey, has been working with Southern Water at Lidsey sewage treatment plant (LSTP) to improve fecal coliform removal efficiency in the tertiary stages of the treatment process. The present research involves the application of a continuous field assessment operating with different flow conditions, with a view to identifying which variables affect the bacteriological quality of the effluent from the LSTP maturation ponds, and optimising the engineering design and system operation under natural conditions. Lidsey sewage treatment plant is located in southern England, near the town of Bognor Regis (Figure 3). It consists of a conventional treatment plant with tertiary treatment by 3 maturation ponds.

The tertiary stages of the plant entail three parallel maturation ponds of similar geometry and dimensions (15 m x 122 m). These were termed the North, Central and South ponds (Figure 3). The geometry of the South pond was modified by installing baffles parallel to the flow pattern, to obtain a pond with three channels, and was named the "channel" pond. The two control ponds (North and Central) were collectively called the "open" pond.

In this investigation, we studied the relationship among several important parameters that should be considered when trying to better the bacteriological quality of effluent coming from maturation ponds.

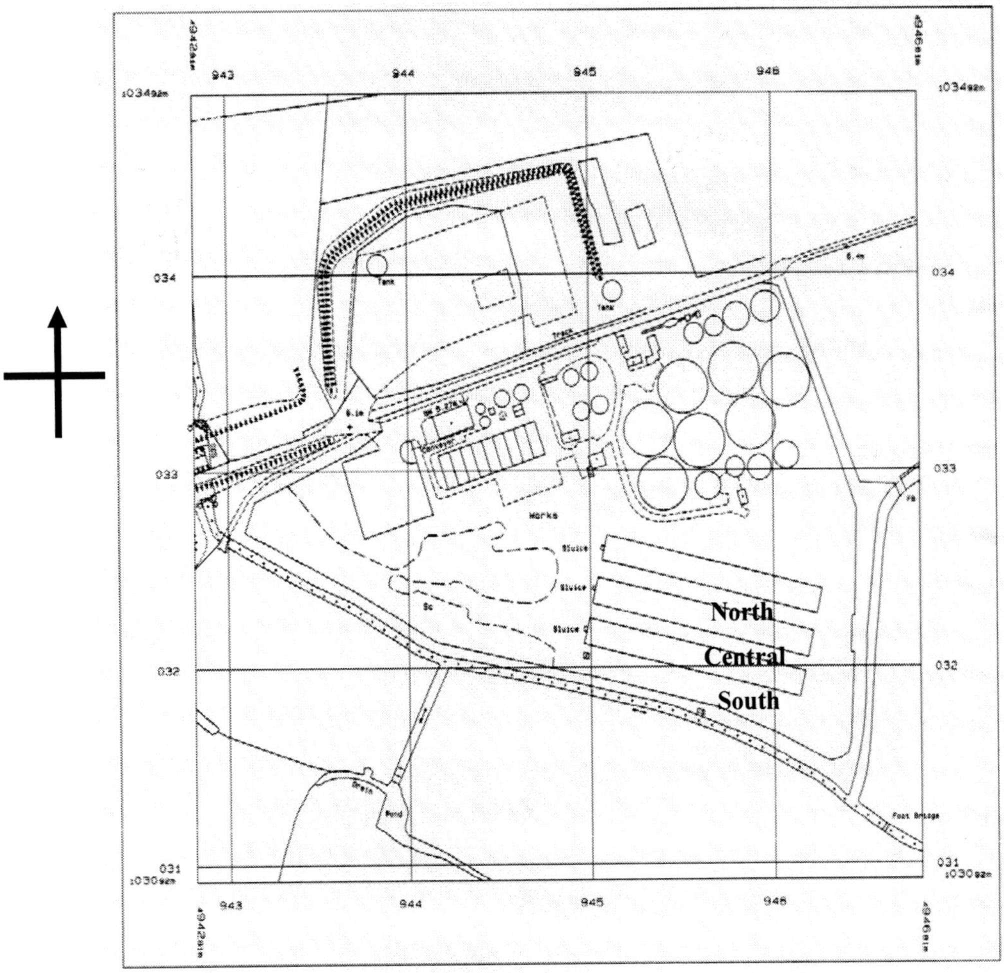

Figure 3. Plan view of Lidsey sewage treatment plant.

The experiment was designed to determine the following simultaneously in the channel pond:

- Temperature, pH, turbidity, DO, ammonium in the South pond (channel pond) outlet. The Grant/logger was brought online, registering values at 30-minute intervals.
- Samples were taken at the inlet and outlet of the channel pond, and FC were determined.
- Flow was measured manually at the inlet and outlet, to determine nominal retention time.
- A tracer study was developed to determine hydraulic retention time.
- Finally, manual records were kept of light intensity, using a light meter during the period that the samples were collected.

This experiment was pursuing several objectives, including confirmation of the results obtained at the Godalming pilot-scale study. The Statistical Package for the Social Sciences (SPSS) was applied, using the stepwise estimation method –a method of selecting variables in the regression model that starts with selecting the best predictor for the dependent variable.

The following parameters were used for multi-regression: FC removal efficiency as the dependent variable, Temperature, pH, Ammonium (NH_4), DO, light intensity, and Nominal retention time.

Some observations about the conditions during data collection should be explained, before reporting on the multi-regression reports for the Lidsey experiment:

The experiment was at full-scale with continuous flow, unlike the pilot-scale batch experiment, where the flow pattern was plug flow. It must also be noted that, in a batch experiment, there is no difference between nominal and hydraulic retention time.

In a continuous flow experiment, nominal retention time is different from hydraulic retention time (Yánez 1993, Lloyd *et al.* 2002, Bracho 2003).

The multi-regression results revealed three possible models (Table 2).

Statistical evidence revealed that Model 3 was most robust, where three independent variables (DO, ammonium, and light intensity) were involved simultaneously. These variables, however, do not impact FC removal when they are isolated. Logically, these three variables depend on photosynthesis, a natural process governed by presence of sunlight.

Table 2. Model summary from multiregression analysis

Model	R- squared	Adjusted R square	Std. error of estimate.	$P < 0.05$
1	0.365[a]	0.330	3.1705	0.005
2	0.665[b]	0.625	2.37391	0.000
3	0.857[c]	0.830	1.59755	0.000

[a] Predictors: Constant, DO.
[b] Predictors : Constant, DO, Ammonium.
[c] Predictors: Constant, DO, Ammonium, Light.

The multi-regression equation for FC removal for continuous flow was obtained for a retention time between 3 and 4.5 days, with temperature oscillating between 13.5°C and 15°C. This expression is given by:

$$FC\,(\%) = 63.548 + 5.130\,(DO) + 1.149\,(A) + 5.45\times10^{-3}\,(I) \qquad (3)$$

Where:

DO = dissolved oxygen mg/l,
A = ammonium (mg/l), and
I = light intensity (1000 x *lux*).

3. Factors Controlling FC Removal in the Batch Experiment in Godalming and in the Channel Maturation Pond in Lidsey

Note that operating conditions were different for the pilot-scale experiment at Godalming and the full-scale experiment at Lidsey, so the comparison was not completely just. The first test was for a batch experiment, where total control of retention time was established. Besides, in a batch experiment, bacteriological quality of the inlet has a unique value, i.e., there is greater precision in the FC removal measured.

The second experiment, developed at Lidsey sewage treatment plant, used continuous flow with constant variations in flow and influent quality. In addition, the full-scale pond included some vegetation on the bottom and along the walls. Thus, presence of vegetation may have affected treatment by causing some modifications to the system's hydraulic and biological behavior. However, a good comparison may still be made between both the pilot- and full-scale experimental results, to take full advantage of both studies and indicate which parameters contribute most to FC removal.

The pilot-scale experiment clearly revealed that retention time was the main parameter for FC removal, regardless of presence or absence of sunlight, but FC removal efficiency (6%) improved substantially in presence of sunlight. The other parameters monitored (pH, temperature, ammonium, etc.) had no significant influence on FC removal.

The full-scale experiment was developed with very little variation of retention time, so this variable was statistically blocked. When the effect of retention time is eliminated, other parameters become statistically significant, such as DO, ammonium and light intensity. Both experiments, then, offered complementary information to help understand FC removal mechanisms in Lidsey.

Sarikaya and Saatci (1987) included a study with a tracer (salt) in their investigation, but gave no details about the hydraulic results. They only concluded "*...in our study, the effect of retention time on the K values was found to be small*". They then indicated that retention time was kept constant in all three ponds, thereby eliminating its effect in the analysis of the influence of pond depth. Undoubtedly, if the effect of retention time is blocked or eliminated, the effect of sunlight would then become the first parameter that affects FC removal, such as occurred in Lidsey. On the contrary, retention time would become the main parameter in FC removal, like in the pilot-scale experiment at Godalming.

3.1. Relationship of Exposure to Sunshine vs. FC Removal with Flow Type in Lidsey Sewage Treatment Plant

For this experiment, values of sunshine hours and temperature were found in the Internet (Bognor Regis station), and associated with FC removal in the open (Central pond) and the channel pond (South pond).

Fecal coliform removal efficiency in the channel pond was directly related to the number of sunshine hours (Figure 4). Flow regime in this pond was near the plug type (d = 0.074), where all the elements in the water have the same retention time (plug flow, d = 0). This means that treatment is favored, because the water elements are exposed to sunlight for the same amount of time. This confirms the importance of flow type and its relationship to sunlight exposure. It also confirms the evidence presented in section one, on the pilot batch scale.

There was greater variation between sunshine hours and faecal coliform removal efficiency in the open pond (Central) (Figure 5), because the type of flow for that period was "completely" mixed (d = ∞). In general, ponds which are supposed to be completely mixed produce short circuits (i.e. fluid velocities are higher than the mean velocity), and dead zones (when fluid velocity equals zero). These problem factors are reduced with plug flow. The studies with Rhodamine WT carried out in this research offer the following evidence:

a) When the flow is "completely" mixed, the tracer begins to appear one hour after it is injected into the water. This means there is a fraction of water that exits untreated, because of the strong short circuits caused by jet flow (high velocity at the surface of the fluid).

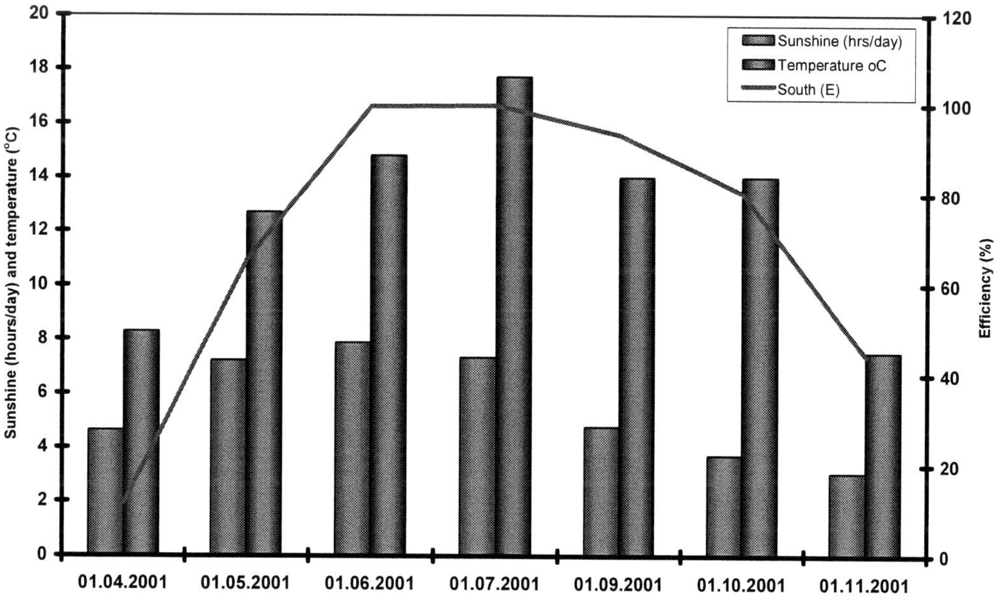

Figure 4. Average sunshine (hours/day), temperature (°C), and fecal coliform efficiency (%) in the channel pond (South pond).

b) When the regime is close to plug flow, as in the case of the channel pond (Figure 4), the tracer begins to appear up to 17.5 hours after injection (Bracho *et al.* 2006a), with longer residence in the pond. This results in better treatment, because channels retard jet flow. In other words, water that usually leaves the pond quickly, due to short circuits, remains at least 44.47% of the mean hydraulic retention time, causing increased exposure of water to natural disinfection, at least during this time. This process does not occur in the open pond (central). The results confirm the importance of plug flow and its relationship with sunshine hours/day.

3.2. Determining Chlorophyll a in Lidsey Sewage Treatment Plant

Algae are an important feature and characteristic of healthy operating wastewater pond treatment systems. They contain chlorophyll and exhibit true photosynthesis, utilizing light as an energy source for cell synthesis. It is via photosynthesis that simple, stable inorganic compounds are converted into energy-rich matter (algal cells) and oxygen. This process depends on pond environmental conditions conducive to growth and development of healthy algal communities (Frederick 1995).

Low pH (7-8) and DO (5-9 mg/l) values were reported during the research in Lidsey (2000-2002). This did not help to clarify faecal coliform removal mechanisms in maturation ponds proposed by Curtis *et al.* (1992a) or by researchers that reported high pH (>9) and DO as being responsible for FC removal (Him *et al.* 1980, Pearson *et al.* 1987a, b, 1996). Also, there was no evidence of significant amounts of algae or chlorophyll *a* in the pond.

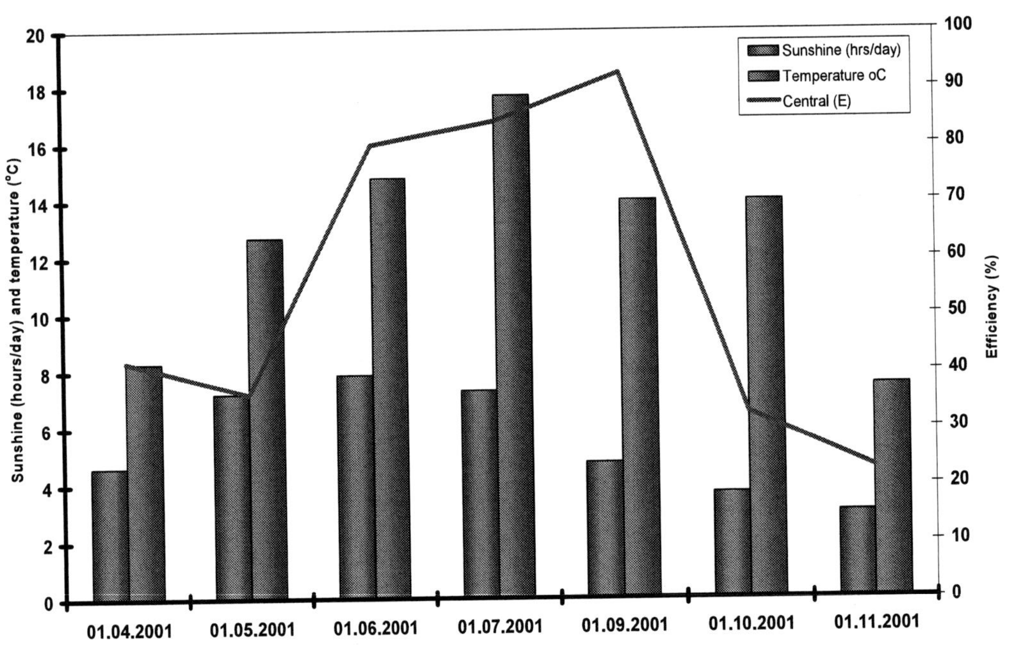

Figure 5. Average sunshine (hours/day), temperature (°C), and fecal coliform efficiency (%) in the open pond (Central pond).

3.2.1. Chlorophyll a

Samples were taken at the inlet and outlet of the open and channel ponds during July 2002, and certain pond samples were analyzed for chlorophyll *a* to determine productivity (Table 3). The results showed that chlorophyll concentration was very low at the three monitoring points, because algae concentrations in maturation ponds usually range from 500-2000 μg chlorophyll *a* per liter (Mara and Pearson 1998).

Table 3. Chlorophyll *a* concentration at Lidsey

	Inlet pond (μg/l)	North outlet pond (μg/l)	South outlet pond (μg/l)
Average (μg/l)	11.65	5.55	5.80

A sample was taken at six internal points in the channel pond, two points in each channel at a depth of 50 cm (Figure 6). A sample was also taken at the inlet and outlet. The samples in channel A were taken at 50 m and 100 m from the inlet and were identified as A_1 and A_2. Samples from channels B and C were taken at points adjacent to channel A and were identified as B_1, B_2, C_1 and C_2. Attempts to collect samples from the bottom were abandoned, due to too much mud.

Results are shown in Table 4. During sampling a film of algae was observed on certain parts of the inner walls of channel B. The samples in this channel were taken near the walls. In sample B_1, a large quantity of organisms was observed.

This indicated that the ecosystem in channel B was not uniform throughout and that it was different from the ones in channels A and C. A great quantity of floating and sunken material was seen in channel B, and could be diatoms, but samples of this material must be identified.

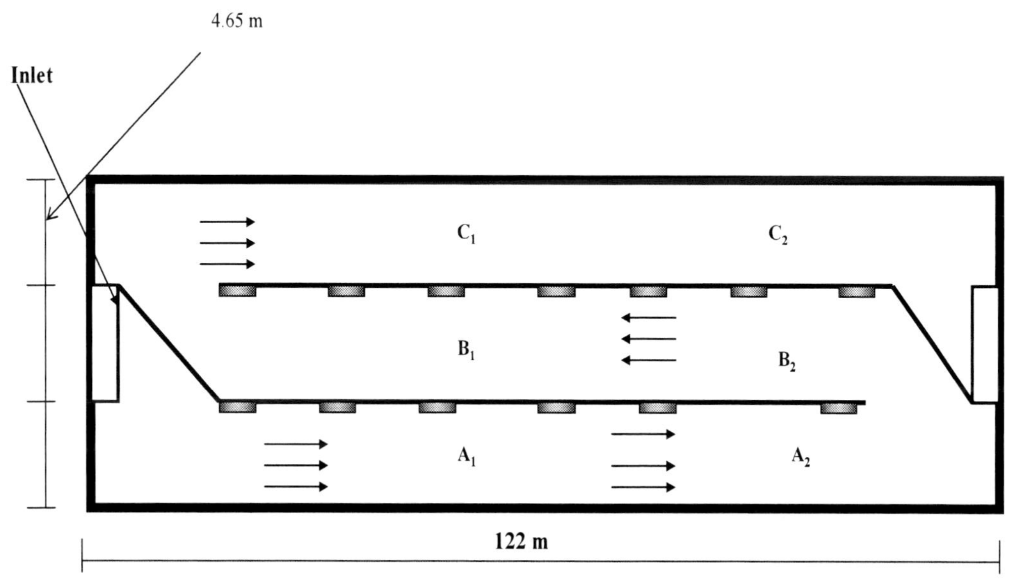

Figure 6. Internal points monitored in the channel pond at Lidsey (not to scale).

Table 4. Values for the physico-chemical parameters of the samples taken in the South pond (channel) at Lidsey

Parameter	Points monitored.							
	A_1	**A_2**	**B_1**	**B_2**	**C_1**	**C_2**	**Inlet**	**Outlet**
pH	7.30	7.26	7.20	7.27	7.17	7.25	7.20	7.20
Temperature °C	20.4	20.8	20.7	20.3	20.7	20.9	19.7	19.5
Chlorophyll (μg/l)	0.35	0	100	66	118.9	1.04	0	2.78
SS (mg/l)	7	8	116	110	3	11	15	1
COD (mg/l)	36	35	102	46	38	40	42	40
NH_4	0.45	0.77	0.78	1.14	0.75	1.06	0.78	1.02
Turbidity (NTU)	6	1	1	1	1	1	6	1

The highest chlorophyll *a* concentrations were detected at points B_1 and B_2 (100 μg^1 and 66 μg^1, respectively), and at point C_1 (118 μg^1), but they were all less than 500 μg^1, and defined by Mara and Pearson (1998) as typical for maturation ponds. SS were also high at points B_1 (116 mg/l) and B_2 (110 mg/l), but not at point C_1 (3 mg/l) (Table 4). The highest COD concentration was 102 mg/l, recorded at point B_1.

Chlorophyll *a* behavior in channel B was atypical when compared with A and C. This may be due to a different hydraulic behavior, which can be explained as follows: The walls on both sides of the central channel were built with rectangular polyurethane. This geometry differs from that of channels A and C due to trapezoidal geometric that includes a side with polyurethane and the other natural land. The geometric configuration of channel B differs from that of channels A and C, the hydraulic behavior may be different, thus favoring algae growth.

Alabaster *et al.* (1991) reported chlorophyll *a* concentrations between 59-3178 μg/l, in a facultative pond in Kenya. Pearson *et al.* (1987c), in Lourdes Portugal, documented mean chlorophyll *a* of 154 μg/l in primary facultative ponds; an average of 1,227 μg/l in secondary facultative ponds; the maturation average was higher at 1,454 μg/l. These investigations were done with WSP systems and are different from the Lidsey case, where maturation ponds are used as tertiary treatment, after conventional treatment in which a high proportion of nutrients have been removed. Algal cell reproduction takes place after 24 hours, i.e. retention time in Lidsey is not enough to generate massive algal growth.

Summarizing, every possible effort was made to link FC removal with the removal mechanism proposed by Curtis *et al.* (1992a). However, high pH values were not recorded at Lidsey, which would incline in favor of the proposal of Davies-Colley *et al.* (1999) "*Sunlight inactivation of* Escherichia coli *is strongly dependent on DO and also increases strongly with pH > 8.5.*" At lower pH, sunlight inactivation is independent of WSP constituents and damage is mainly by UV-B. The evidence points to sunlight as a strong bactericidal agent. This was manifested in Lidsey due to the transparency of the water, especially in June and July, months with maximum sunshine hours/day and longer sunshine periods.

Table 5. Monthly average for faecal coliform removal in an open and channel pond, from March to July 2002

	Raw sewage	Inlet pond	North Open pond	South Channel pond
March			Q = 6.45 T = 10.99	Q = 7.18 T = 10.99
Fecal coliform geomean (cfu/100 ml)	6.45×10^{6}	6.6×10^{4}	4.31×10^{4}	1.56×10^{4}
Fecal coliform removal (%)		98.99	34.65	76.34
April			Q = 5.81 T= 13.9	Q = 7.53 T= 13.9
Fecal coliform geomean (cfu/100 ml)	9.8×10^{6}	5.48×10^{4}	8.75×10^{3}	7.85×10^{2}
Fecal coliform removal (%)		99.44	84.04	98.57
May			Q = 7.88 T = 14.24	Q = 8.61 T= 14.24
Fecal coliform geomean (cfu/100 ml)	9.38×10^{6}	6.8×10^{4}	2.09×10^{4}	5.34×10^{3}
Fecal coliform removal (%)		98.27	69.48	92.19
June			Q = 9.68 T = 17.4	Q = 12.16 T = 17.4
Fecal coliform geomean (cfu/100 ml)	1.06×10^{7}	6.59×10^{4}	1.42×10^{4}	4.44×10^{3}
Fecal coliform removal (%)		99.38	78.51	93.27
July			Q = 13.83 T = 17.4	Q = 15.89 T = 17.4
Fecal coliform geomean (cfu/100 ml)	1.18×10^{7}	1.42×10^{5}	3.1×10^{4}	2.35×10^{4}
Fecal coliform removal (%)		98.8	77.7	83.46

Q = flow l/s; T = temperature °C.

3.3. Effect of Geometric Configuration on Fecal Coliform Removal in Lidsey

Pond geometry affects the flow pattern, and in turn, the retention time. Square shaped ponds tend to produce complete mixing, whereas long rectangular ponds are characterized by plug flow. Importance of flow patterns with respect to treatment efficiency was treated on a pilot-scale by Camp (1946), who showed that plug flow was the ideal flow for contaminant removal. Marais (1974) also concluded that plug flow was the ideal flow pattern for bacteria removal and proposed the use of ponds in series to produce plug flow. Bracho 2003 and

Bracho *et al.* (2006a, b) reported that plug flow may almost be achieved by installing baffles in the pond, allowing FC removal to increase as high as 50% on a pilot-scale. Results are given in Table 5.

Both ponds operated simultaneously under the same temperature conditions, but flow rate in the channel pond was greater than in the open pond. Thus, the channel pond was always more efficient. On the other hand, FC removal increased in both ponds during months with longer days, except in April, the month with brightest sunlight accompanied with low precipitation. We also observed in March, month with the shortest light period, that 50% more FC was removed in the channel pond than in the open pond. In the channel pond, the baffle configuration increases the distance the water has to travel, as well as reducing dead spaces and short circuits, thus increasing retention time due to the change to plug flow (Lloyd *et al.* 2002, Bracho *et al.* 2006a). The baffle configuration permits better hydraulic behavior, permitting maximum natural disinfection of the water, without resorting to construction of additional ponds that need more available space.

Conclusions

- It is recognized that FC removal mechanisms are complex to analyze. Even more complex is the analysis of the individual effect of each parameter because their contributions to FC removal are not isolated. Interaction exists between parameters, and was detected quite clearly in the statistical analysis carried out in the case of Lidsey.
- In previous investigations, Curtis (1990) reported that bacteria in maturation ponds can be removed by a photo-oxidation process. The statistical model for FC removal obtained for Lidsey involves dissolved oxygen, ammonium and light intensity, so the possibility of photo-oxidation cannot be discarded. But it cannot be asserted either, because in the photo-oxidation process Curtis (1990) includes humic substances (yellow gilvin) that were not determined in Lidsey. However, Chemical Oxygen Demand (COD), an indicator of the presence of organic material, was measured, but the values detected were low (Table 4).
- Davies-Colley *et al.* (1999) said that solar rays cause direct damage to bacterial cells. This is perhaps what happened in Lidsey, and may explain why turbidity plays such an important role, because it limits sunlight penetration, a parameter mentioned by Alkaan *et al.* (1995) and indirectly mentioned by Mayo (1989), Sarikaya and Saatci (1987), Moeller and Calkins (1980), and Lian *et al.* (1998).
- In the pilot experiment, it was demonstrated that retention time was the most important variable in FC removal –accounting for 85% FC removal. Light intensity, on the other hand, plays a secondary role, with 6% more FC removal being attributed to presence of sunlight.
- Based on evidence gathered in this investigation, there are parameters that affect FC removal, such as retention time and sunlight, and that, in turn; these parameters are affected by physical, chemical, biological and environmental parameters that interact

with each other. Mentioning them all is beyond the scope of this investigation, but some of the important ones are given below:

- Retention time: In general existing maturation ponds require engineering interventions to improve their flow pattern and change them from completely mixed or dispersed flow to close plug flow. This would guarantee that all elements in the water have the same retention time and same exposure to sunlight for their natural disinfection.
- Turbidity: Natural disinfection requires that the disinfecting agent (sunlight, in this case) penetrate throughout the water column. Undoubtedly, this does not occur in the majority of WSP systems. It could, however, occur frequently in maturation ponds located after conventional treatment, where nutrients and SS have been removed by treatment prior to entering the pond.
- Control of excessive algal growth: Ponds with very long retention times may acquire massive algal growth. This condition would unfavorably limit sunlight penetration into the pond, an undesirable factor in maturation ponds located after conventional treatment.
- Algal multiplication occurs after 24 hours. The logical thing would be to implement an optimum hydraulic retention time for the treatment, one that would allow a minimal concentration of algae to:
  - Maintain an adequate concentration of DO within the system, as well as an adequate natural habitat for the treatment
  - Guarantee water transparency, to allow free entry of sunlight.
  - Obtain maximum FC removal efficiency.

- It may be mentioned, with strong evidence, that channel ponds are most appropriate for:
  - Controlling, handling or manipulating hydraulic pond behavior, thus providing a series of benefits directly related to sunlight intensity. These two parameters constitute the principal binomial for FC removal in maturation ponds as tertiary treatment after conventional wastewater treatment.
  - There is no technology to control sunlight intensity, because sunlight is a natural phenomenon peculiar to the environmental conditions of each region, but channel ponds may be operated intelligently, using minimum retention time for days with maximum periods of sunlight (boreal summer) and longer retention times for days with minimum periods of sunlight (boreal winter). Great success may be obtained with intelligent management, and sunlight periods (not light intensity) could be controlled indirectly.
  - Besides, channel ponds are the most adequate technology for rehabilitating existing maturation ponds or to be included in any new designs, because they require less land.
  - However, this study has demonstrated that where turbidity in the liquid column is low (< 2), high efficiency of FC removal can be achieved in absence of significant photosynthetic activity and at relatively lower temperatures (< 19 oC). We summarize here in Figure 7 the principal factors relating to light which control FC removal based on Curtis' (1990) contribution.

- However, it is also worthwhile to provide a general hierarchical view of FC removal factors, to stress their relative importance as shown in Figure 8.

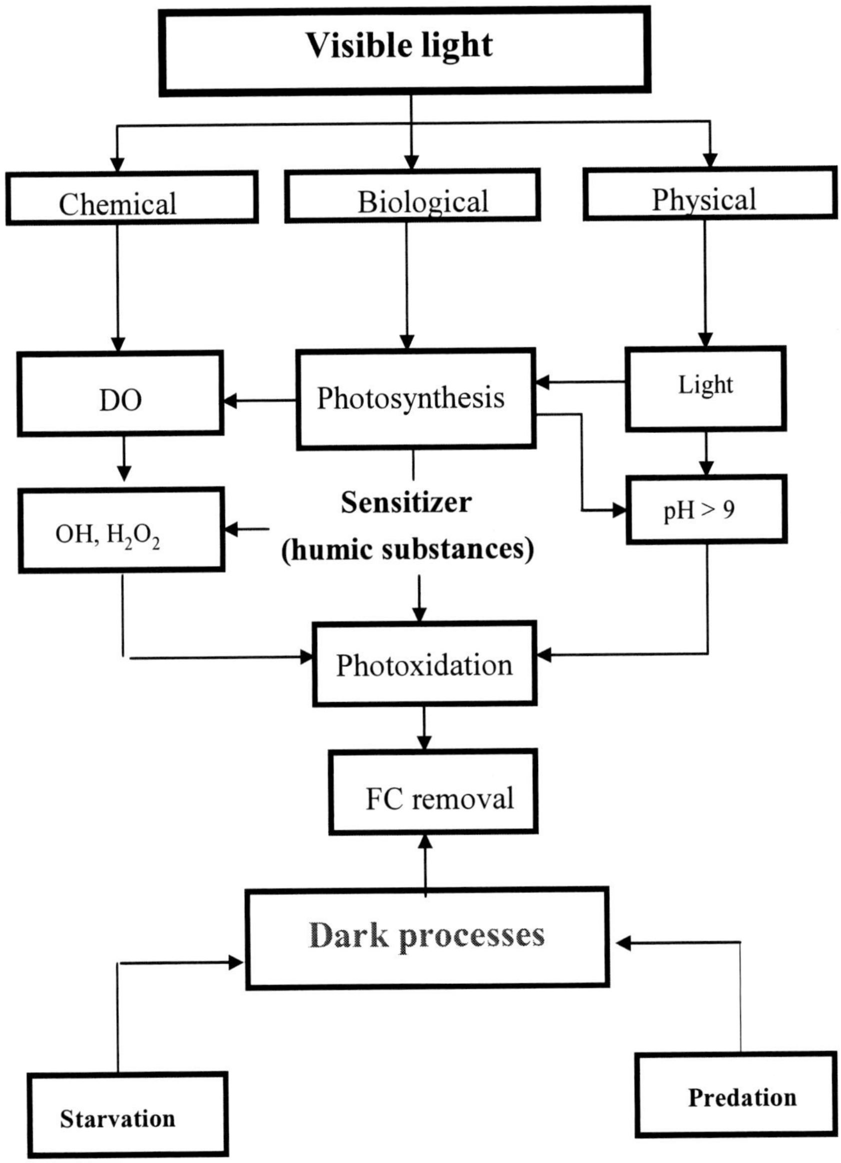

Figure 7. Factors controlling FC removal in maturation ponds after WSP in tropical climate, according to Curtis (1990); modified by the authors.

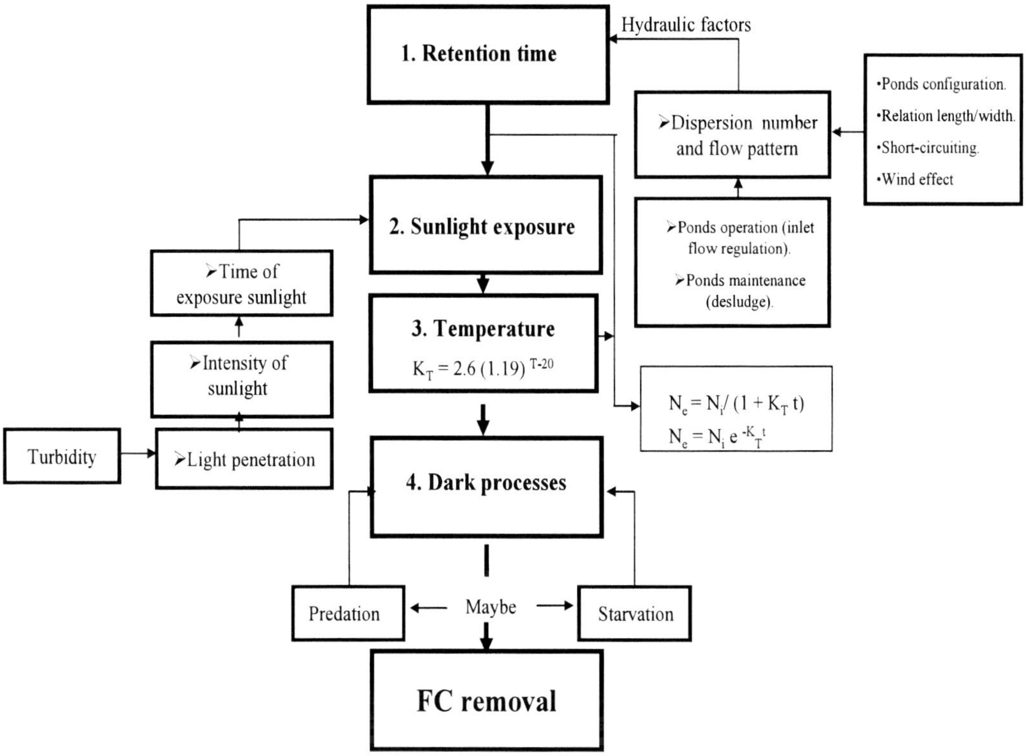

Figure 8. Factors controlling FC removal in low turbidity maturation ponds after conventional treatment.

References

Alabaster, G. P., Mills, S. W., Osebe, S. A., Thitani, W. N., Pearson, H. H., Mara, D. D., & Muiruri, P. (1991). Combined treatment of domestic and waste stabilisation pond systems in Kenya. *Water Sci.* Tech. 24:43-52.

Alkan, U., Elliot, D. J., & Evison, M. (1995). Survival of enteric bacteria in relation to simulated solar radiation and other environmental factors in marine waters. *Water Research* 29:2071-2081.

Bitton, G. (1999). *Wastewater microbiology* (Second Edition). New York-USA: Wiley and Sons, Inc.

Bracho N. R. (2003). *Optimisation of faecal coliform removal performance in three tertiary maturation ponds*. PhD Thesis, University of Surrey, Guildford-England.

Bracho, N., Lloyd, B., & Aldana, G. (2006a). Optimisation of hydraulic performance to maximize faecal coliform removal in maturation ponds. Water Research 40:1677-1685.

Bracho, N., Lloyd, B., & Aldana, G. (2006b). Re-habilitación de una laguna de estabilización utilizando bafles. *Revista Ciencia* 14:309-319.

Camp, T. (1946). Sedimentation and the design of settling tanks. *American Society Civil Engineer* (ASCE) 111:895-958.

Curtis, T. P. (1990). *Mechanisms of removal of faecal coliforms from waste stabilisation ponds.* PhD Thesis, University of Leeds, Leeds-England.

Curtis, T. P., & Mara, D. D. (1994). The effect of sunlight mechanisms for the die-off faecal coliform bacteria in waste stabilisation ponds. *Research Monographs in Tropical Public Health Engineering*, No. 1.

Curtis, T.P ., Mara, D. D., & Silva, S.A. (1992a). Influence of pH, oxygen, and humic substances on ability of sunlight to damage faecal coliforms in waste stabilization ponds. *Applied Environmental Microbiology* 58:1335-1343.

Curtis, T. P., Mara, D. D., & Silva, S. A. (1992b). The effect of sunlight on faecal coliforms in ponds: implications for research and design. *Water Sci. Tech.* 26:7-8, 1729-1738.

Davies-Colley, R. J., Donnison, M., Speed, D., Ross, C., & Nagels, J. (1999). Inactivation of faecal indicator micro-organisms in waste stabilisation ponds: Interactions of environmental factors with sunlight. *Water Research* 33:1220-1230.

Droste, R. (1997). *Theory and practice of water and wastewater treatment.* New York-USA: John Wiley & Sons, Inc.

Him, J., Vijamaa, H., & Raevuori, M. (1980). The effect of physiochemical, phytoplacton and seasonal factors on faecal indicators bacteria in Northern brackish water. *Water Research* 14:279-286.

Lian, Y., Cheung, R. Y., Everitt, S., & Wong, H. (1998). Reclamation of wastewater for polyculture of freshwater fish: Wastewater treatment in ponds. *Water Research* 32:1864-1880.

Lloyd, B., Vorkas, C., & Guganesharajah, K. (2002). Reducing hydraulic short-circuiting in maturation ponds to maximize pathogen removal using channels and wind breaks. *5th International IWA Specialist Conference on Waste Stabilisation Ponds,* New Zealand , Vol. 2:445-458.

Mara, D. D., & Pearson, H. W. (1998). *Design manual for waste stabilisation ponds in Mediterranean Countries.* Leeds, England: Lagoon Technology International Ltd.

Marais, G. V. R. (1974). Faecal bacterial kinetics in stabilisation ponds. *J. Env. Eng. Div. ASCE* 100(EE1):119-139.

Mayo, A. (1989). Effect of pond depth on bacterial mortality rate. *J. Environmental Engineering* 115:964-977.

Moeller, J. R., & Calkins, J. (1980). Bactericidal agents in wastewater lagoons and lagoon design. *J. Water Poll. Cont. Fed.* 52:2442-2451.

Muttamara, S., & Puetpaiboon, U. (1997). Roles of baffles in waste stabilisation ponds. *Water Sci. Tech.* 35:275-284.

Pearson, H. W., Mara, D. D., Cawley, L. R., Arridge, H. A., & Silva, S. A. (1996). The performance of an innovative tropical experimental waste stabilization pond system operating at high organic loading. *Water Sci. Tech.* 33:63-73.

Pearson, H. W., Mara, D. D., Mills, S. W., & Smallman, D. J. (1987a). Factors determining algal populations in waste stabilisation ponds and influence of algae on pond performance. *Water Sci. Tec.* 19:131-140.

Pearson, H. W., Mara, D. D., Mills, S. W., & Smallman, D. J. (1987b). Physico-chemical parameters influencing faecal bacterial survival in waste stabilisation ponds. *Water Sci. Tech.* 19:145-152.

Pearson, H. W., Mara, D. D., Smallman, D. J., & Mills, S. W. (1987c). Parameters influencing faecal coliform survival in waste stabilization ponds. *Water Sci. Tech.* 19:145-152.

Sarikaya, H. Z., Saatci, A. M, & Abdulfattah, A. F. (1987). Effect of pond depth bacterial die-off. *J. Environmental Engineering* 113:1350-1361.

Sauce, F. (1978). Interaction des algues et des autres micro-organismes dans les milieux pollués. *Ind. Aliment Agric.* :1239-1243.

Yánez, F. (1993). *Lagunas de estabilización. Teoría diseño, evaluación y mantenimiento.* Cuenca, Ecuador: Imprenta Monsalve.

In: Progress in Environmental Microbiology
Editor: Myung-Bo Kim, pp. 223-234
ISBN: 978-60021-940-5

Chapter VIII

Antagonistic Effect of Microbially-Treated Mixture of Agro-Industrial Wastes and Inorganic Insoluble Phosphate to *Fusarium* Wilt Disease

N. Vassilev[1*], *M. Fenice*[2], *E. Jurado* [1], *A. Reyes*[1], *I. Nikolaeva*[3], *M. Vassileva*[1]

[1]Department of Chemical Engineering, Faculty of Sciences, University of Granada, Granada-18071, Spain
[2]Department of Agrobiology and Agrochemistry, University of Tuscia, Italy
[3]Department of Public Technology, Malardalen University, Vasteras, Sweden

Abstract

A microcosm studies was carried out to determine the effect of *Aspergillus niger*-treated mixture of two agro-industrial waste (AIW) materials, dry olive wastes and sugar beet press mud, on tomato (*Lycopersicon esculentum*) plants grown in a soil inoculated with *Fusarium oxysporum* f. sp. *lycopersici* (*Fol*). These waste materials were selected as they constitute a major environmental problem especially for Mediterranean countries. Agro-wastes were treated in conditions of solid-state fermentation in the presence of Morocco apatite (RP) and further applied at a rate of 50 g/kg soil. Soil-plant systems were additionally inoculated or not with the arbuscular mycorrhizal (AM) fungus *Glomus intraradices*. Plant growth and nutrition, symbiotic developments and soil enzymatic activities were stimulated in *Fol*-free soil supplemented with treated agro-wastes and significantly greater in treatments where AM fungus was introduced compared with the non-amended control. The introduction of *Fol* into the soil-plant system adversely influenced all studied parameters. AM fungus alone reduced the effect of the plant pathogen on tomato plants and significantly decreased (by $2.3x10^3$) the number of *Fol*-CFU compared with the non-mycorrhizal control grown in non-amended soil. Higher levels of pathogen control were achieved in *A. niger*/AIW/RP-amended mycorrhized plant-soil system. In separate, *in vitro* studies, *A. niger* demonstrated a strong suppressive effect on *Fol* with a 5-fold reduction of the pathogen colony diameter compared to the control. Production of siderophore-like metabolites was also detected in chrome azurol S plate assays with *A. niger*.

Keywords: *Aspergillus niger*; Sugar beet wastes; Dry olive wastes; Rock phosphate; Solid-state fermentation product; Biocontrol; Soil-plant system; *Glomus intraradices*; Fungal pathogen

Introduction

The direct application of insoluble rock phosphates (RP) is increasing globally because of the rapid expansion in the biological-organic agriculture and the need of inexpensive phosphate. Accordingly, the efficacy of free and immobilized cells of phosphate solubilizing microorganisms (PSM) applied in RP solubilization in fermentation systems or as biofertilizers in establishing plant cover and enhancing crop yields is well documented in several review papers (Kucey et al., 1989; Vassilev et al., 2001; Vassilev and Vassileva, 2003). Particularly fungal microorganisms were widely accepted as excellent phosphate solubilizers (Whitelaw, 2000). Their application is considered as an environmentally-mild alternative in substituting highly polluting chemical RP processing (Goldstein, 2000) and/or in preventing frequent applications of soluble forms of inorganic phosphate fertilizers which often results in P leaching to the groundwater thus causing eutrophication of natural water reservoirs (Del Campillo et al., 1999).

A number of *in vitro* and *in vivo* studies have shown the ability of PSM to release metabolites such as organic acids which through their hydroxyl and carboxyl groups chelate the cations bound to phosphate, the latter being converted to soluble forms (Sagoe et al., 1998). In any case, metabolizable C compounds must be applied to the microbes to ensure their growth, organic acid production, and, simultaneously, RP solubilization. In general, the interaction of minerals such as phosphate rocks with organic matter and microorganisms has an enormous impact on biodiversity, global climate, biological productivity, human nutrition and the toxicity of metal pollutants (Huang, 2002).

During the last decade, we have developed a biotechnological scheme for RP solubilization by fungal microorganisms grown on agro-industrial wastes (Vassilev et al., 1995). Introduction of the resulting fermentation product, containing mineralized organic matter, soluble phosphate and fungal mycelium, into soil-plant systems in greenhouse and field conditions resulted in a significant plant growth enhancement, higher level of mycorrhization and increased soil enzyme activities (Vassilev et al., 1996; 1998; Vassileva et al., 1999; Cereti et al., 2004; Medina et al., 2004; Vassilev et al., 2006). The same approach

* Corresponding Author: nbvass@yahoo.com

was successfully proved in revegetation and bioremediation strategies (Caravaca et al., 2004; Medina et al., 2005). Bearing in mind that micronutrients found in natural phosphates such as zinc and cooper, are known to suppress pathogens (Duffy and Defago, 1999) and the effect of mineralized organic matter and PSM on phytopathogenic fungi (Ros et al., 2005), the further step in our investigations was to test the potential biocontrol properties of the above mentioned fermentation product.

As most of the soil-borne pathogens are fungi, the two main methods for disease control currently available in crop production are repeated applications of fungicides and the use of cultivars resistant or tolerant to the pathogens. However, the overuse of chemicals to combat plant diseases has caused soil pollution and had harmful effect on human health. On the other hand, cultivars resistant to pathogens have been developed but their use is limited, especially in fruit and vegetable crops. During the last years, a large number of fungal microorganisms have been reported as antagonists of soil-borne fungal pathogens (Kiss, 2003).

The aim of this work was to study the effect of an acid-producing strain of *Aspergullus niger* grown on agro-industrial wastes (AIW) in the presence of insoluble inorganic phosphate (rock phosphate, RP) and further introduced into a soil-plant-phytopathogen fungus system.

Materials and Methods

Fermentation Stage

Fermentation process details and analytical methods have been published in previous articles (Vassilev et al., 1995; 1996; 2006; Vassileva & Vassilev 1999). However, we will briefly mention the most important methodological points of the fermentation experiment.

The strain of *Aspergillus niger* NB2 used in this study had previously been selected as producing citric acid on complex substrates including lignocellulosic materials (Vassilev et al., 1995). It was maintained on potato-dextrose agar slants at 4° C. For inoculum preparation, *A. niger* was grown on a slant at 30° C for 7 days and spores were scraped in sterile distilled water enriched with 0.1 ml Tween 80.

Sugar beet wastes (sugar beet press mud) were obtained from the local fabric for sugar production (Azucarera de Jaen, Spain). Dry olive wastes were kindly provided by "COLGRA", Spain. Portions of 15 g of the solid wastes (dry olive wastes:sugar beet wastes, 1:0.5, w/w), ground to pass a 2-mm-pore screen, were placed in 250-ml Erlenmeyer flasks and mixed with medium strength Czapek-Dox mineral salt solution at a ratio of 1:1 (solid particles:liquid phase, w/v). All flasks were sterilized by autoclaving at 120° C for 30 min and inoculated with spore suspension of *A. niger* (1.2 x 10^7 spores/flask). Morocco low-grade rock phosphate (12.8 % P, 1 mm mesh) was sterlized separately and added at a rate of 3 g/l (0.15 g per flask) prior spore inoculation. Fermentation experiments were carried out at 30° C for 15 days. Characteristics of the resulting product, which was further used in soil-plant experiments, was pH, 3.8; titratable acidity, 62 mmol/l (83% of which was determined as citric acid); electrical conductivity (1:10), 1023 μS/cm; total P, 0.83%; total N, 0.7%; water soluble C, 1210 μg/g.

Soil-Plant Experiment

The treatments (five replications each) used in this experiment were as follows: (i) (C) control, soil without amendments; (ii) (+AM), control + *Glomus intraradices*; (iii) (+AM+*A. niger*/AIW/RP), control amended with dry olive/sugar beet wastes treated with *A. niger* in the presence of rock phosphate + *G. intraradices*. The fermentation products obtained at the fermentation stage were mixed at a rate of 5 % with a steam-sterilized (100° C/1h, 3 consecutive days) soil-sand mixture (1:1; w/w) and left for equilibration for 3 weeks at room temperature. The soil used for the experiments was the top 0-20 cm of Granada province (Spain) soil, with pH 8.1 in a 1:1 soil-water ratio, organic carbon 0.46%, 2.1 mg/kg N, 1.7 mg/kg P ($NaHCO_3$-extractable P). The soil texture was 358 g/kg sand, 436 g/kg silt, 205 g/kg clay and 12 g/kg organic matter. When necessary, soil was inoculated with *Fusarium oxysporum* f. sp. *lycopersici*. The inoculum of the pathogen, maintained on potato-dextrose agar slants, was prepared after 7-day cultivation in flasks containing 100 ml of sterile Czapek-Dox liquid medium as described by De Cal et al. (2000). Each pot received 50 ml of a microconidial suspension to reach a final concentration of 4.5×10^5 microconidia of the pathogen per g of soil.

Surface sterilized (70% ethanol, 2 min; 0.5% sodium hypochlorite, 2 min; sterile distilled water, rinsed 4-5 times) seeds of tomato (*Lycopersicon esculentum*) were sown in sterilized soil. Two weeks after germination the seedlings were transplanted in pots (three seedlings per pot) containing 300 g of amended or non-amended soil according to the treatments i-iii. The seedlings were inoculated (treatments ii and iii) with *in vitro* produced spores of *G. intraradices* Schenck & Smith as described by Vimard et al. (1999).

Experiments were performed under controlled conditions in a climate room at 21 to 28° C, 16h light, 50% relative humidity. Water loss from field capacity was replaced daily by top watering.

Analyses

Plants were harvested after 7 weeks and the shoot and root fresh and dried weights were recorded. P contents were determined by the dry ashing digestion method of Jones et al. (1991) followed by the molybdo-vanado determination as described by Lachica et al. (1973). The nitrogen concentration was assayed by using the Kjeldahl digestion method following a modification of Jones et al. (1991).

Roots were carefully washed and the percentage of root length colonized by *G. intraradices* was determined by the gridline intersect method (Giovannetti and Mosse, 1980) after staining with 0.05 % Trypan blue in lactophenol (v/v) (Phillips and Hayman, 1970).

Rhizosphere soil samples (closely associated with the plant roots) were collected, stored at 4° C, and subsequently assayed for acid phosphatase and β-glucosidase activity using as substrates *p*-nitrophenyl disodium phosphate (PNPP, 0.115 M) and *p*-nitro-phenyl-□-D-glucopyranoside (PNG, 0.05M), respectively (Naseby and Lynch, 1997; Masciandaro et al., 1994). The amount of *p*-nitrophenol formed was determined spectrophotometrically at 398 nm (Tabatabai and Bremner, 1969).

Severity of symptoms was determined after 10, 20, 30 and 49 days using the following index: A, plant without symptoms; B, lower leaves yellow; C, lower leaves dead and some upper leaves wilted; D, lower leaves dead and upper leaves wilted; E, dead plant. Rhizospheric population of *F. oxysporum* f. sp. *lycopersici* was determined using a dilution plate method. Ten grams of air-dried rhizosphere soil were shaken for 1h in a 250-ml Erlenmeyer flask containing 100 ml of 0.2% agar-water. One ml of the suspension from each sample (100-fold dilution) was spread on Petri dishes containing selective medium. The plates were further incubated for 7 days and the number of fungal colony-forming units per gram soil was determined by colony counting.

In vitro Agar Plate Bioassays

A modified chrome azurol S (CAS)-agar plate assay was carried out according to Milagres et al. (1999) to test the ability of *A. niger* to produce siderophores. *A. niger* was also assayed for inhibition of *F. oxysporum* on potato-dextrose agar. 10-mm disks of fresh culture of the pathogen fungus grown for 7 days at 28° C were cut out and placed in the center of a 9-cm Petri plate with PDA. Spores of *A. niger* were further inoculated on either side of the *F. oxysporum* disk at the corner of the plate. Fungal growth for individual (*F. oxysporum*, control) and dual (*F. oxysporum*+*A. niger*) fungal cultures was observed daily during 7 days at 28° C. All in vitro experiments were carried out in triplicate.

New Duncan's multiple range tests was used where appropriate to test for significant difference.

Results

Microscopic observation of plant roots showed that only AM-inoculated plants were root colonized (Table 1). The percentage of AM root length colonization was 16 % higher in pathogen-non-infected treatments amended with the fermentation product, compared with treatments where only *G. intraradices* was introduced into soil. In this latter case, the fungal pathogen significantly reduced AM colonization while no significant difference was observed in treatments amended with microbially-treated agro-wastes and RP compared with pathogen-non-infected mycorrhizal plants.

Table 1. Mycorrhizal colonization of tomato plants grown in amended or non-amended soil infected or not with *Fusarium oxysporum*

Treatment	AM colonization (%)	
	- *F. oxysporum*	+ *F. oxysporum*
+AM	43*b*	31*b*
+AM+*A. niger*/AIW/RP	59*a*	53*a*

Column values followed by the same superscript letters are not significantly different (P<0.05) using Duncan's multiple range test

The second goal of this set of experiments was to assess growth and nutrient (nitrogen, N and phosphate, P) uptake of tomato plants under different conditions of cultivation (Table 2 and 3). After 7-weeks culture, fresh and dry weights of shoots and roots of tomato plants non-infected with *F. oxysporum* were significantly higher in treatments where the arbuscular mycorrhizal fungus was introduced compared to the non-mycorrhizal control. However, the greatest increase in plant weight was registered in treatments amended with the fermentation mixture containing *A. niger*, treated AIW, and partially solubilized RP. Similarly, mineral N and P contents in tomato shoots and roots were significantly greater in treatments where *G. intraradices* was introduced into the soil-plant system compared with the control. These parameters were further enhanced in the treatments amended with microbially-treated AIW and RP. It should be noted, however, that particularly the contents of P, expressed as a percentage of tomato shoot dry weight, showed no significant differences between mycorrhizal plants and mycorrhizal plants +*A. niger*/AIW/RP which was possibly due to a dilution effect.

The introduction of *F. oxysporum* into the soil-plant system adversely affected the plant growth and development. This effect was more pronounced in the control treatment when the fresh plant biomass of 3.1 g was 35% lower compared with 4.75 g obtained in the control non-infected with *F. oxysporum*. *G. intraradices* significantly reduced the effect of *F. oxysporum* and fresh biomass (5.26 g) of mycorrhizal plants was 69.6 % higher than that recorded in the pathogen-infected control treatments. The highest plant growth and nutrient uptake among all pathogen-infected treatments were measured in soil-plant systems amended with *A. niger*/AIW/RP and inoculated with *G. intraradices*. In this treatment, differences between levels of plant growth, nutrient (particularly P) uptake and root mycorrhization of plants grown in the presence and absence of the pathogen were found to be insignificant. It should be noted, that the introduction of fermentation mixture improved soil enzyme activities involved in the P (acid phosphatase) and C (β-glucosidase) cycles compared with the same biochemical parameters in the control and +AM treatments (Table 4).

Table 2. Fresh weight (FW), dry weight (DW), nitrogen (N) and phosphate (P) content of tomato shoots in soil amended or non-amended with *Aspergillus niger* (*A.n.*) treated agro-industrial wastes (AIW) in the presence of rock phosphate (RP)

Parameter	FW (g)	DW (g)	N (mg)	N* (%)	P (mg)	P* (%)
- *Fusarium oxysporum*						
Control	4.75^{bc}	0.44^{c}	22.1^{d}	5.0	0.9^{bc}	0.21
+AM	6.91^{b}	0.7^{b}	32.9^{c}	4.7	1.2^{b}	0.17
+AM+ *A.n.*/AIW/RP	14.5^{a}	1.53^{a}	65.8^{a}	4.3	2.8^{a}	0.18
+ *Fusarium oxysporum*						
Control	3.1^{c}	0.28^{d}	8.71^{e}	3.1	0.61^{d}	0.21
+AM	5.26^{b}	0.51^{bc}	20.4^{d}	4.0	0.97^{bc}	0.19
+AM+ *A.n.*/AIW/RP	13.4^{a}	1.31^{a}	49.8^{b}	3.8	2.34^{a}	0.18

* % of DW.

Column values followed by the same superscript letters are not significantly different ($P<0.05$) using Duncan's multiple range test.

Table 3. Fresh weight (FW), dry weight (DW), N and P content of tomato roots in soil amended or non-amended with microbially-treated AIW in the presence of rock phosphate

Parameter	FW (g)	DW (g)	N (mg)	N* (%)	P (mg)	P* (%)
- *Fusarium oxysporum*						
Control	1.34^{c}	0.11^{c}	2.53^{c}	2.3	0.2^{c}	0.18
+AM	1.71^{b}	0.15^{b}	3.0^{bc}	1.99	0.28^{b}	0.19
+AM+ *A.n.*/AIW/RP	2.53^{a}	0.22^{a}	4.16^{a}	1.73	0.59^{a}	0.24
+ *Fusarium oxysporum*						
Control	0.93^{d}	0.1^{c}	2.0^{d}	2.0	0.16^{c}	0.16
+AM	1.36^{c}	0.13^{bc}	2.5^{c}	1.95	0.23^{bc}	0.18
+AM+AIW/RP/*A.n.*	2.1^{ab}	0.19^{a}	3.4^{b}	1.78	0.41^{a}	0.21

* % of DW.

Column values followed by the same superscript letters are not significantly different (P<0.05) using Duncan's multiple range test.

Under these conditions, it was found that the presence of *G. intraradices* significantly reduced the effect of *F. oxysporum* on tomato plants accompanied by a significant decrease of 2.3×10^3 in the number of CFU compared with the control treatment (Table 5). However, the AM symbiosis only partly limited the negative effect of the pathogen. Further introduction of *A. niger*/AIW/RP into soil demonstrated higher capacity to control the pathogen effect and only 6×10^2 CFU were found per gram soil at the end of the experiment.

Table 4. Phosphatase and β-glucosidase activity in rhizosphere of tomato plants grown in amended or non-amended soil infected with *Fusarium oxysporum*

Treatment	Phosphatase (μg PNF per g per h)	β-glucosidase (μg PNF per g per h)
Control	23.65*c*	36.5*c*
+AM	63.84*b*	59.89*b*
+AM+*A.n.*/AIW/RP	117.32*a*	102.54*a*

Values within a column followed by same letter are not significantly different (P<0.05) as determined by Duncan's multiple range test

Table 5. Observations of disease severity and number of *Fusarium oxysporum* colony-forming units (CFU) in the rhizosphere of tomato plants grown in amended or non-amended soil inoculated or not with *Glomus intraradices* (AM)

Treatment	Days/severity of symptoms				*F. oxysporum*
+*F. oxysporum*	10	20	30	45	(log CFU/g soil)
Control	A	C	D	E	3.83
+AM	A	B	B	C	3.64
+AM+*A.n.*/AIW/RP	A	A	A/B	B	2.77

In a separate *in vitro* experiment, *A. niger* showed a strong suppressive effect on *F. oxysporum* with a 5-fold reduction of the pathogen colony diameter ($14^{\pm}0.6$ mm) compared with the control ($72^{\pm}2.1$mm) after a 7-day cultivation. In addition, *A. niger* produced a rapid (after 2 days of growth) color-change reaction from blue to purplish-red in chrome azurol S-agar plate assay which evidenced siderophore-like substance production.

Discussion

It is now well established that biological control of fungal pathogens offers a potential alternative to chemical fungicides. A fungal biocontrol agent can prevent or suppress the pathogen by mechanisms involving antibiosis, parasitism, and competition for space and/or nutrition although the antagonistic effect can be based on more than one mode of action (Kiss, 2003).

In this work, the role of natural microflora in the suppressiveness of *F. oxysporum* was concisely avoided in order to determine the effect of *G. intraradices* alone and in combination with the fermentation product containing *A. niger*/AIW/partially solubilized RP.

Arbuscular mycorrhizal fungi are accepted as an important part of biocontrol fungi (Johansson et al., 2004) as they provoke increased resistance to certain wilt and root rot pathogens including *Fusarium* (Filion et al., 1999; Dar et al., 1997) due to an increased nutritional status of the host plant and/or releasing unspecified substances. These statements were confirmed in our study as *G. intraradices* not only significantly improved the growth and nutrient (N and P) uptake of tomato plants but in the presence of *F. oxysporum*, although at lower root mycorrhizal colonization (in comparison with the treatment free of pathogen), simultaneously decreased the pathogen population density (CFU/g rhizosphere soil) as compared with the non-mycorrhizal control. Improved pathogen performance is not uncommon in P deficient soil (as used in this study), where the addition of AM fungi produces vigorous plants better able to support pathogens (Borowitcz, 2001). In addition, poorer pathogen performance may be due to competition between pathogens and AM fungi for plant metabolites or infection sites (Traquair, 1995). Although fungal pathogens reduced root colonization by AM fungi, the latter were shown to provoke protection through increased hydrolytic enzyme activity including that directly involved in the regulation of the symbioses (Lambais and Mehdy, 1995; Pozo et al., 1998).

These effects of *G. intraradices* were further enhanced in soil-plant systems amended with *A. niger*/AIW/RP fermentation product. Availability of P was enhanced by the arbuscular mycorrhizal fungus which may be attributed to the overall increase in surface area and phosphatase activity of extraradical hyphae. The addition of microbially-treated organic matter to soil has a positive effect on soil fertility and increases its physical, chemical and biochemical characteristics as reported in previous studies using fermentation product with similar characteristics (Vassilev et al., 1996; Medina et al., 2004). Several studies assessed ecologically sustainable options for controlling fungal pathogens by combined protective actions of organic soil amendments and biocontrol microorganisms (Cotxarrera et al., 2002; Bardin et al., 2004).

In this study, results of *in vitro* experiments with *A. niger* showed a significant suppressive effect on *F. oxysporum* although the exact mechanisms of pathogen inhibition were not investigated. However, the biocontrol properties of *A. niger* could be attributed to siderophore production demonstrated in plate assays. In general, elaboration of ferrichrome-type siderophores by strains of genus *Aspergillus* has been observed which gives a competitive advantage (particularly for iron) over other microorganisms (Renshaw et al., 2002). Moreover, the presence of soluble phosphate obtained as a result of RP *A. niger*-mediated solubilization, most likely stimulated the production of siderophores as reported by Gram (1996). In addition, *A. niger* is known as a producer of a wide range of hydrolytic enzymes as confirmed in soil conditions by this study. This fungal microorganism can be considered as a biocontrol agent as recently suggested by other authors who demonstrated the biocontrol properties of P-solubilizing bacteria and fungi (Khan and Khan, 2002). Therefore, it was not surprising that *A. niger*/AIW/RP amendment reduced the capacity of *F. oxysporum* to harm the tomato plants by significantly reducing fungal-pathogen growth.

Conclusion

A solid-state fermentation process seems to be the preferred mass production method for obtaining large quantities of biocontrol agents (Larena et al., 2004). Direct applications of the resulting fermented products in soil results in a rapid mortality of plant pathogens (Adams et al., 2002). In our study, the fermented material additionally contains partially solubilized soluble P which is important bearing in mind the soil characteristics in the Mediterranean region. As agriculture moves toward more biological and environmentally mild methods, the combined effect of dual application of *A. niger*/AIW/RP and arbuscular mycorrhizal fungus *G. intraradices* on suppressing *F. oxysporum* should be further studied in natural conditions to critically assess a possible use in a large scale.

Acknowledgements

This work was supported by Project MEC CTM2005-06955, EC Project "Waste to Products", and R&C Programme (CICYT-MEC).

References

Adams, T.T., Eitman, M.A. & Hanel, B.M. (2002). Solid state fermentation of broiler litter for production of biocontrol agents. *Bioresource Technology, 82*, 33-41.

Bardin, S.D., Huang, H.C. & Moyer, J.R. (2004). Control of *Pythium* damping-off of sugar beet by seed treatment with crop straw powders and a biocontrol agent. *Biological Control, 29*, 453-460.

Borowitcz, V. (2001). Do arbuscular mycorrhizal fungi alter plant-pathogen relations? *Ecology, 82*, 3057-3068.

Caravaca, F., Alguacil, M.M., Vassileva, M., Diaz, G. & Roldan, A. (2004). AM fungi inoculation and addition of microbially treated dry olive cake-enhanced afforestation of a desertified Mediterranean site. *Land Degradation and Development, 15*, 153-161.

Cereti, C.F., Rossini, F., Federici, F., Quarantino, D., Vassilev, N. & Fenice, M. (2004). Reuse of microbially treated olive mill wastewater as fertilizer for wheat (*Triticum durum* Desf.). *Bioresource Technology, 91*, 135-140.

Cotxarrera, L., Trillas-Gay, M.I., Steinberg, C. & Alabouvette, C. (2002). Use of sewage sludge compost and *Trichoderma asperellum* isolates to suppress Fusarium wilt of tomato. *Soil Biology and Biochemistry, 34*, 467-476.

Dar, Gh. Hassan, Zargar, M.Y. & Beigh, G.M. (1997). Biocontrol of *Fusarium* root rot in the common bean (*Phaseolus vulgaris* L.) by using symbiotic *Glomus mosseae* and *Rhizobium leguminosarum*. *Microbial Ecology, 34*, 74-80.

De Cal, A., Garcia-Lopez, R., Melgarejo, P. (2000). Induced resistance by *Penicillium oxalicum* against *Fusarium oxysporum* f. sp. *lycopersici*: Histological studies of infected and induced tomato stems. *Phytopatology, 90*, 260-268.

Del Campillo, S.E., Van der Zee, S.E.A.T.M. & Torrent, J. (1999). Modelling long-term phosphorus leaching and changes in phosphorus fertility in excessively fertilized acid sandy soils. *European Journal of Soil Science, 50*, 391-399.

Duffy, B.K.; Defago, G. Trace mineral amendments in Agriculture for optimizing the biocontrol activity of plant-associated bacteria. In: P Berthelin, M Huang, JM Bollag, F Andreux editors. *Effect of mineral-organic-microorganism interactions on soil and freshwater environments*. New York: Kluwer Academic/Plenum Publishers, 1999, 295-304.

Filion, M., St-Arnaud, M. & Fortin, J.A. (1999). Direct interaction between the arbuscular mycorrhizal fungus *Glomus intraradices* and different rhizosphere microorganisms. *New Phytologist, 141*, 525-533.

Giovannetti, M. & Mosse, B. (1980). An evaluation of techniques for measuring vesicular-arbuscular mycorrhizal infection in roots. *New Phytologist, 84*, 489-500.

Goldstein, A.H. (2000). Bioprocessing of rock phosphate ore: essential technical considerations for the development of a successful commercial technology. IFA Technical Conference, New Orleance, pp. 1-21, http://goldsteinlab.alfred.edu/publications.html.

Gram, L. (1996). The influence of substrate on siderophore production by fish spoilage bacteria. *Journal of Microbiological Methods, 25*, 199-205.

Huang, P.M. Foreseeable impacts of soil mineral-organic component-microorganism interactions on society ecosystem health. In: A Violante, PM Huang, J-M Bollag and L Gianfreda, editors. *Soil Mineral-Organic Matter-Microorganism Interactions and Ecosystem Health*. Boston: Elsevier, 2002, 163-176.

Johansson, J.F., Paul, L.R. & Finlay, R.D. (2004). Microbial interactions in the mycorrhizosphere and their significance for sustainable agriculture. *FEMS Microbiology Ecology, 48*, 1-13.

Jones, J.B., Wolf, B. & Mills, H.A. Plant analysis handbook. Athens: Micro-Macro Publishing, 1991, 195-203.

Khan, M.R. & Khan, S.M. (2002). Effects of root-dip treatment with certain phosphate solubilizing microorganisms on the fusarium wilt of tomato. *Bioresource Technology, 85*, 213-215.

Kiss, L. (2003). A review of fungal antagonists of powdery mildews and their potential as biocontrol agents. *Pest Management Science, 59*, 475-483.

Kucey, R.M.N., Jansen, H.H. & Legett, M.E. (1989). Microbially mediated increases in plant-available phosphorus. *Advances in Agronomy, 42*, 198-228.

Lachica, M,, Aguilar, A. & Yanez, J. (1973). Metodos analiticos en la Estacion Experimental del Zaidin. *Anales en Edafologia y Agrobiologia, 32*, 1033-1047.

Lambais, M.R. & Mehdy, M.C. (1995). Differential expression of defense-related genes in arbuscular mycorrhiza. *Canadian Journal of Botany, 73*, 533- 540.

Larena, I., De Cal, A. & Melgarejo, P. (2004). Solid substrate production of *Epicoccus nigrum* conidia for biological control of brown rot on stone fruits. *International Journal of Food Microbiology, 94*, 161-167.

Masciandaro, G., Ceccanti, B. & Garcia, C. (1994). Anaerobic digestion of straw and piggery wastewater: II. Optimization of the process. *Agrochimica, 3*, 195-203.

Medina, A., Vassilev, N., Alguacil, M., Roldan, A. & Azcon, R. (2004). Increased plant growth, nutrient uptake, and soil enzymatic activities in a desertified Mediterranean soil amended with treated residues and inoculated with native mycorrhizal fungi and a plant growth-promoting yeast. *Soil Science, 169*, 260-270.

Medina, A., Vassilev, N., Barea, J.M., Azcon, R. (2005). Application of *Aspergillus niger*-treated agrowaste residue and *Glomus mossae* for improving growth and nutrition of *Trifolium repens* in a Cd-contaminated soil. *Journal of Biotechnology, 116*, 369-378.

Milagres, A.M.F., Manchuca, A. & Napoleao, D. (1999). Detection of siderophore production from several fungi and bacteria by a modification of chrome azurol S (CAS) agar plate assay. *Journal of Microbiological Methods, 37*, 1-6.

Naseby, D.C. & Lynch, J.M. (1997). Rhizosphere soil enzymes as indicators of perturbation caused by a genetically modified strain of *Pseudomonas fluorescens* on wheat seeds. *Soil Biology and Biochemistry, 29*, 1353-1362.

Phillips, J.M. & Hayman, D.S. (1970). Improved procedure for clearing roots and staining parasitic and vesicular arbuscular mycorrhizal fungi for rapid assessment of infection. *Transactions of the British Mycological Socety, 55*, 159-161.

Pozo, M.J., Azcon-Aguilar, C. & Dumas-Gaudot, E. (1998). Chitosanase and chitinase activities of tomato roots during interactions with arbuscular mycorrhizal fungi or *Phytophthora parasitica. Journal of Experimental Botany, 49*, 1729-1739.

Renshaw, J., Robson, G., Trinci, P.J., Wiebe, M.G., Livens, F.R., Collison, D. & Taylor, R.J. (2002). Fungal siderophores: structures, functions and applications. *Mycological Research, 106*, 1123-1142.

Ros, M., Hernandez, M.T., Garcia, C., Bernal, A. & Pascual, J.A. (2005). Biopesticide effect of green compost against fusarium wilt on melon plants. *Journal of Applied Microbiology, 98*, 845-854.

Sagoe, C.I., Ando, T., Kouno, K. & Nagaoka, T. (1998). Relative importance of protons and solution calcium concentration in phosphate rock dissolution by organic acids. *Soil Science and Plant Nutrition, 44*, 617-625.

Tabatabai, M.A. & Bremner, J.M. (1969). Use of *p*-nitrophenol phosphate in assay of soil phiosphatase activity. *Soil Biology and Biochemistry, 1*, 301-307.

Traquair, J.A. (1995). Fungal biocontrol of root diseases: endomycorrhizal suppression of cylindrocarpon root rot. *Canadian Journal of Botany, 73*, S89-S95.

Vassileva, M. & Vassilev, N. (1999). Rock phosphate solubilization by *Aspergillus niger* on olive cake-based medium and its further application in soil-plant system. *World Jounal of Microbiology and Biotechnology, 14*, 281-284.

Vassilev, N. & Vassileva, M. (2003). Biotechnological solubilization of rock phosphate on media containing agro-industrial wastes. *Applied Microbiology and Biotechnology, 61*, 435-440.

Vassilev, N., Baca, M., Vassileva, M. & Franco, I. (1995). Rock phosphate solubilization by *Aspergillus niger* grown on sugar beet waste medium. *Applied Microbiology and Biotechnology, 44*, 546-549.

Vassilev, N., Franco, I. & Vassileva, M. (1996). Improved plant growth with rock phosphate solubilized by *Aspergillus niger* grown on sugar beet wastes. *Bioresource Technology, 55*, 237-241.

Vassilev, N., Medina, A. & Vassileva, M. (2006). Microbial solubilization of rock phpsjhate on media containing agro-industrial wastes and effect of the resulting products on plant growth and P uptake. *Plant and Soil, 287*, 77-84.

Vassilev, N., Vassileva, M., Fenice, M. & Federici, F. (1998). Fetilizing effect of microbially treated olive mill wastewaters on *Trifolium* plants. *Bioresource Technology, 66*, 133-137.

Vassilev, N., Vassileva, M., Fenice, M. & Federici, F. (2001). Immobilized cell technology applied in solubilization of insoluble inorganic (rock) phosphate. *Bioresource Technology, 79*, 263-271.

Vimard, B., St-Arnaud, M., Furlan, V. & Fortin, J.A. (1999). Colonization potential of in vitro- produced arbuscular mycorrhizal fungus spores compared with a root-segment inoculum from open pot culture. *Mycorrhiza, 8*, 335-338.

Whitelaw, M.A. (2000). Growth promotion of plants inoculated with phosphate-solubilizing fungi. *Advances in Agronomy, 69*, 99-151.

In: Progress in Environmental Microbiology
Editor: Myung-Bo Kim, pp. 235-244
ISBN: 978-60021-940-5

Chapter IX

Fluorescence In Situ Hybridization (FISH) in Aquatic Bacteria

Ilias Tirodimos * ***and Malamatenia Arvanitidou***
Laboratory of Hygiene, Medical School, Aristotle University of Thessaloniki, Thessaloniki, Greece

Abstract

Present methods for the detection of micro-organisms in the aquatic environment are slow, inefficient and often unreliable. Furthermore, direct microscopic enumeration has shown that numbers of bacteria capable of forming colonies on "nonselective" media are usually several orders of magnitude fewer than numbers actually present and metabolically active in freshwater and marine environments. The development of molecular techniques, in particular in situ hybridization offers the potential for rapid and specific assays. The fluorescent in situ hybridization (FISH) technique uses rRNA-targeted fluorescent nucleic acid probes, and over the last decade has become an important tool for microbial ecologists. In short, cells are fixed (i.e., they are not viable anymore and the status quo of their DNA and RNA is preserved), permeabilized to facilitate access of the probe to the target site and then hybridized with nucleic acid probes. The probes are labeled with a fluorochrome, and the samples can then be analysed by epifluorescence. The classic FISH technique rely on (usually 16S) rRNA as a probe target. In this mini-review we will summarise methods and applications for FISH in aquatic environment and we will highlight in particular their advantages and limitations.

Keywords: Fluorescent in situ Hybridization (FISH), Aquatic Bacteria

---

* Corresponding author: Ilias Tirodimos, Laboratory of Hygiene, Medical School, Aristotelian University of Thessaloniki, 54124 Thessaloniki, Greece. Tel.#2310999142; E-mail: iltirodimos@yahoo.gr

I. Introduction

Direct microscopic enumeration has shown that numbers of bacteria capable of forming colonies on "nonselective" media are usually several orders of magnitude fewer than numbers actually present and metabolically active in freshwater and marine environments. This phenomenon is described as "great plate count anomaly" and has been known to microbiologists for generations [1]. It has been estimated that as little as 0.3% of bacteria in soil and <0.1% in marine water is culturable.

Identification and detection of microbial indicators, like coliforms and *E.coli*, are the most commonly performed tests within the water industry. At present microbiological water quality is determined using one of a number of methods including membrane filtration or most probable number methods [2]. However, these methods require many hours for a result to be obtained. There is a need for rapid tests which will detect and correctly identify faecal indicators and other pathogens, so that water providers can respond quickly to bacteriological failures of water quality and thereby help protect public health. It should be recognized also, that during membrane filtration, there are many bacterial species known to be able to pass through 0.45-mm membranes, some of which may well be opportunistic pathogens, such as various mycobacteria (hence the recommendation to use 0.2-mm membranes).

Furthermore, water may be the source of bacterial pathogens such as *Legionella* spp. and *Campylobacter* spp. [3, 4]. In the long run, the ability to determine whether these pathogens are present or absent in a sample and the level of contamination is the goal of any screening technique. Since many indicator bacterial numbers do not correlate well with pathogen levels, direct detection of pathogens is most desirable. However, often these important microbial populations seem to be difficult to cultivate. For example, the environmental pathogen *Legionella pneumophila* which is the etiologic agent of Legionnaire's disease, normally inhabits aquatic environments or wet soil, usually surviving as intracellular parasites of amoebae and ciliates [5]. Thus, isolation and reliable culturing of *Legionella* on selective medium is fastidious, especially because the bacterium is able to form viable but not culturable cells which cannot be cultured without previous passage through host cells, e.g. amoebae.

During the last decade the use of molecular methods have supplied the means for examining microbial diversity and detecting specific organisms in aquatic environment without the need for cultivation [6, 7]. Fluorescence in situ hybridization (FISH) uses gene probes with a fluorescent marker, typically targeting the 16S rRNA. Concentrated and fixed cells are permeabilized and mixed with the probe. Incubation temperature and addition of chemicals can influence the stringency of the match between the gene probe and the target sequence (Image 1). Since the signal of a single fluorescent molecule within a cell does not allow detection, target sequences with multiple copies in a cell have to be selected (e.g. there are 10^2- 10^4 copies of 16S rRNA in active cells). Probes can be designed to be complementary to species-, group-, or kingdom- specific target sites. Generally, a probe targeting the *Bacteria* domain is used in combination with more specific probes. The most widely used probe for aquatic ecosystems is probably the eubacterial probe EUB338 [8]. After hybridization, the microbial communities are typically examined by epifluorescent microscopy [9].

II. Tools and Technology

i) Fluorescence Microscopy

After hybridization, fluorescence conferred by rRNA- targeted oligonucleotide probes can be analyzed by epifluorescence microscopy. Fluorescence is the luminescence of a substance when it is excited by radiation. Most biological molecules or structures do not fluorescence on their own, so they must be linked with fluorescent molecules (or fluorochromes) in order to create specific fluorescent probes.

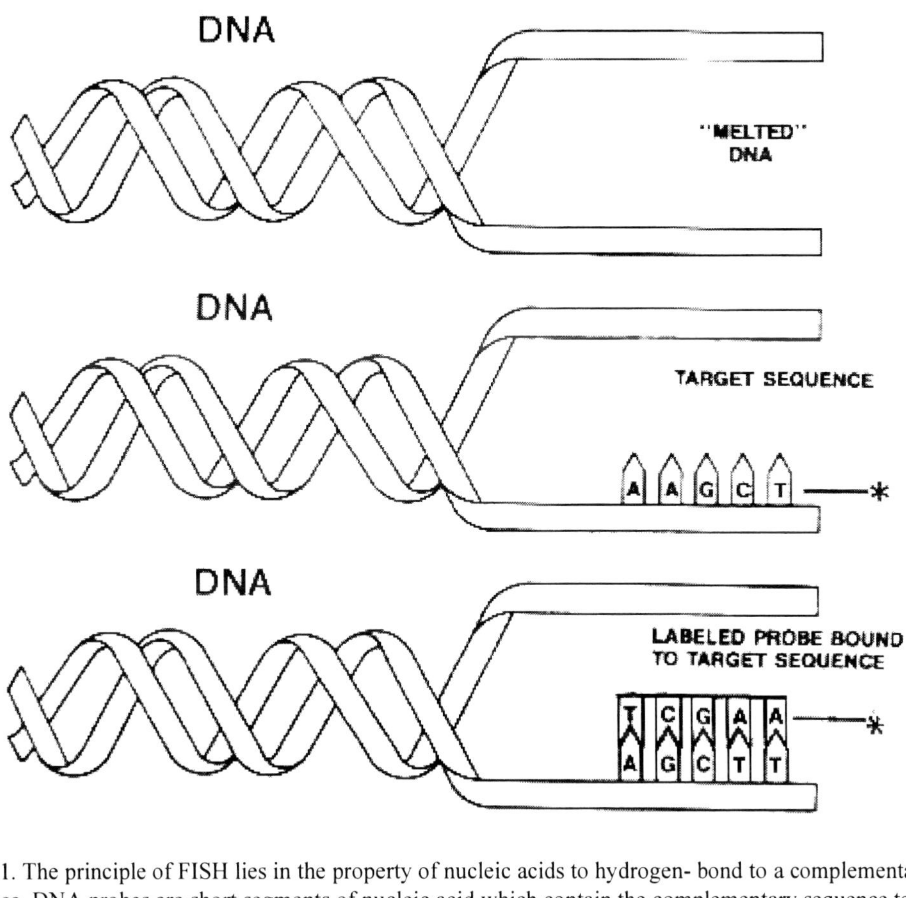

Image 1. The principle of FISH lies in the property of nucleic acids to hydrogen- bond to a complementary sequence. DNA probes are short segments of nucleic acid which contain the complementary sequence to a target sequence of interest. An appropriately designed and tagged probe will identify those nucleic acid molecules which contain the target sequence. The specific nature of the hydrogen-bonding allows the selection to occur in the presence of a heterogeneous population of DNA or RNA molecules.

Fluorescence of a substance is seen when the molecule is exposed to a specific wavelength of light (excitation wavelength or spectrum) is always of a higher wavelength. To view this fluorescence in the microscope, several light filtering components are needed. Specific filters are needed to isolate the excitation and emission wavelengths of a fluorochrome. A bright light source with proper wavelengths for excitation is also needed.

For normal fluorescence applications, this is a mercury vapor arc burner. Mercury arc burners are very bright lamps with a limited lifetime and require some maintenance and care to make sure that they are producing the brightest possible light beam for fluorescence excitation [10].

A dichroic beam splitter or partial mirror which reflects lower wavelengths of light and allows higher wavelengths to pass is also required. A beam splitter is required because the objective acts as a condenser lens for the excitation wavelength as well as the objective lens for emission. One only wishes to see the light emitted from the fluorochrome and not any of the excitation wavelength. This epi- illumination type of light path is required to create a dark background so that the fluorescence can be easily seen. The wavelength at which a beam splitter allows the higher wavelengths to pass must be set between the excitation and emission wavelengths of any given fluorochrome so that excitation light is reflected and emission light is allowed to pass through it (Image 2).

Filter sets must be made to correspond to the excitation and emission characteristics of a given fluorochrome. Some sets are made to allow visualization of two or even three fluorochromes simultaneously.

ii) Ribosomal RNA (rRNA) as the Target

The general approach to environmental studies has been to target discrete regions of the rRNAs or their genes for hybridization to group- and species- specific oligonucleotide probes. Extensive sequencing of the ribosomal RNAs (16S- and 23S- like rRNAs) has been particularly informative. An average bacterial 16S rRNA molecule has a length of 1,500 nucleotides, and 23S rRNA molecules are around 3,000 nucleotides. When fully or almost analyzed (at least >1,000 nucleotides should be determined), both molecules contain sufficient information for reliable phylogenetic analyses. Sequence divergence among these individual rRNAs has defined the outline of a natural classification of microorganisms. These data can also be used to design hybridization probes for studies in water microbiology [11].

A single *Escherichia coli* cell has between 10^4 and 10^5 ribosomes and, consequently, as many copies of 16S- and 23S- like rRNAs. Thus, the rRNA is a naturally amplified target for hybridization probes. Furthermore, the cellular rRNA content is not constant for a certain species but directly correlated with the growth rate of the microbial cells. This could be exploited to gain insights into the physiological activity of individual cells.

However, the use of rRNA as the target for oligonucleotide probes may have problems, as we mention in paragraph IV.

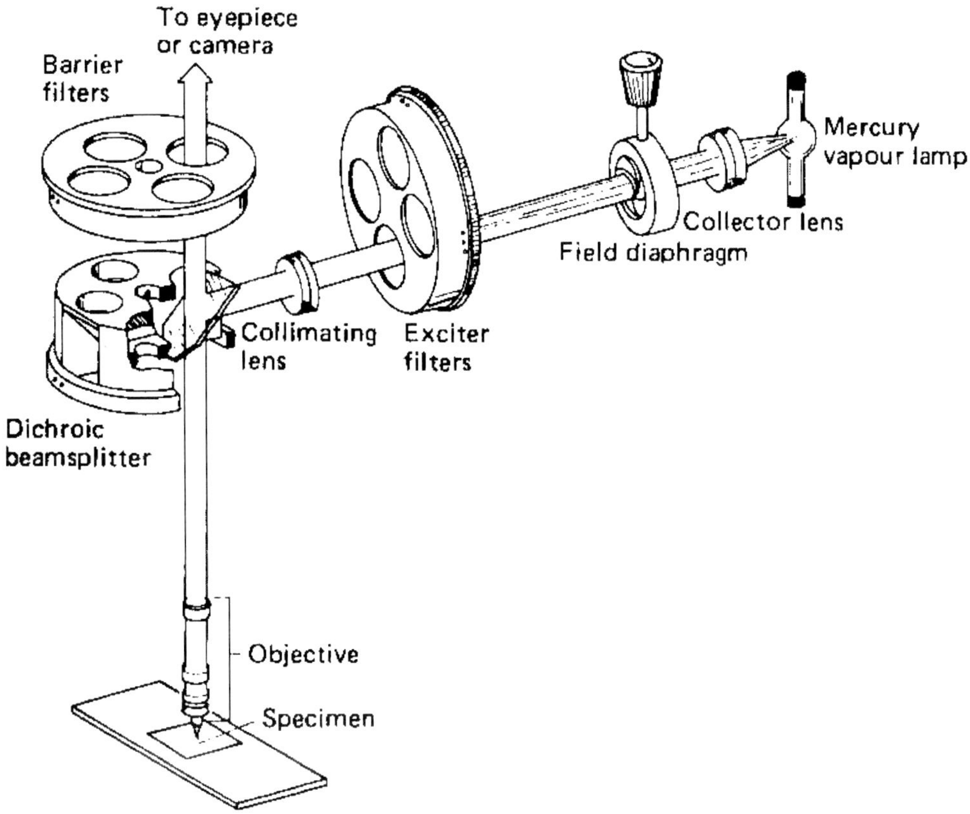

Image 2. Incident light fluorescence: mercury vapour lamp provides rich desirable ultraviolet radiation sources. Collector lens directs light to the exciter filter, which transmits selected wavelengths. Dichroic beam-splitter reflects virtually all of the desirable short wavelength, exciting light down through the objective to the specimen. Fluorochromes in specimen react to excitation wavelengths and emit longer wavelengths of visible fluorescence up through the system to the eyepiece. The barrier filter passes the emmited light from the specimen and blocks out unwanted background illumination.

iii) Fluorescent Dyes and Stains

Fluorescent dyes with affinities for all of the major macromolecules occurring within microbial cells are commercially available. Variables of practical importance to fluorescence include the intrinsic properties of the fluorophore: its excitation and emission spectra, molar absorbance coefficient, quantum yield, quantum efficiency and photostability.

In the early nineties, the most commonly used dyes for FISH in microbiology were fluorescein and rhodamine- derivatives (e.g. FITC, FLUOS and TRITC). Their low fluorescence intensity per mole fluorochrome, and their sensitivity to pH and bleaching limits their application, especially in the case of cells with low abundance of rRNA targets [12]. Fluorescent dyes with high quantum yields and extinction coefficients such as the cyanine series Cy are increasingly being used to overcome these problems. The latter are superior to the classical fluorescein dyes, because they result in significantly brighter staining and are very stable to photo-bleaching. However, the use of high- performance fluorochrome such as

Cy3 has a tendency to increase the background signal caused by cross- reaction with other specimens, inorganic particles and detritus. Improving the hybridization and wash conditions and the adjustment of stringency can reduce this drawback.

Hybridization buffers contain reagents to maximize nucleic acid duplex formation and inhibit nonspecific binding of probes. Formamide was used to lower the optimal hybridization temperature to minimize cell damage and encourage specific probe binding. Detergents such as SDS and heterologous nucleic acids inhibit background signals due to charge or nonspecific interactions with nucleic acids. Prehybridization without a probe in the hybridization buffer proved to be crucial to avoid higj background fluorescence.

III. Applications of In Situ Hybridization in Aquatic Bacterial Pathogens

As we mentioned in Chapter I, identification and detection of coliforms and *E.coli* are the most commonly performed tests within the water industry. Fluorescently labeled probes were used effectively to demonstrate the presence of *E.coli* in water samples [13, 14].

The genus *Campylobacter* includes the species *C.jejuni* and *C.coli*, which are the most common human enteric pathogens causing acute bacterial diarrhea worldwide. Drinking water is widely regarded as the main source of infection due to the presence of those organisms as part of the intestinal flora of many animals. Isolation of campylobacters may require about 4 to 5 days due to slow growth and lack of a suitable selective medium. Both 16S and 23S rRNA sequence data were used to design oligonucleotide probes for FISH analyses concerning Campylobacter strains [3, 15, 16].

Pseudomonas spp. is a typical psychrophillic bacterium that can be found in rivers, soil and other wet locations. The species *Ps.aeruginosa* is a major opportunistic pathogen in both nosocomial and community infections. FISH under optimized hybridization and washing conditions specifically detected *Pseudomonas* spp. [17].

The environmental pathogen *Legionella pneumophila*, which is the etiologic agent of Legionnaire's disease, normally inhabits aquatic environments or wet soil, usually surviving as intracellular parasites of amoebae and ciliates. Development of legionellosis has been attributed to the inhalation of viable organisms in fine aerosols into the lung, in which they invade the alveolar macrophages and other phagocytic cells. Isolation and reliable culturing of *Legionella* on selective medium is fastidious, especially because the bacterium is able to form viable but not culturable cells which cannot be cultured without previous passage through host cells, e.g. amoebae. Oligo-nucleotide probes were designed against regions of the 16S rRNA sequence and tested by in situ hybridization to strains of *L.pneumophila* [18, 19, 20].

Helicobacter pylori colonizes the gastric mucosa of approximately half the world population. Most infected individuals are asymptomatic but in 10-20% of carriers chronic infection is associated with the development of gastric diseases ranging from ulcer to adenocarcinoma. Various studies have provided evidence suggesting untreated and contaminated water supplies as a possible environmental source. Furthermore, there is evidence that *H.pylori* may have a higher tolerance than *E.coli* to common disinfectants used in distribution systems so aiding its survival in treated municipal systems. Work performed with various types of water (freshwater and seawater) using the FISH technique enabled researchers to confirm the presence of *H.pylori* cells in analyzed samples [21, 22].

The oligonucleotide probes for the above genera and species that have been used are listed in Table 1.

Other environmental bacteria and protozoa like *Cryptosporidium*, have been identified by rRNA FISH [13, 23].

IV. Problems with the Method

Despite its importance in microbiology, the standard FISH technique still suffers from many limitations.

Some basic limitations have to do with the target molecule. Apart from permeability issues, the main reason for weak signals with the classic rRNA targeted FISH is the low ribosome content found in slowly growing or metabolically inactive cells in environmental samples. Problems of this nature have led to a number of different approaches for signal amplification being developed in recent years. A modification of the technique, known as TSA-FISH (Tyramide Signal Amplification of FISH), boosts the fluorescent signal from hybridized cells 20-40 times fold versus background [24]. Another improvement of the method is the CARD-FISH (Catalyzed Reporter Deposition- FISH). Briefly, samples are concentrated on membrane filters, which are then embedded in low-gelling-point agarose, before permeabilizing the cells using lysozyme [25]. It has also been reported that a fluorescently labeled peptide nucleic acid (PNA) probe, targeted a site of the 16S rRNA with a very low accessibility for oligonucleotide probes, nevertheless confers a bright signal to its target cell [26]. As PNA pseudopeptides have an uncharged polyamide backbone, hybridization can be performed at low salt concentrations and high temperatures. These conditions significantly decrease the stability of the rRNA secondary structure, thereby leading to a higher accessibility of otherwise difficult target sites. Furthermore, microbiologists move away from rRNA and instead target other nucleic acids present in lower copy numbers, such as mRNA, plasmids or even single copy genes [27]. This trend is further aided by the increasing amount of available sequence data gathered from multiple genome projects.

Table 1. Probes used for rRNA FISH of bacterial species and genera

Probes	Sequences (5'→ 3')	Target organisms
EUB338	GCT-GCC-TCC-CGT-AGG-AGT	Eubacteria
CAMP4	GGT-AAG-CTA-CTA-AGA-GCG	*Campylobacter* spp.
PSEUD	GAT-CCG-GAC-TAC-GAT-CGG-TTT	*Pseudomonas* spp.
LEGPNE1	ATC-TGA-CCG-TCC-CAG-GTT	*Legionella pneumophila*
HPY	CTG-GAG-AGA-CTA-AGC-CCT-CC	*Helicobacter pylori*

Another problem may be that the 16S rRNA may be too well-conserved to discriminate between closely related populations. Different species may have almost identical 16S rRNA sequences. In such cases, the 23S rRNA may be useful. It is approximately twice as long and contains several highly variable regions. Another limitation originates from the fact that the rRNA diversity has only been partially described. There may be unknown microorganisms which are phylogenetically members of a probe target group, but do not contain a perfectly matching target site. This problem often occurs when designing group- specific probes. The solution to the problem of quantifying populations in complex microbial communities lies to the effort that goes into maintaining and enlarging high-quality rRNA databases and into the continuous development of tools for rational probe design.

A number of environmental factors also influence FISH performance. Systems that have little resource limitation, such as laboratory cultures, biofilms and sewage sludge, tend to yield a high percentage of hybridized cells. Freshwater on average appear to yield somewhat lower hybridization rates than marine samples in general. However, there is an absence of a clear relationship between the perceived productivity of a system and the FISH signal, which may result from the fact that a system's trophic status and the observed bacterial growth rate and biomass are often relatively uncoupled [28].

The most important practical limitation for the wide application of rRNA- targeted nucleic acid probes in water microbiology is the lack of automation. Data on the identification and the abundance of in situ stained microorganisms in complex water samples are mostly obtained by time- consuming manual analysis and microscopic counting. The development of a semiquantitative oligonucleotide chip containing two or three specific probes for every rRNA sequence known, will significantly promote the existing FISH technique.

V. Conclusions

The microbial community of water systems is examined from different points of view. One of them, is from a health perspective: researchers are interested in determining the identity and level of pathogens. This helps the epidemiologists track the source and spread of disease and also the engineers optimize the treatment conditions for elimination of hazardous organisms from the systems. However, until two decades ago the gathering of the information involved a substantial amount of time and effort. Advances in molecular biology in the past 20 years have resulted in a number of new detection methods that depend on the recognition of specific gene sequences. Such methods are usually rapid and can be tailored to detect specific strains of organisms on the one hand or groups of organisms on the other. The methods such as FISH, have a substantial potential for future applications in the field of water hygiene.

However, these methods are still not sufficiently sensitive to detect the low concentrations of pathogens considered important in water: many FISH experiments have been done with water samples spiked with a known amount of bacteria. Hence, there is still a research need to both increase the sensitivity of FISH and to develop routine approaches to where these organisms concentrate- e.g. biofilm-based detection methods or improved concentration methods for large- volume water samples.

Apart from the development of ultrasensitive FISH technology, a next desirable step would be an efficient automation to increase the amount of samples that can be analyzed, maybe by optimizing a microarray based approach.

References

Amann, R. and Ludwig, W. (2000). Ribosomal RNA-targeted nucleic acid probes for studies in microbial ecology. *FEMS Microbiol. Rev*. 24: 555-565.

Amann, R., Ludwig, W. and Schleifer, K.-H. (1995). Phylogenetic Identification and In Situ Detection of Individual Microbial Cells without Cultivation. *Microbiol. Rev.* 59: 143-169.

Amann, R.I., Binder, B.J., Olson, R.J., Chisholm, S.W., Devereux, R. and Stahl, D.A. (1990). Combination of 16S rRNA-targeted oligonucleotide probes with flow cytometry for analyzing mixed microbial populations. *Appl. Environ. Microbiol*. 56: 1919-1925.

Amann, R.I., Krumholz, L. and Stahl, D.A. (1990). Fluorescent-Oligonucleotide Probing of Whole Cells for Determinative, Phylogenetic, and Environmental Studies in Microbiology. *J. Bacteriol*. 172: 762-770.

Biegala, I., Not, F., Vaulot, D. and Simon, N. (2003). Quantitative assessment of picoeukaryotes in the natural environment by using taxon-specific oligonucleotide probes in association with tyramide signal amplification- fluorescence in situ hybridization and flow cytometry. *Appl. Environ. Microbiol.* 69: 5519-5529.

Böckelmann, U., Manz, W., Neu, T.R. and Szewzyk, U. (2000). Characterization of the microbial community of lotic organic aggregates ('river snow') in the Elbe River of Germany by cultivation and molecular methods. *FEMS Microbiol. Ecol*. 33: 157-170.

Bouvier, T. and Del Giorgio, P.A. (2003). Factors influencing the detection of bacterial cells using fluorescence in situ hybridization (FISH): a quantitative review of published reports. *FEMS Microbiol. Ecol*. 44:3-15.

Bouvier, T.C. and Del Giorgio, P.A. (2002). Compositional changes in free-lining communities along a salinity gradient in two temperature estuaries. *Limnol. Oceanogr.* 47: 453-470.

Brand, B.C. and Hacker, J.(1996). The biology of *Legionella* infection. In: S.H.E. Kaufmann (ed.): Host response to intracellular pathogens, pp. 291-312. R.G. Landes Company, Austin, Tex.

Brehm-Stecher, B.F. and Johnson, E.A. (2004). Single-Cell Microbiology: Tools, Technologies, and Applications. *Microbiol. and Mol. Biol. Rev*. 68: 538-559.

Declerck, P., Verelst, L., Duvivier, L., Van Damme, A. and Ollevier, F. (2003). A detection method for *Legionella* spp. in (cooling) water: fluorescent in situ hybridization (FISH) on whole bacteria. *Water Sci. Technol*. 47: 143-146.

Dutil, S., Tessier, S., Veillette, M., Laflamme, C., Mériaux, A., Leduc, A., Barbeau, J. and Duchaine, C. (2006). Detection of *Legionella* spp. by fluorescent *in situ* hybridization in dental unit waterlines. *J. Appl. Microbiol*. 100: 955-963.

Fricker, C.R. (2000). From media to molecules: new approaches to the detection of micro-organisms in water. *Water, Air, and Soil Pollution* 123: 35-41.

Garcia-Armisen, T. and Servais, P. (2004). Enumeration of viable *E.coli* in rivers and wastewaters by fluorescent in situ hybridization. *J. Microbiol. Methods* 58: 269-279.

Grimm, D., Merkert, H., Ludwig, W., Schleifer, K.-H., Hacker, J. and Brand, B.C. (1998). Specific Detection of *Legionella pneumophila*: Construction of a New 16S rRNA-Targeted Oligonucleotide Probe. *Appl. Environ. Microbiol.* 64: 2686-2690.

Horan, N. (2003). Faecal indicator organisms. In: D. Mara and N. Horan (eds): The Handbook of Water and Wastewater Microbiology, pp. 105-112. Academic Press, London.

Kenzaka, T., Yamaguchi, N., Tani, K. and Nasu, M. (1998). rRNA-targeted fluorescent in situ hybridization analysis of bacterial community structure in river water. *Microbiology* 144: 2085-2093.

Kitaguchi, A., Yamaguchi, N and Nasu, M. (2006). Simultaneous enumeration of viable Enterobacteriaceae and *Pseudomonas* spp. within three hours by multicolor fluorescence in situ hybridization with vital staining. *J. Microbiol. Methods* 65: 623-627.

Lehtola, M.J., Loades, C.J. and Keevil, C.W. (2005). Advantages of peptide nucleic acid oligonucleotides for sensitive site directed 16S rRNA fluorescence in situ hybridization (FISH) detection of *Campylobacter jejuni*, *Campylobacter coli* and *Campylobacter lari*. *J. Microbiol. Methods* 62: 211-219.

Moreno, Y., Botella, S., Alonso, J.L., Ferrús, M.A., Hernández, M. and Hernández, J. (2003). Specific Detection of *Arcobacter* and *Campylobacter* Strains in Water and Sewage by PCR and Fluorescent In Situ Hybridization. *Appl. Environ. Microbiol.* 69: 1181-1186.

Moreno, Y., Ferrús, M.A., Alonso, J.L., Jiménez, A. and Hernández, J. (2003). Use of fluorescent in situ hybridization to evidence the presence of *Helicobacter pylori* in water. *Water Res.* 37: 2251-2256.

Moreno, Y., Hernández, M., Ferrús, M.A., Alonso, J.L., Botella, S., Montes, R. and Hernández, J. (2001). Direct detection of thermotolerant campylobacters in chicken products by PCR and in situ hybridization. *Res. Microbiol.* 152: 577-582.

Pernthaler, A. and Amann, R. (2004). Simultaneous Fluorescence In Situ Hybridization of mRNA and rRNA in Environmental Bacteria. *Appl. Environ. Microbiol.* 70: 5426-5433.

Pernthaler, A., Pernthaler, J. and Amman, R.I. (2002). Fluorescence in situ hybridization and catalyzed reporter deposition for the identification of marine bacteria. *Appl. Environ. Microbiol.* 68: 3094- 3101.

Piqueres, P., Moreno, Y., Alonso, J.L. and Ferrús, M.A. (2006). A combination of direct viable count and fluorescent in situ hybridization for estimating *Helicobacter pylori* cell viability. *Res. Microbiol.* 157: 345-349.

Wagner, M., Horn, M. and Daims, H.(2003). Fluorescence in situ hybridization for the identification and characterization of prokaryotes. *Curr. Opin. Microb.* 6: 302-309.

Worden, A.Z., Chisholm, S.W. and Binder, B.J. (2000). In situ hybridization of *Prochlorococcus* and *Synechococcus* (marine cyanobacteria) spp. with RRNA-targeted peptide nucleic acid probes. *Appl. Environ. Microbiol.* 66, 284-289.

www. Fluorescence Microscopy (PDF). Robert Bagnell. UNC

Index

A

B

C

D

E

F

G

H

I

K

L

M

N

O

P

Q

R

S

T

U

V

W

X

Y

Z